Advanced Concepts and Principles of Atmospheric Sciences

Volume I

Advanced Concepts and Principles of Atmospheric Sciences
Volume I

Edited by **Smith Paul**

R CALLISTO REFERENCE

New York

Published by Callisto Reference,
106 Park Avenue, Suite 200,
New York, NY 10016, USA
www.callistoreference.com

Advanced Concepts and Principles of Atmospheric Sciences: Volume I
Edited by Smith Paul

International Standard Book Number: 978-1-63239-013-4 (Hardback)

Printed in the United States of America.

Contents

Preface

The term 'Atmospheric Sciences' refers to the study of atmosphere. This subject deals with the study of the impact and effects of different systems on the atmosphere and vice versa. It wouldn't be wrong to say that this very term is a broad category, under which a lot of different disciplines are covered. Numerous different areas of study fall under the category of atmospheric science such as meteorology, environmental soil sciences, etc. Some of the prominent names in this field include Léon Teisserenc de Bort and Richard Assmann.

This field uses principles from physics, chemistry, biology, computer science and other scientific areas. There is a specific area of study that falls under atmospheric science, known as climatology which deals with the study of atmospheric changes, for both long term and short term duration. Aeronomy under atmospheric sciences focuses on the study of layers of atmosphere, mostly upper layers of atmosphere, where ionization and dissociation occurs. This field has a broad scope and also covers planetary sciences, which includes the studies of planets and solar systems. Numerous different varieties of instruments are required in the field of Atmospheric Sciences, to study different aspects of the atmosphere. Lasers, rocketsondes, radiosondes, satellites and weather balloons are some among the various instruments which are involved in this field of study.

I wish to thank all the authors for their efforts and time that they have given to this project. Without their dedication and timely submissions, this publication wouldn't have been possible. I must also acknowledge the team at the publishing house, who have done a tremendous job with this book. Last but not the least, I wish to thank my family and friends, who have supported me in my life through everything.

Editor

Ambient Ozone Exposure in Czech Forests: A GIS-Based Approach to Spatial Distribution Assessment

I. Hůnová,[1] J. Horálek,[1] M. Schreiberová,[1] and M. Zapletal[2, 3]

[1] *Ambient Air Quality Department, Czech Hydrometeorological Institute, 14306 Prague, Czech Republic*
[2] *Faculty of Philosophy and Science, Silesian University at Opava, 74601 Opava, Czech Republic*
[3] *Ekotoxa s.r.o.-Centre for Environment and Land Assessment, 74601 Opava, Czech Republic*

Correspondence should be addressed to I. Hůnová, hunova@chmi.cz

Academic Editor: Shaibal Mukerjee

Ambient ozone (O_3) is an important phytotoxic pollutant, and detailed knowledge of its spatial distribution is becoming increasingly important. The aim of the paper is to compare different spatial interpolation techniques and to recommend the best approach for producing a reliable map for O_3 with respect to its phytotoxic potential. For evaluation we used real-time ambient O_3 concentrations measured by UV absorbance from 24 Czech rural sites in the 2007 and 2008 vegetation seasons. We considered eleven approaches for spatial interpolation used for the development of maps for mean vegetation season O_3 concentrations and the AOT40F exposure index for forests. The uncertainty of maps was assessed by cross-validation analysis. The root mean square error (RMSE) of the map was used as a criterion. Our results indicate that the optimal interpolation approach is linear regression of O_3 data and altitude with subsequent interpolation of its residuals by ordinary kriging. The relative uncertainty of the map of O_3 mean for the vegetation season is less than 10%, using the optimal method as for both explored years, and this is a very acceptable value. In the case of AOT40F, however, the relative uncertainty of the map is notably worse, reaching nearly 20% in both examined years.

1. Introduction

Ambient ozone (O_3) is a widely studied air pollutant [1]. Due to its unsaturated chemical structure, it is highly reactive and contributes to the oxidative power of atmosphere, essential for scavenging many pollutants from the atmosphere [2]. It has important negative impacts on both human health and the environment as acknowledged in numerous studies [3]. Moreover, due to its absorption-radiation abilities, O_3 is an important greenhouse gas [4, 5], and there are significant interactions between O_3 and climate change [6].

Current ambient O_3 levels have increased by approximately two times as compared to those measured over a century ago [7]. Although the magnitude and origin of the hemispheric O_3 trends are still not completely understood [8], there are indications that background O_3 levels over the midlatitudes of the Northern Hemisphere have continued to rise over the past three decades within the range of approximately 0.5–2% per year [9]. A significant contribution to O_3 levels both in Europe and North America originates in East Asia as a result of its dynamic development regarding population growth and increased fossil fuel consumption [10].

For the above reasons, the detailed knowledge of spatial distribution of ambient O_3 levels is becoming increasingly important. To develop a reliable, accurate, and continuous air pollutant surface predicting the values at locations without measurements is an essential task which we frequently encounter in environmental and health-related studies. This benefit of O_3 mapping stands out when viewed alongside the increasingly limited financial resources available for costly ambient air quality monitoring networks.

There are a wide range of techniques available for spatial interpolation, the advantages and limitations of which are widely discussed in the scientific literature [11]. In principle these techniques are classified as deterministic (the nearest neighbor and polynomial regression) or stochastic (geostatistical approaches as kriging and cokriging). The difference between these two is that the geostatistical methods use the spatial correlation structure and allow a prediction variability estimate to assess, under certain conditions, prediction accuracy. In between the deterministic and stochastic methods, there are a wide range of radial basis functions or splines.

The quality of maps of air pollution depends mainly on the quality of the input data measured at the stations, on the number of measuring sites, and also on their spatial distribution [12, 13]. Air pollution measurements, particularly those from online permanent monitoring used in routine monitoring networks, are very costly and so the number of sites is generally very limited. The number of required sites depends obviously on the type of air pollutant and on the representativeness of the measuring site. The representativeness is closely related to proximity of emission sources and topography; more stations are needed in complex terrain in contrast to flat [14]. When developing a surface for pollutants with high spatial variability (due to the importance of their local emission sources), for example, PM, benzo(a)pyrene, or toxic metals, more sites are needed. In contrast, pollutants like O_3, with a more regional character, depending mostly on regional phenomena such as meteorology and long-range or regional air pollution transport need a less-dense monitoring network.

Maps of ambient O_3, in context of its impacts on environment in rural areas, produced by different approaches were published for different regions: United Kingdom [15], Sierra Nevada in California [16–18], and the Carpathians in Europe [19]. Across the EU, mapping of background O_3 at a fine spatial scale (1×1 km) was carried out [20], as well as mapping of exposure index AOT40 at 2×2 km grid resolution [21, 22]. Across the Czech Republic, the method of [15] was applied for ozone deposition mapping [23].

In the Czech Republic (CR), ambient O_3 levels are elevated [24, 25], limit values over vast regions are frequently exceeded, and phytotoxic potential is high [26–28]. The aim of this paper is to compare the different spatial interpolation techniques and to recommend the optimal approach for producing a reliable map of O_3 with respect to its phytotoxic potential.

2. Methods

2.1. Ambient Ozone Data.
We used real-time O_3 levels measured at sites in the nationwide air quality monitoring network by UV absorbance, a reference method as declared by the EC [29]. The ozone analyzers used were the Thermo Environmental Instruments (TEI), M49. The samplers were placed ca 2 m above ground. Standard procedures for quality control and quality assurance [29] were applied. We considered O_3 seasonal means (April–September) and the exposure index AOT40 for forests—AOT40F [30], calculated according to [31]. The input data were 1 h mean concentrations. Data capture required for calculations of seasonal means was 75%. The AOT40 index as a cumulative variable is very sensitive to the quality of measured data [32, 33] and obviously also to missing values. More stringent requirements are, thus, needed for calculation of AOT40 as compared to the seasonal mean. In cases when we had less

than 90% of hourly O_3 concentrations for the period of April–September for calculating the AOT40F, we corrected it according to [34] so as to prevent underestimation of the O_3 exposure.

With respect to the aim of this study to assess O_3 exposure for forests, urban sites were not considered. From a total of 55 sites monitoring real-time O_3 concentrations across the CR, we accounted only for rural sites as specified by EoI [35]. The rural sites according to EoI are the sites with no important emission sources nearby and which are assumed to be affected only by long-range or regional air pollution transport. Thus, the representativeness of such sites is considerable, mostly in tens to hundreds of kms. The selection of sites resulted in 24 rural sites distributed unevenly across the CR (Figure 1), with more sites at border mountain areas as compared to the interior of the country. Considering that the area of the CR is 79 000 km^2, the sampling density was approximately one monitor per 3 292 km^2.

2.2. Spatial Interpolation.
Maps for mean vegetation season O_3 concentrations and for exposure index AOT40F for forests were prepared. For spatial interpolation, we used 24 rural sites run by the CHMI. In addition to the Czech sites, we also used data from selected measuring sites in Germany and Poland to improve the interpolation near border areas (Figure 1). Measuring sites in Slovakia are too distant so they cannot be used. Data from Austrian sites situated near the border were not available. The maps were prepared using ArcGIS Geostatistical Analyst [36] on a grid of 1×1 km resolution. A 25 m resolution DEM was used.

For spatial interpolation, eleven methods were used (Table 1). These methods are adequately referenced in scientific literature, so we comment on them only very briefly. In principle, we use three different interpolation methods using measured data only and two basic methods which combine measured and supplementary data, with four subvariants.

2.2.1. Interpolation Methods Using Measured Data Only

(A) Inverse Distance Weighted Method. The inverse distance weighted method (IDW) is likely to be the most frequently used deterministic method [37]. For interpolation we used

$$\hat{Z}(s_0) = \frac{\sum_{i=1}^n Z(s_i)/d_{0i}^k}{\sum_{i=1}^n 1/d_{0i}^k}, \qquad (1)$$

where $\hat{Z}(s_0)$ is the interpolated value of concentration in the point s_0, $Z(s_i)$ is the measured value of concentration in the ith point, $i = 1, \dots, n$, d_{0i} is the distance between the interpolated point and the ith point with measurement, and n is the number of sites used for interpolation.

(B) Radial Basis Functions Method. Radial basis functions (RBF) interpolate the measured value while minimizing the total curvature of the surface. The interpolation is described by

$$\hat{Z}(s_0) = \sum_{i=1}^n w_i \cdot \Phi(d_{0i}) + w_{n+1}, \qquad (2)$$

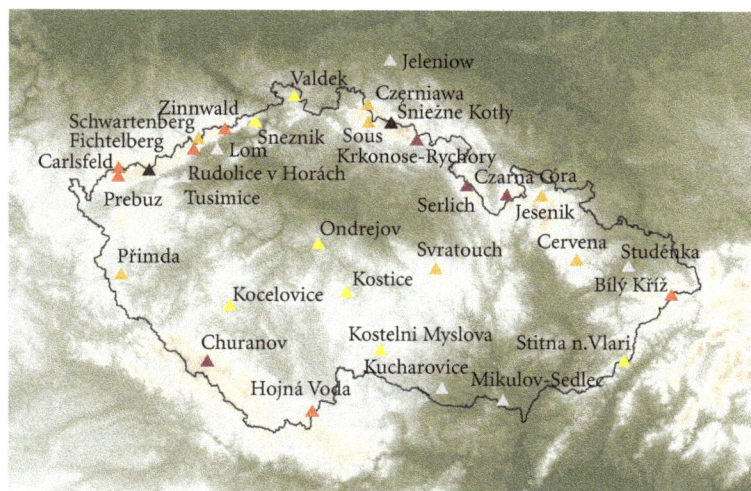

FIGURE 1: Sites with online monitoring of ambient O_3 concentrations ranked according to altitude.

TABLE 1: Methods used for interpolation.

Method	Acronym
Interpolation of measured data by inverse distance weighting	IDW
Interpolation of measured data by radial basis functions	RBF
Interpolation of measured data by ordinary kriging	OK
Linear regression of measured data and altitude	LR
Linear regression of measured data and altitude with subsequent interpolation of its residuals by IDW	LR + res_IDW
Linear regression of measured data and altitude with subsequent interpolation of its residuals by RBF	LR + res_RBF
Linear regression of measured data and altitude with subsequent interpolation of its residuals by OK	LR + res_OK
Interpolation of mean afternoon O_3 concentration minus regression of mean O_3 afternoon increment with altitude [15]	ALR
Interpolation of mean afternoon O_3 concentration minus regression of mean O_3 afternoon increment with altitude [15] with subsequent interpolation of its residuals by IDW	ALR + res_IDW
Interpolation of mean afternoon O_3 concentration minus regression of mean O_3 afternoon increment with altitude [15] with subsequent interpolation of its residuals by RBF	ALR + res_RBF
Interpolation of mean afternoon O_3 concentration minus regression of mean O_3 afternoon increment with altitude [15] altitude with subsequent interpolation of its residuals by OK	ALR + res_OK

where $\Phi(x)$ is a specific RBF function, d_{0i} is the distance between the interpolated point and the ith point with measurement, $w_1, \ldots, w_{n+1}$ are the weighting parameters, and n is the number of surrounding sites used for interpolation.

Although calculation of the RBF and estimation of its parameters is rather complicated, the computation is simple and fast. The parameters $w_1, \ldots, w_{n+1}$ are obtained from the system of equations given by

$$\sum_{j=1}^{n} w_j \Phi\left(d_{ij}\right) + w_{n+1} = Z(s_i), \quad i = 1, \ldots, n,$$
$$\sum_{j=1}^{n} w_j = -w_{n+1}. \tag{3}$$

A more detailed description is given by [36]. Coyle et al., 2002, [15] applied this interpolation technique within his approach for ambient O_3 mapping for Great Britain.

(C) Ordinary Kriging. Ordinary kriging is a geostatistical interpolation method. It considers the statistical model:

$$Z(s) = \mu + \varepsilon(s), \quad s \in D, \qquad (4)$$

where μ represents the constant mean structure of the concentration field, $\varepsilon(s)$ is a smooth variation plus measurement error (both zero-mean), and D is the examined area.

The interpolation is performed according to the equation

$$\hat{Z}(s_0) = \sum_{i=1}^{n} \lambda_i Z(s_i), \qquad \sum_{i=1}^{n} \lambda_i = 1, \qquad (5)$$

where $\hat{Z}(s_0)$ is the interpolated value of concentration in the point s_0, $Z(s_i)$ is the measured value of concentration in the ith point, $i = 1,\ldots,n$, n is the number of surrounding sites used for interpolation, and $\lambda_1, \ldots, \lambda_n$ are the weights assumed based on a semivariogram.

The weights λ_i are derived from a semivariogram $\gamma(\cdot)$ in order to minimize the mean square error. The explicit calculation is achieved by the system of equations given by

$$-\sum_{j=1}^{n} \lambda_j \gamma\left(s_i - s_j\right) + \gamma(s_0 - s_i) - m = 0, \quad i = 1,\ldots, n,$$

$$(6)$$

$$\sum_{i=1}^{n} \lambda_i = 1.$$

Kriging is a commonly used standard method. For construction of an O_3 surface, it was used, for example, by [38–40].

2.2.2. Interpolation Methods Using Both Measured and Auxiliary Data. The methods described in Section 2.2.1 were used only for the interpolation of the measured O_3 concentrations. Apart from these methods, there are others using well-correlated physical relationships between concentrations and other characteristics, for which more complex spatial information is known. The simplest approaches are the linear regression models without spatial interpolation; more complicated are different combinations of linear regression and spatial interpolation.

(A) Linear Regression Model without Spatial Interpolation. The basic linear regression model equation considered is

$$Z(s) = c + a_1 * X_1(s) + a_2 * X_2(s) + \cdots, \qquad (7)$$

where $X_i(s)$ are different supplementary parameters at the point s, for $i = 1, 2,\ldots,c$, and $a_1, a_2,\ldots$, are the parameters of the linear regression model.

In our case, altitude is used as the auxiliary parameter.

(B) Linear Regression Model Followed by the Spatial Interpolation of Residuals. The interpolation is estimated according to

$$\hat{Z}(s_0) = c + a_1 \cdot X_1(s_0) + a_2 \cdot X_2(s_0) + \cdots + \eta(s_0), \qquad (8)$$

where $\hat{Z}(s_0)$ is the estimated value of the O_3 concentration at the point s_0, $X_1(s_0), X_2(s_0), \ldots, X_n(s_0)$ are the n number of individual auxiliary variables at the point s_0, $c, a_1, a_2, \ldots, a_n$ are the n selected parameters of the linear regression model calculated at the points of measurement, and $\eta(s)$ is the spatial interpolation of the residuals of the linear regression model at the points of measurement.

The output of a dispersion model, altitude, meteorological variables (temperature, relative humidity, global radiation) may be among the auxiliary characteristics. For preparing an O_3 surface, this method was used, for example, by Loibl et al., 1994 [41], who used relative altitude as the auxiliary characteristics, Horálek et al., 2008 [42], who used model EMEP, altitude, and global radiation as the auxiliary characteristics, and Abraham and Comrie, 2004 [43].

In this study we used the altitude as the sole auxiliary characteristic. The major reason was that preliminary analysis of our data showed the best association between O_3 concentrations and altitude. Inclusion of meteorological variables did not bring any further benefit to our model. The assumption of linear distribution of O_3 with altitude was tested prior to the regression analysis.

Different interpolation methods, as described in Section 2.2.1, can be used for interpolation of residuals.

(C) Interpolation of Mean Afternoon Concentration Minus Regression of Mean Afternoon Increment with Altitude. Ozone concentrations show diurnal variation. Next to this, mean afternoon increment (i.e., the difference between the mean afternoon concentration and the mean whole-day concentration) is strongly related to altitude; see [15]. Coyle introduced the mapping method in which this regression relation is combined with the spatial interpolation of the afternoon concentration; that is,

$$\hat{Z}(s_0) = \rho(s_0) - R_\Delta(s_0), \qquad (9)$$

where $\rho(s_0)$ is the spatial interpolation of the mean afternoon concentrations at the point s_0 and $R_\Delta(s_0)$ is the regression relation of the increment Δ based on altitude at the point s_0.

While Coyle uses minimum curvature function (i.e., one of the RBF function) for the interpolation of afternoon values, we use ordinary kriging, as it shows generally better results; see bellow.

(D) Interpolation of Mean Afternoon Concentration Minus Regression of Mean Afternoon Increment with Altitude Followed by the Spatial Interpolation of Residuals. The variant of the method C is the addition of the interpolation of its residuals to the results of this method. As for the method B, different interpolation methods as introduced under Section 2.2.1 are used.

2.3. Uncertainty of Maps. We used cross-validation analysis for the assessment of uncertainty of the map. Cross-validation compares a value measured at a monitoring site with an estimated value based on interpolation of values measured at other sites. The root mean square error (RMSE)

TABLE 2: RMSE values (ppb) from cross-validation: comparison for different interpolation techniques for O_3 means, 2007-2008 vegetation seasons.

	Seasonal O_3 mean (ppb)			
	2007	2008	Average	
	40.15	38.87	39.51	
Interpolation technique	RMSE (ppb)			
	2007	2008	Average	Ranking
IDW	5.23	4.97	5.10	11
RBF	4.92	4.69	4.80	10
OK	4.73	4.67	4.70	9
LR	3.45	3.30	3.38	5
LR + res_IDW	3.49	3.48	3.48	8
LR + res_RBF	3.40	3.37	3.38	6
LR + res_OK	**3.31**	**3.15**	**3.23**	1
ALR	3.43	3.38	3.40	7
ALR + res_IDW	3.44	3.22	3.33	4
ALR + res_RBF	3.41	3.15	3.28	3
ALR + res_OK	**3.39**	**3.07**	**3.23**	1

TABLE 3: RMSE values (ppb.h) from crossvalidation: comparison for different interpolation techniques for AOT40F, 2007-2008 vegetation seasons.

	AOT40F (ppb.h)			
	2007	2008	Average	
	17570	15202	16386	
Interpolation technique	RMSE (ppb.h)			
	2007	2008	Average	Ranking
IDW	3695	3105	3400	11
RBF	3556	3023	3289	10
OK	3364	2961	3163	9
LR	3438	2841	3140	8
LR + res_IDW	3335	2897	3116	7
LR + res_RBF	3271	2844	3058	6
LR + res_OK	**3169**	**2827**	**2998**	1
ALR	3222	2842	3032	3
ALR + res_IDW	3274	2809	3041	4
ALR + res_RBF	3254	2852	3053	5
ALR + res_OK	**3207**	**2850**	**3029**	2

of the map, calculated according to (10), was used as the uncertainty criterion. RMSE should be as small as possible:

$$\text{RMSE} = \sqrt{\frac{1}{N}\sum_{i=1}^{N}\left(Z(s_i) - \hat{Z}(s_i)\right)^2}, \quad (10)$$

where RMSE is the root mean square error of the whole map, $Z(s_i)$ is the measured concentration at the ith site, $i = 1, \ldots, N$, $\hat{Z}(s_i)$ is the concentration at the ith site estimated from concentrations measured at other sites, $i = 1, \ldots, N$, and N is the number of measuring sites.

3. Results

The RMSE values comparing the different interpolation techniques both for O_3 seasonal means and AOT40F are presented in Tables 2 and 3. Considering the average from the vegetation seasons of 2007 and 2008, our results indicate that the optimal interpolation approaches are LR+res_OK and ALR+res_OK. In the case of O_3 seasonal means, the rankings for both methods were exactly the same, while for AOT40F LR+res_OK slightly outperformed ALR+res_OK. All three interpolation techniques alone—IDW, RBF, and OK—gave much worse results in comparison to linear regression of

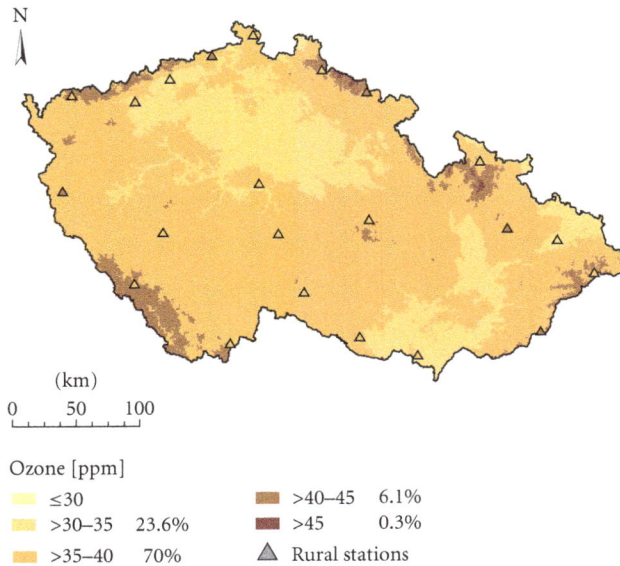

FIGURE 2: Spatial interpolation of mean O_3 concentrations in the 2008 vegetation season (ppb), interpolation technique LR + res_OK, grid resolution 1 × 1 km.

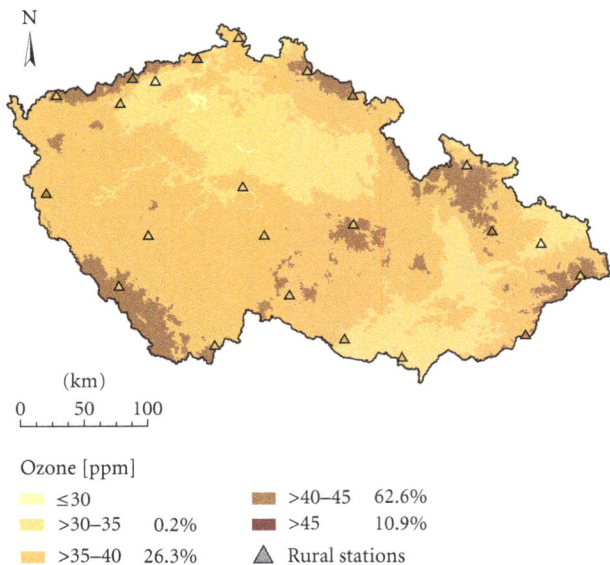

FIGURE 3: Spatial interpolation of mean O_3 concentrations in the 2008 vegetation season (ppb), interpolation technique ALR + res_OK, grid resolution 1 × 1 km.

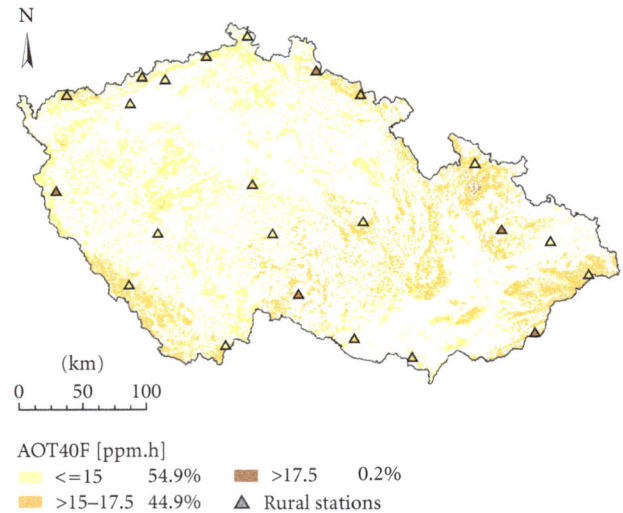

FIGURE 4: Spatial interpolation of exposure index AOT40F for the 2008 vegetation season (ppb.h), interpolation technique LR + res_OK, grid resolution 1 × 1 km, forested areas.

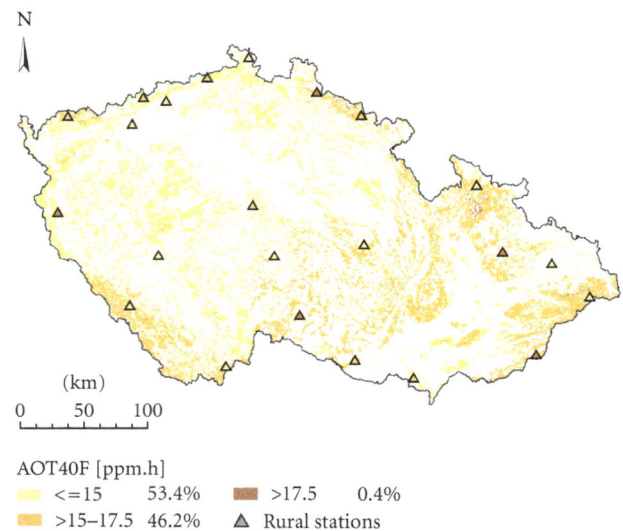

FIGURE 5: Spatial interpolation of exposure index AOT40F for the 2008 vegetation season (ppb.h), interpolation technique ALR + res_OK, grid resolution 1 × 1 km, forested areas.

measured data and altitude with subsequent interpolation of its residuals. This held both for O_3 seasonal means and AOT40F but was more pronounced for the O_3 seasonal means.

The relative uncertainty of the map of mean O_3 for the vegetation season was 8% for LR+res_OK and ALR+res_OK methods for both explored years. This is a thoroughly acceptable value. Even though the IDW method ranking indicated that it was the worst interpolation approach, the relative uncertainty of the map of mean O_3 for the vegetation season was 13%. In the case of AOT40F, however, the relative uncertainty of the map was notably worse. For LR+res_OK, ranking as the best approach, its relative uncertainty values were 18% in 2007 and 19% in 2008, while for IDW, assessed as the worst approach, its values were 21% for 2007 and 20% for 2008.

Figures 2 and 3 show the spatial interpolation of mean O_3 concentrations in the 2008 vegetation season prepared by interpolation techniques LR+res_OK and ALR+res_OK which ranked best in the comparison (Table 2). The two approaches resulted in maps which are very similar and exhibited only minor differences. The same applies for Figures 4 and 5 which show the spatial variability of AOT40F values in the 2008 vegetation season prepared by interpolation techniques LR+res_OK and ALR+res_OK which ranked best in

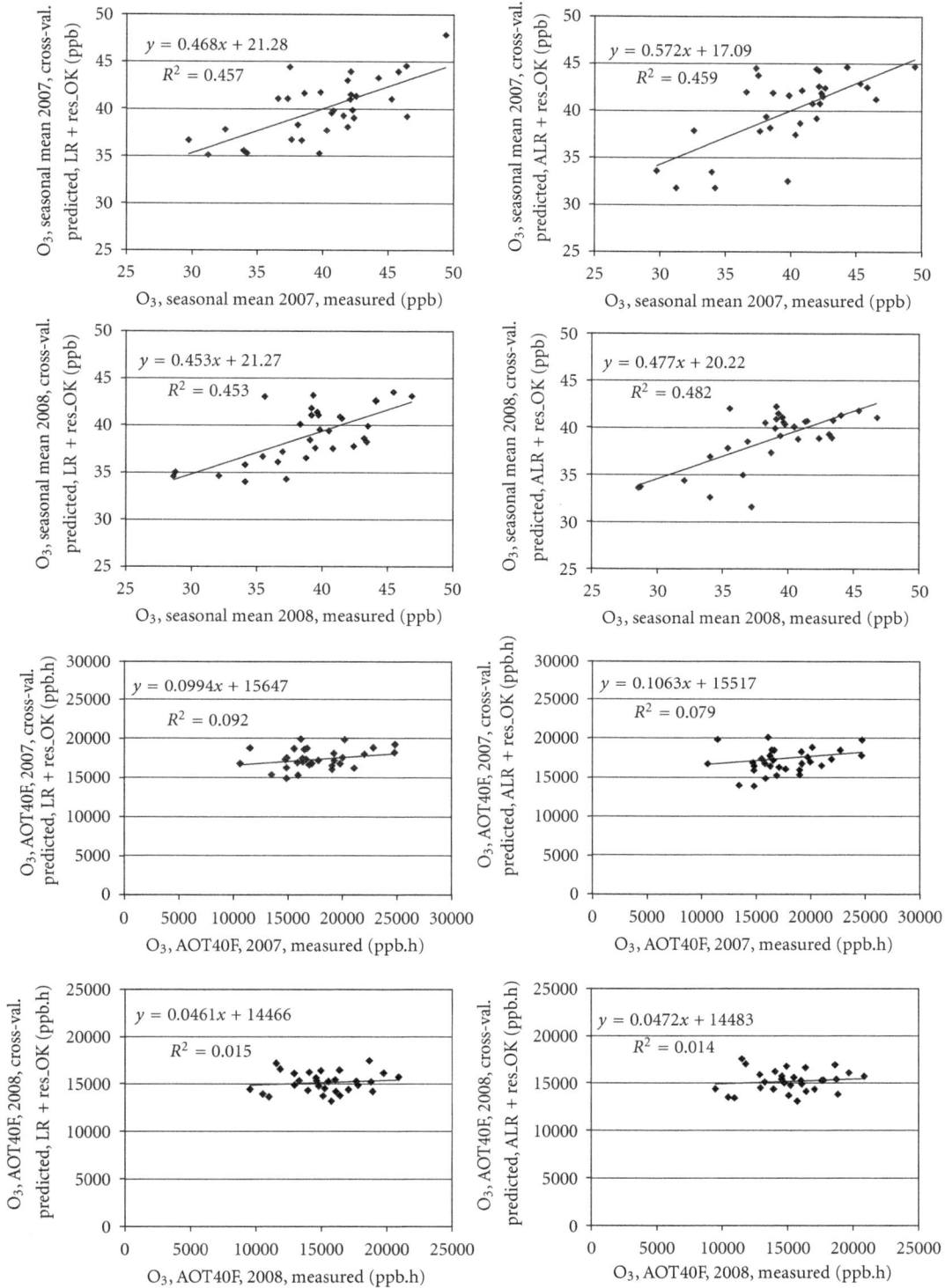

FIGURE 6: Scatter plots showing the relationship of cross-validation interpolated and measured values for the interpolation techniques LR + res_OK (left) and ALR + res_OK (right) for (a) seasonal mean O_3 concentrations (b) exposure index AOT40F.

the comparison (Table 3). The relationships between the results of these two interpolation approaches are presented by scatter plots (Figure 6). Notably better results were obtained for O_3 seasonal means.

4. Discussion

To produce a reliable air pollutant map to predict values in regions without measurements is an essential yet challenging task. Generally, dense monitoring networks are expensive but give a precise picture of spatial variability of a given phenomenon. Sparse sampling and monitoring networks, however, although less expensive, may miss significant spatial features of the studied phenomenon [12]. To make a real monitoring network denser, it is possible, for example, to add virtual measuring sites to improve the quality of interpolation [44]. Currently, however, no rigorous methodology for the determination of the number of monitoring sites sufficient/adequate to develop a reliable air pollutant surface is available [45]. Familiarity with the terrain and various phenomena that could affect the air pollutant concentrations and distributions are among the most important issues [12].

For mapping purposes, a number of techniques are available. There are substantial qualitative differences between the maps derived using different interpolation techniques as shown, for instance, by [46] for the maps of NO_2. The assessment of performance of the different techniques is extremely important. The maps derived by different interpolation techniques may be compared and evaluated by using the objective criteria, such as cross-validation [42]—when we omit one site in interpolation process and predict its values based on the rest of the sites and then compare the predicted and measured values.

Presented maps are applicable merely for rural areas. The obvious limitation of the maps is the low number of measuring sites which are unevenly distributed across the country. The spatial distribution of sites has strong historic connotations. Originally the measuring sites were located preferentially to more polluted regions, and they still remain in their original setting to observe the long-time trends.

Our results show that using the auxiliary data, in our case the dependence of O_3 concentrations on altitude in particular, significantly improves the overall quality of the resulting map (see Tables 2 and 3). Meteorology was not factored into the linear regression models as the preliminary analysis of our data showed that including meteorological variables did not bring any further benefit to our model. The likely reason is fairly low number of O_3 measuring sites. In near future we intend to use the Eulerian photochemical dispersion model CAMx [47] as auxiliary characteristics. The preliminary results seem to be promising. Next to this, it can be stated that the methods using ordinary kriging in its spatial interpolation part show the best results.

If we compare the two methods which ranked as the best, the LR+res_OK (i.e., linear regression of measured values with altitude followed by the interpolation of its residuals using ordinary kriging) approach slightly outperforms ALR+res_OK (i.e., the Coyle's approach [15] followed by the interpolation of its residuals using ordinary kriging) or gives comparable results. ALR+res_OK, however, is much more complicated and demanding for computation and, thus, less practical for application.

We can reasonably assume that notably worse results of mapping of AOT40F as compared to mean O_3 concentration for a vegetation season (see Figure 6) are due to more random/incidental factors affecting AOT40. Moreover, exposure index AOT40 is less robust characteristic as compared to mean O_3 concentrations [33].

Presented maps show the high-resolution O_3 spatial patterns which can be used for assessment of O_3 effects on vegetation. Exposure maps in particular are useful for indication of areas with high O_3 phytotoxic potential for forests and were already used across the Czech forests earlier [26]. Spatial patterns of O_3 seasonal means are useful for estimation of O_3 deposition as published for the Czech coniferous and deciduous forests by [23] and for estimation of stomatal flux.

5. Conclusions

We developed reasonable continuous surfaces for ambient O_3 vegetation season mean concentrations and AOT40F using eleven interpolation approaches. The comparison based on RMSE indicates that linear regression between measured O_3 data and altitude with subsequent interpolation of its residuals outperforms the interpolation techniques IDW, radial basis functions, and ordinary kriging significantly. This holds for both O_3 seasonal means and AOT40F, and, in the case of O_3 seasonal means, this feature is more pronounced as compared to AOT40F. Considering all different aspects, including the results of cross-validation analysis and the demandingness of computation, linear regression of O_3 data and altitude with subsequent interpolation of its residuals by ordinary kriging can be recommended as the optimal approach out of the eleven spatial interpolation techniques examined. Notably better results in mapping were obtained for mean seasonal O_3 concentrations in comparison to exposure index AOT40F.

Acknowledgments

This study was funded by the Ministry of Environment of the Czech Republic MŽP (SP/1b7/189/07). The data on ambient ozone were provided by the Czech Hydrometeorological Institute. We appreciate the assistance of Linton Corbet who revised the English and commented on the final version of the paper.

References

[1] J. N. Cape, "Surface ozone concentrations and ecosystem health: past trends and a guide to future projections," *Science of the Total Environment*, vol. 400, no. 1–3, pp. 257–269, 2008.

[2] J. H. Seinfeldt and S. N. Pandis, *Atmospheric Chemistry and Physics*, Wiley-Interscience, New York, NY, USA, 1998.

[3] B. S. Felzer, T. Cronin, J. M. Reilly, J. M. Melillo, and X. Wang, "Impacts of ozone on trees and crops," *Comptes Rendus—Geoscience*, vol. 339, no. 11-12, pp. 784–798, 2007.

[4] O. N. Singh and P. Fabian, *Atmospheric Ozone: A Millenium Issue*, EGU Special Publication Series 1. Copernicus, Berlin, Germany, 2003.

[5] IPCC, *Climate Change 2007. Synthesis Report*, IPCC, Geneva, Switzerland, 2007.

[6] I. S. A. Isaksen, *Ozone-Climate Interactions*, Air Pollution Research Report No. 81. EC, Brussels, Belgium, 2003.

[7] A. Volz and D. Kley, "Evaluation of the Montsouris series of ozone measurements made in the nineteenth century," *Nature*, vol. 332, no. 6161, pp. 240–242, 1988.

[8] J. E. Jonson, D. Simpson, H. Fagerli, and S. Solberg, "Can we explain the trends in European ozone levels?" *Atmospheric Chemistry and Physics*, vol. 6, no. 1, pp. 51–66, 2006.

[9] R. Vingarzan, "A review of surface ozone background levels and trends," *Atmospheric Environment*, vol. 38, no. 21, pp. 3431–3442, 2004.

[10] D. J. Jacob, J. H. Crawford, M. M. Kleb et al., "Transport and chemical evolution over the pacific (TRACE-P) aircraft mission: design, execution, and first results," *Journal of Geophysical Research D*, vol. 108, no. 20, article 8781, 2003.

[11] N. Cressie, *Statistics for Spatial Data*, John Wiley & Sons, New York, NY, USA, 1993.

[12] W. Fraczek, A. Bytnerowicz, and A. Legge, *Optimizing a Monitoring Network for Assessing Ambient Air Quality in the Athabasca Oil sands Region of Alberta Canada. Alpine space—Man & Environment 7: Global Change and Sustainable Development in Mountain Regions*, Innsbruck University Press, 2003.

[13] L. Wu and M. Bocquet, "Optimal redistribution of the background ozone monitoring stations over France," *Atmospheric Environment*, vol. 45, pp. 772–783, 2011.

[14] D. L. Peterson, "Monitoring air quality in mountains: designing an effective network," *Environmental Monitoring and Assessment*, vol. 64, no. 1, pp. 81–91, 2000.

[15] M. Coyle, R. I. Smith, J. R. Stedman, K. J. Weston, and D. Fowler, "Quantifying the spatial distribution of surface ozone concentration in the UK," *Atmospheric Environment*, vol. 36, no. 6, pp. 1013–1024, 2002.

[16] W. Fraczek, A. Bytnerowicz, and M. J. Arbaugh, "Use of geostatistics to estimate ambient ozone patterns," in *Ozone Air Pollution in the Sierra Nevada—Distribution and Effects on Forests*, A. Bytnerowicz, M. J. Arbaugh, and R. Alonso, Eds., pp. 215–247, Developments in Environmental Science 2, Elsevier, Amsterdam, The Netherlands, 2003.

[17] E. H. Lee, "Use of auxiliary data for spatial interpolation of ambient ozone patterns," in *Ozone Air Pollution in the Sierra Nevada—Distribution and Effects on Forests*, A. Bytnerowicz, M. J. Arbaugh, and R. Alonso, Eds., pp. 165–194, Developments in Environmental Science 2, Elsevier, Amsterdam, The Netherlands, 2003.

[18] H. K. Preisler and S. Schilling, "Use of nonparametric local regression to estimate surface ozone patterns over space and time," in *Ozone Air Pollution in the Sierra Nevada—Distribution and Effects on Forests*, A. Bytnerowicz, M. J. Arbaugh, and R. Alonso, Eds., pp. 195–214, Developments in Environmental Science 2, Elsevier, Amsterdam, The Netherlands, 2003.

[19] A. Bytnerowicz, B. Godzik, W. Fraczek et al., "Distribution of ozone and other air pollutants in forests of the Carpathian Mountains in central Europe," *Environmental Pollution*, vol. 116, no. 1, pp. 3–25, 2002.

[20] R. Beelen, G. Hoek, E. Pebesma, D. Vienneau, K. de Hoogh, and D. J. Briggs, "Mapping of background air pollution at a fine spatial scale across the European Union," *Science of the Total Environment*, vol. 407, no. 6, pp. 1852–1867, 2009.

[21] J. Horálek, B. Denby, P. de Smet et al., "Spatial mapping of air quality for European scale assessment," ETC/ACC Technical paper 2006/6, 2007, http://acm.eionet.europa.eu/docs/ETCACC_TechnPaper_2006_6_Spat_AQ.pdf.

[22] J. Horálek, J. Fiala, P. Kurfürst, B. Denby, P. de Smet, and F. de Leeuw, Spatial assessment of PM10 and ozone concentrations in Europe (2005) EEA Technical report No 1/2009, 2009, http://acm.eionet.europa.eu/reports/EEA_TR_1_2009_Spatial_AQ_assessment_2005.

[23] M. Zapletal and P. Chroust, "Ozone deposition to a coniferous and deciduous forest in the Czech Republic," *Water, Air, and Soil Pollution*, vol. 7, no. 1–3, pp. 187–200, 2007.

[24] I. Hůnová, "Ambient air quality for the territory of the Czech Republic in 1996-1999 expressed by three essential factors," *Science of the Total Environment*, vol. 303, no. 3, pp. 245–251, 2003.

[25] I. Hůnová, J. Šantroch, and J. Ostatnická, "Ambient air quality and deposition trends at rural stations in the Czech Republic during 1993-2001," *Atmospheric Environment*, vol. 38, no. 6, pp. 887–898, 2004.

[26] I. Hůnová, H. Livorová, and J. Ostatnická, "Potential ambient ozone impact on ecosystems in the Czech Republic as indicated by exposure index AOT40," *Ecological Indicators*, vol. 3, no. 1, pp. 35–47, 2003.

[27] I. Hunová, R. Novotný, H. Uhlířová et al., "The impact of ambient ozone on mountain spruce forests in the Czech Republic as indicated by malondialdehyde," *Environmental Pollution*, vol. 158, no. 7, pp. 2393–2401, 2010.

[28] I. Hůnová, L. Matoušková, R. Srněnský, and K. Koželková, "Ozone influence on native vegetation in the Jizerske hory Mts. of the Czech Republic: results based on ozone exposure and ozone-induced visible symptoms," *Environmental Monitoring and Assessment*, vol. 183, no. 1–4, pp. 501–515, 2011.

[29] EC, Directive 2008/50/EC of the European Parliament and of the Council of 21 May 2008 on ambient air quality and cleaner air for Europe, OJEC L 152, 2008.

[30] J. Fuhrer, L. Skärby, and M. R. Ashmore, "Critical levels for ozone effects on vegetation in Europe," *Environmental Pollution*, vol. 97, no. 1-2, pp. 91–106, 1997.

[31] UN/ECE, Mapping Manual Revision, UNECE convention on long-range transboundary air pollution, Manual on the Methodologies and Criteria for Modelling and Mapping Critical Loads and Levels and Air Pollution Effects, Risks and Trends, 2004, http://www.icpmapping.org.

[32] J. P. Tuovinen, "Assessing vegetation exposure to ozone: properties of the AOT40 index and modifications by deposition modelling," *Environmental Pollution*, vol. 109, no. 3, pp. 361–372, 2000.

[33] M. Sofiev and J. P. Tuovinen, "Factors determining the robustness of AOT40 and other ozone exposure indices," *Atmospheric Environment*, vol. 35, no. 20, pp. 3521–3528, 2001.

[34] EC, "Directive 2002/3/EC of the European Parliament and of the Council of 12 February 2002 relating to ozone in ambient air," *OJEC*, no. L 67, pp. 14–30, 2002.

[35] EC, "Council Decision 97/101/EC of 27 January 1997 establishing a reciprocal exchange of information and data from networks and individual stations measuring ambient air pollution within the Member States," *Official Journal of the European Communities*, no. L 35/14, 1997.

[36] K. Johnston, J. Ver Hoef, K. Krivoruchko, and N. Lucas, *Using ArcGIS Geostatistical Analyst*, Environmental Systems Research Institute, Redlands, Calif, USA, 2001.

[37] E. H. Isaaks and R. M. Srivastava, *An Introduction to Applied Geostatistics*, Oxford University Press, Oxford, UK, 1989.

[38] L. S. Casado, S. Rouhani, C. A. Cardelino, and A. J. Ferrier, "Geostatistical analysis and visualization of hourly ozone data," *Atmospheric Environment*, vol. 28, no. 12, pp. 2105–2118, 1994.

[39] A. S. Lefohn, H. P. Knudsen, and L. R. McEvoy, "The use of kriging to estimate monthly ozone exposure parameters for

the Southeastern United States," *Environmental Pollution*, vol. 53, no. 1–4, pp. 27–42, 1988.

[40] A. G. Hjellbrekke, "Ozone measurements 1998," EMEP/CCC-Report 5/2000, Norwegian Institute for Air Research, 2000.

[41] W. Loibl, W. Winiwarter, A. Kopsca, J. Zueger, and R. Baumann, "Estimating the spatial distribution of ozone concentrations in complex terrain," *Atmospheric Environment*, vol. 28, no. 16, pp. 2557–2566, 1994.

[42] J. Horálek, P. de Smet, F. de Leeuw, B. Denby, P. Kurfurst, and R. Swart, "European air quality maps for 2005 including uncertainty analysis," ETC/ACC Technical Paper 2007/7, 2008, http://acm.eionet.europa.eu/docs/ETCACC_TP_2007_7_SpatialAQmapping2005_annual_interpolations.pdf.

[43] J. S. Abraham and A. C. Comrie, "Real-time ozone mapping using a regression-interpolation hybrid approach, applied to Tucson, Arizona," *Journal of the Air and Waste Management Association*, vol. 54, no. 8, pp. 914–925, 2004.

[44] A. L. Beaulant, G. Perron, J. Kleinpeter, C. Weber, T. Ranchin, and L. Wald, "Adding virtual measuring stations to a network for urban air pollution mapping," *Environment International*, vol. 34, no. 5, pp. 599–605, 2008.

[45] G. Hoek, R. Beelen, K. de Hoogh et al., "A review of land-use regression models to assess spatial variation of outdoor air pollution," *Atmospheric Environment*, vol. 42, no. 33, pp. 7561–7578, 2008.

[46] C. D. Lloyd and P. M. Atkinson, "Increased accuracy of geostatistical prediction of nitrogen dioxide in the United Kingdom with secondary data," *International Journal of Applied Earth Observation and Geoinformation*, vol. 5, no. 4, pp. 293–305, 2004.

[47] ENVIRON, "User s guide to the Comprehensive Air Quality model with extensions (CAMx) version 5.10," 2009, http://www.camx.com.

Amazon Rainforest Exchange of Carbon and Subcanopy Air Flow: Manaus LBA Site—A Complex Terrain Condition

Julio Tóta,[1] David Roy Fitzjarrald,[2] and Maria A. F. da Silva Dias[3]

[1] *Universidade do Estado do Amazonas, Manaus, AM, Brazil*
[2] *State University of New York, Albany, NY, USA*
[3] *Universidade de São Paulo, Sao Paulo, SP, Brazil*

Correspondence should be addressed to Julio Tóta, totaju@gmail.com

Academic Editors: J. Dodson and C. Ottle

On the moderately complex terrain covered by dense tropical Amazon Rainforest (Reserva Biologica do Cuieiras—ZF2—02°36′17.1″ S, 60°12′24.4″ W), subcanopy horizontal and vertical gradients of the air temperature, CO_2 concentration and wind field were measured for the dry and wet periods in 2006. We tested the hypothesis that horizontal drainage flow over this study area is significant and can affect the interpretation of the high carbon uptake rates reported by previous works at this site. A similar experimental design as the one by *Tóta et al.* (2008) was used with a network of wind, air temperature, and CO_2 sensors above and below the forest canopy. A persistent and systematic subcanopy nighttime upslope (positive buoyancy) and daytime downslope (negative buoyancy) flow pattern on a moderately inclined slope (12%) was observed. The microcirculations observed above the canopy (38 m) over the sloping area during nighttime presents a downward motion indicating vertical convergence and correspondent horizontal divergence toward the valley area. During the daytime an inverse pattern was observed. The microcirculations above the canopy were driven mainly by buoyancy balancing the pressure gradient forces. In the subcanopy space the microcirculations were also driven by the same physical mechanisms but probably with the stress forcing contribution. The results also indicated that the horizontal and vertical scalar gradients (e.g., CO_2) were modulated by these micro-circulations above and below the canopy, suggesting that estimates of advection using previous experimental approaches are not appropriate due to the tridimensional nature of the vertical and horizontal transport locally. This work also indicates that carbon budget from tower-based measurement is not enough to close the system, and one needs to include horizontal and vertical advection transport of CO_2 into those estimates.

1. Introduction

The terrestrial biosphere is an important component of the global carbon system. The long-term exchanges estimate of terrestrial biosphere is a challenge and has resulted in ongoing debate [1, 2]. For monitoring long-term net ecosystem exchange (NEE) of carbon dioxide, energy, and water in terrestrial ecosystems, tower-based eddy covariance (EC) techniques have been established worldwide [1].

It is now recognized that the EC technique has serious restrictions for application over complex terrain and under calm and stable nighttime conditions with low turbulence or limited turbulent mixing of air [2–8]. To overcome this problem, the friction velocity (u^*) -filtering approach has been formalized by the FLUXNET committee for the estimation of annual carbon balances [5, 9]. This approach

simply discarded calm night's flux data (often an appreciable fraction of all nights) and replaced them with ecosystem respiration rates found on windy nights [10]. Papale et al. [11] pointed out that this approach itself must be applied with caution and the friction velocity (u^*) corrections threshold is subject to considerable concerns and is very site specific. Miller et al. [10] reported that depending on the u^* threshold value used to correct the flux tower data at Santarem LBA site (Easterly Amazon Region—Brazil), the area can change from carbon sink to neutral or carbon source to the atmosphere.

The transport of CO_2 by advection process has been suggested by several studies as the principle reason for the "missing" CO_2 at night [12–17]. The search for this missing CO_2 has spurred a great deal of research with the goal of explicitly estimating advective fluxes in field experiments

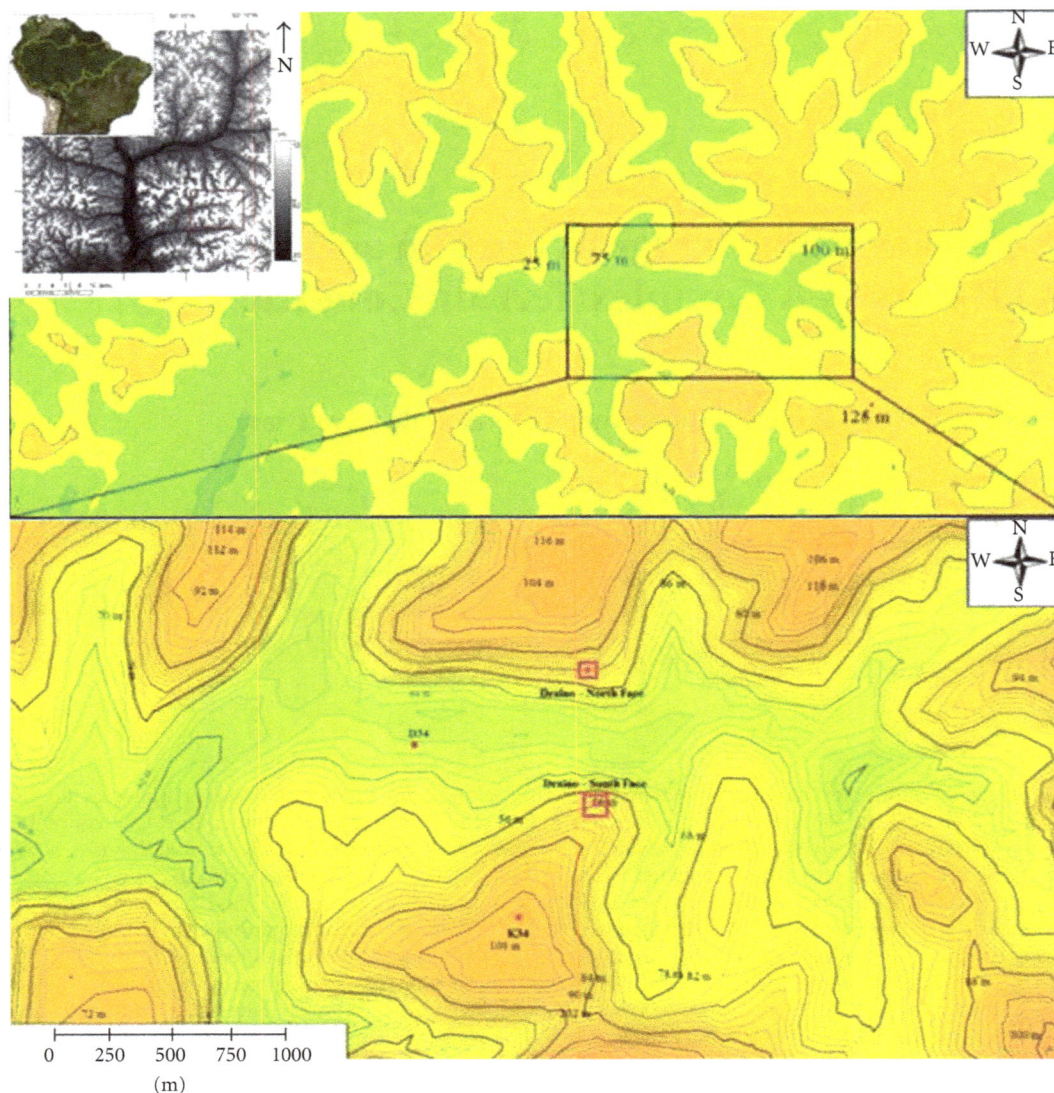

Figure 1: Detailed measurements shows towers' view in the ZF-2 Açu catchment (East-West valley orientation) from SRTM-DEM datasets. The large view in the above panel and below panel: the points of measurements (B34-Valley, K34-Plateau, and subcanopy DRAINO system measurements over slopes in south and north faces (red square).

during the last decade, in order to correct the NEE bias over single tower eddy covariance measurements [8, 15–25].

The complexity of topography and the presence of the valley close to the eddy flux tower have increased the importance to investigating if subcanopy drainage flow account for the underestimation of CO_2 respiration as past studies have asserted [26]. The Manaus LBA site (Central Amazon Region—Brazil) is an example of moderately complex terrain covered by dense tropical forest. The NEE bias is reported by preview works [27, 28, and references therein], and a possible explanation for this is that advection process is happening in that site. This work examines subcanopy flow dynamics and local microcirculation features and how they relate to spatial and temporal distribution of CO_2 on the Manaus LBA Project site. The contribution of exchange of carbon between the atmosphere and the tropical Amazon Rainforest is discussed and correlated with the present work.

2. Material and Methods

2.1. Site Description. The study site is located in the Cuieiras Biological Reserve (54° 58′W, 2° 51′S), controlled by National Institute for Amazon Research (INPA), about 100 km northeast from the Manaus city. At this site, named K34, was implemented a flux tower with 65 m height to monitor long term microclimate, energy, water, and carbon exchanges [29], and various studies have been conducted in its vicinity. The measurements are part of the Large-Scale Biosphere-Atmosphere experiment in Amazonia. Figure 1 presents the study site location including the topographical patterns where the maximum elevation is 120 m and the total area (upper panel) is 97.26 km^2, with distribution of the 31% of plateau, 26% of slope, and 43% of valley [30]. The site area is formed by a topographical feature with moderately complex terrain including a landscape with mosaics of

(a)

(b)

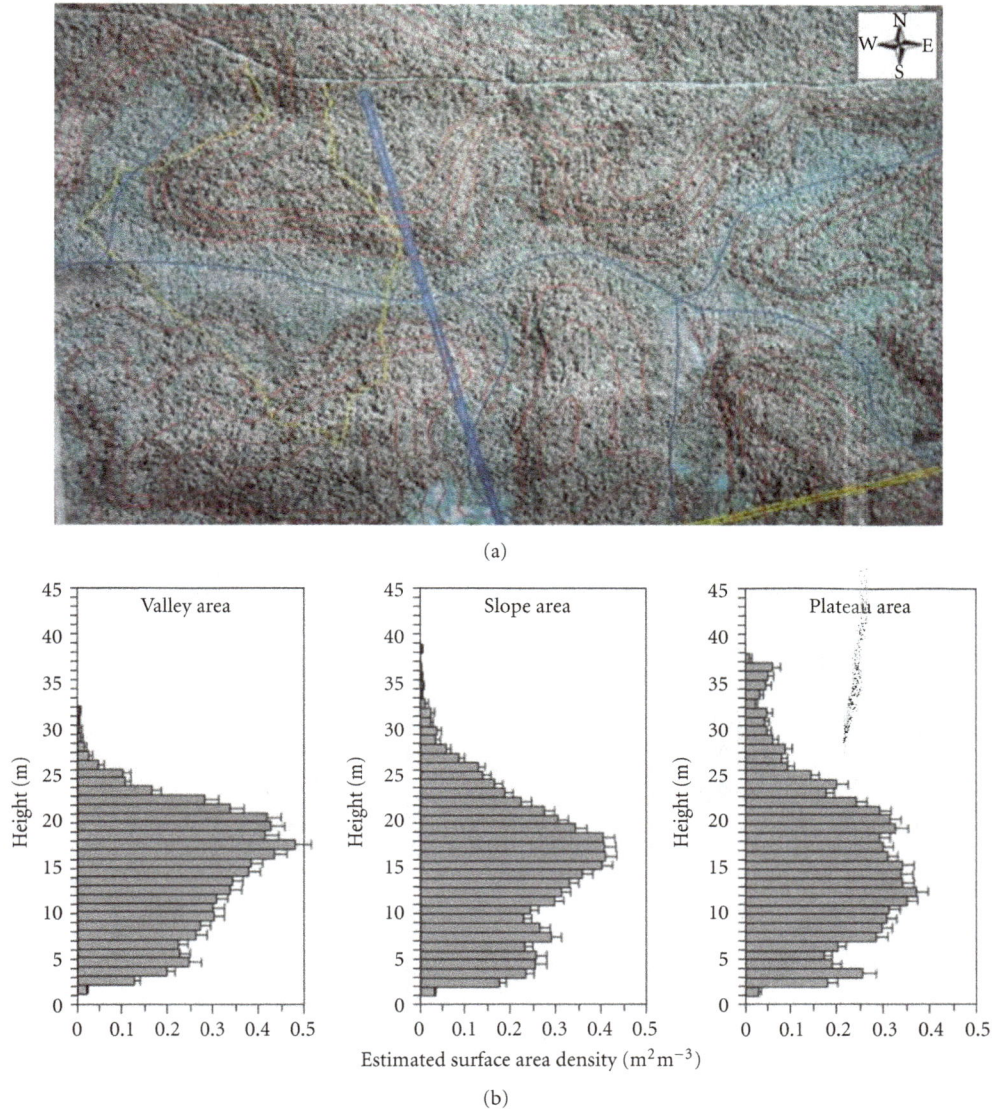

FIGURE 2: (a) IKONOS's image of the site at the Açu catchment with level terrain cotes and vegetation cover and (b) vegetation structure measured from LIDAR sensor over yellow transect in (a). From (a) the valley vegetation (blue color) and vegetation transition to plateau areas (red colors).

plateau, valley and slopes, with elevation differences about 50 m (Figure 1) and with distinct vegetation cover (Figure 2). The eddy flux tower at the Manaus K34 site has footprints that encompass this plateau-valley mosaic.

The vegetation cover on the plateau and slope areas is composed by tall and dense terra firme (nonflood) tropical forest with height varying between 30 to 40 m, maximum surface area density of $0.35 \, m^2 \, m^{-3}$ (Figure 2(b), see also [31], and average biomass of 215 and 492 ton.ha^{-1} [32, 33]).

On the valley area, the vegetation is open and smaller with heights from 15 to 25 m, but with significant surface area density more than the $0.35 \, m^2 \, m^{-3}$ (Figure 2(b)). The soil type on the plateau and slopes area is mainly formed by oxisols (USDA taxonomy) or clay-rich ferralsols ultisols (FAO soil taxonomy), while, on the valley area, waterlogged podzols (FAO)/spodosols (USDA) with sand soil low drained

predominate. Also, in the valley area the presence of small patchy of *campinarana* typical open vegetation with low biomass is also common [34].

The precipitation regime on the site shows wet (December to April) and dry (June to September—less than $100 \, mm \cdot month^{-1}$) periods. The total annual rainfall is about 2400 mm and the average daily temperature is from 26 (April) to 28°C (September). For more detailed information about the meteorology and hydrology of this site, see Waterloo et al. [35] and Cuartas et al. [36].

2.2. Measurements and Instrumentation. The datasets used in this study include a measurement system to monitor airflow above and below the forest, horizontal gradients of CO_2, and the thermal structure of the air below the canopy, named "DRAINO System" (see [8]). The data used in this study

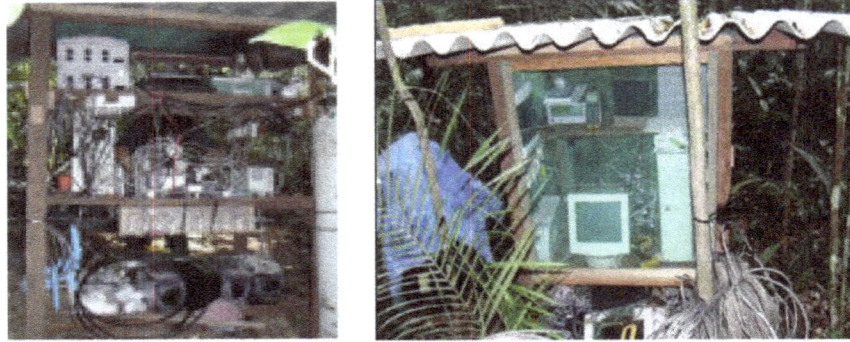

Figure 3: DRAINO measurement system used in Manaus LBA (south face; see also Figure 4).

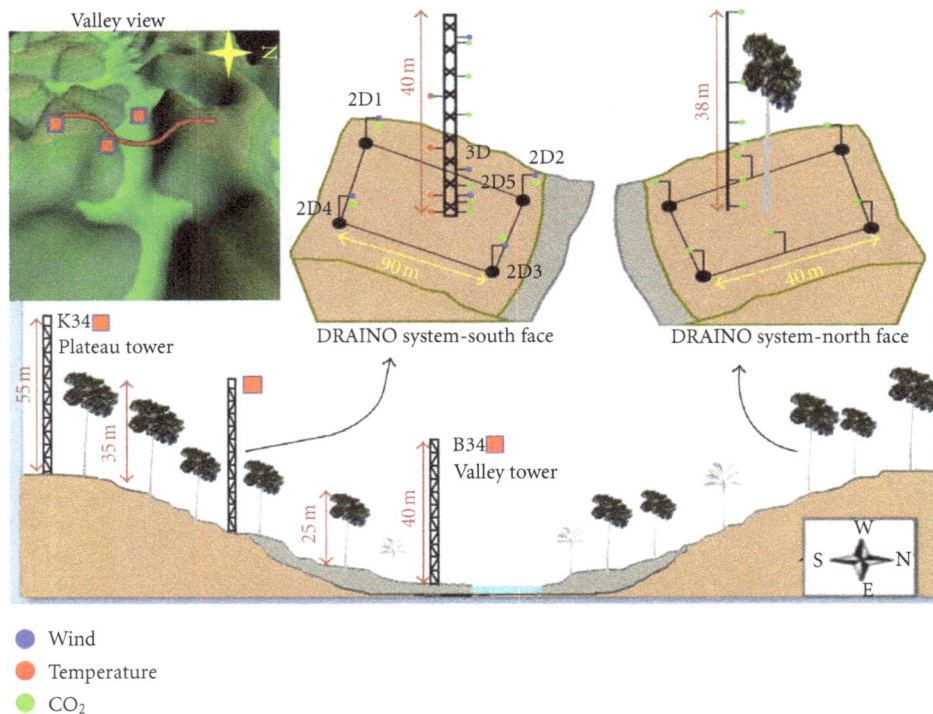

Figure 4: DRAINO measurement system (south and north slope face) implemented at the Manaus LBA site, including topographic view and instrumentation deployed.

were collected during the wet season (DOY 1–151) and the dry season (DOY 152–250) of the year 2006. Complementary information was used from flux tower K34 (LBA tower) on the plateau, and sonic anemometer data collected in the valley flux tower (B34; see [28] for details). The flux tower K34 includes turbulent EC flux and meteorological observations of the vertical profiles of the air temperature, humidity and CO_2/H_2O concentrations, and vertical profile of wind speed, as well as radiation measurements. The fast-response eddy flux data were sampled at 10 Hz and slow response (air temperature and wind profiles) at 30 min average (see [29] for detailed information).

DRAINO Measurement System: Manaus LBA ZF2 site. The DRAINO measurement system used in the Manaus LBA site was similar to that developed by the State University of New York, under supervision of Dr. David Fitzjarrald, and applied at Santarem LBA site, including the same methodological procedures and sampling rates (see [8]). However, due to the terrain complexity, it was modified for the Manaus forest conditions including a long distance power line and duplication of CO_2 observations for different slopes areas (Figure 4). The DRAINO measurement system used in the Manaus LBA site was mounted in an open, naturally ventilated wooden house (Figure 3).

(a) CO$_2$ concentration-3 m

(b) Air temperature at 38 m ($^\circ$C)

(c) Total precipitation (mm)

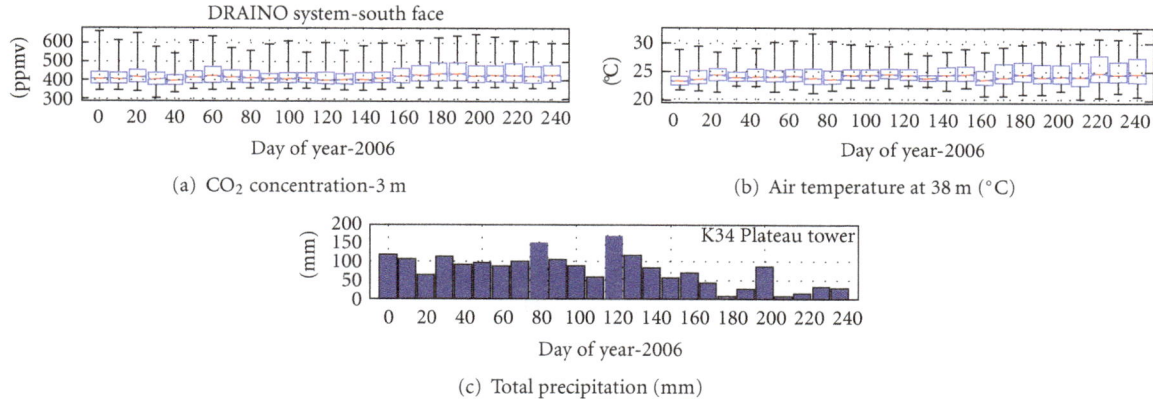

FIGURE 5: 10 days time series of the CO$_2$ concentration (a), air temperature (b) (DRAINO system), and total precipitation (c) (plateau tower).

The system and sensors were deployed (Figure 4) with measurements of air, temperature and humidity (red points), CO$_2$ concentration (green points), and wind speed and direction (blue points), for both south and north faces. The observations of the 3D sonic anemometer were sampled at 10 Hz and all the other parameters (CO$_2$, H$_2$O, air temperature, and humidity) were sampled at 1 Hz (Figure 4).

The acquisition system developed at ASRC was employed [18]. It consists of a PC operating with Linux, an outboard Cyclades multiple serial port (CYCLOM-16YeP/DB25) collecting and merging serial data streams from all instruments in real time, the data being archived into 12-hour ASCII files. At Manaus LBA Site two systems in the both south and north valley slope faces were mounted (Figures 3 and 4).

For each slope face, a single LI-7000 Infrared Gas Analyzer (LI-COR Inc., Lincoln, NE, USA) was used. A multiposition valve (VICI Valco Instrument Co., Inc.) controlled by a CR23x Micrologger (Campbell Scientific, Inc., Logan, Utah, USA), which also monitored flow rates was also used. This procedure minimizes the potential for systematic concentration errors to obtain the horizontal and vertical profiles. Following Staebler and Fitzjarrald [17] and Tóta et al. [8] a similar field calibration was performed during the observations at the Manaus LBA site, including initial instrument intercomparison.

The result was similar to that obtained by Tóta et al. [8], with CO$_2$ mean standard error < 0.05 ppm and mean standard error of about 0.005 m s^{-1} for wind speed measurements. After intercomparison, the sonic anemometers and the CO$_2$ inlet tubes were deployed as shown in Figure 4.

On the south face, the instrument network array (Figure 4 and Table 1) consisted of 6 subcanopy sonic anemometers, one 3-D ATI (Applied Technologies Inc., CO, USA) at 2 m elevation in the center of the grid (named 3D ATI), and 5 SPAS/2Y (Applied Technologies Inc., CO, USA), 2-component anemometers (1 sonic at 6 m in the grid center and 4 sonic along the periphery at 2 m; see Figure 4), with a resolution of 0.01 m s^{-1}. Also, a Gill HS (Gill Instruments Ltd., Lymington, UK) 3-component sonic anemometer was installed above the canopy (38 m). The horizontal gradients of CO$_2$/H$_2$O were measured in the array at 2 m above ground, by sampling sequentially from 4 horizontal points surrounding the main tower location at distances of 70–90 m and from points at 6 levels on the main DRAINO south face tower, performing a 3-minute cycle. On the north face, similar CO$_2$ measurements were mounted including a 6 level vertical profile and 6 points in the array at 2 m above ground, performing a 3-minute cycle.

On both slope faces the air was pumped continuously through 0.9 mm Dekoron tube (Synflex 1300, Saint-Gobain Performance Plastics, Wayne, NJ, USA) tubes from meshed inlets to a manifold in a centralized box. A baseline air flow of 4 LPM from the inlets to a central manifold was maintained in all lines at all times to ensure relatively "fresh" air was being sampled. The air was pumped for 20 seconds from each inlet, across filters to limit moisture effects. The delay time for sampling was five seconds and the first 10 seconds of data were discarded. At the manifold, one line at a time was then sampled using an infrared gas analyzer (LI-7000, Licor, Inc.). To minimize instrument problems, only one LI-7000 gas analyzer sensor, for each slope face, was used to perform vertical and horizontal gradients of the CO$_2$.

3. Results and Discussion

The datasets analyzed in this study were obtained during the periods defined by dry (DOY 1–150 January to June) and wet (DOY 152–250 July to October) seasons of 2006. Figure 5 presents an example of the datasets cover, with 10 days composite statistic, for CO$_2$ concentration and air temperature at south face area of the DRAINO system and the total precipitation on the plateau K34 tower measurements.

The measurements covered almost the entire year of 2006, including dry, wet and the transition from wet to dry season. The air temperature amplitude above canopy on the slope area of the DRAINO system was higher, as expected, in the dry season. A good relationship is observed between CO$_2$ concentration and air temperature with much large amplitudes in the dry season than in the wet season. It probably associates with less vertical mixing during dry than wet season producing much higher subcanopy CO$_2$ concentration and vertical gradient along the forest.

TABLE 1: DRAINO system sensors at ZF2 LBA Manaus site.

Level (m)	Parameter	Instrument
38	$u'\,v'\,w'\,T'$	Gill 3D sonic anemometers
2	$u'\,v'\,w'\,T'$	ATI 3D sonic anemometer
6, 2	$u'\,v'\,w'\,T'$	CATI/2 2D sonic anemometers
2	CO_2 concentration (horizontal array)	LI-7000 CO_2/H_2O analyzer
38, 26, 15, 3, 2, 1	CO_2, H_2O profile (sourth face)	LI-7000 CO_2/H_2O analyzer
35, 20, 15, 11, 6, 1	CO_2, H_2O profile (north face)	LI-7000 CO_2/H_2O analyzer
18, 10, 2, 1	Air temperature and humidity	Aspirated thermocouples

$u'\,v'\,w'$: wind components and T': air temperature fluctuation.

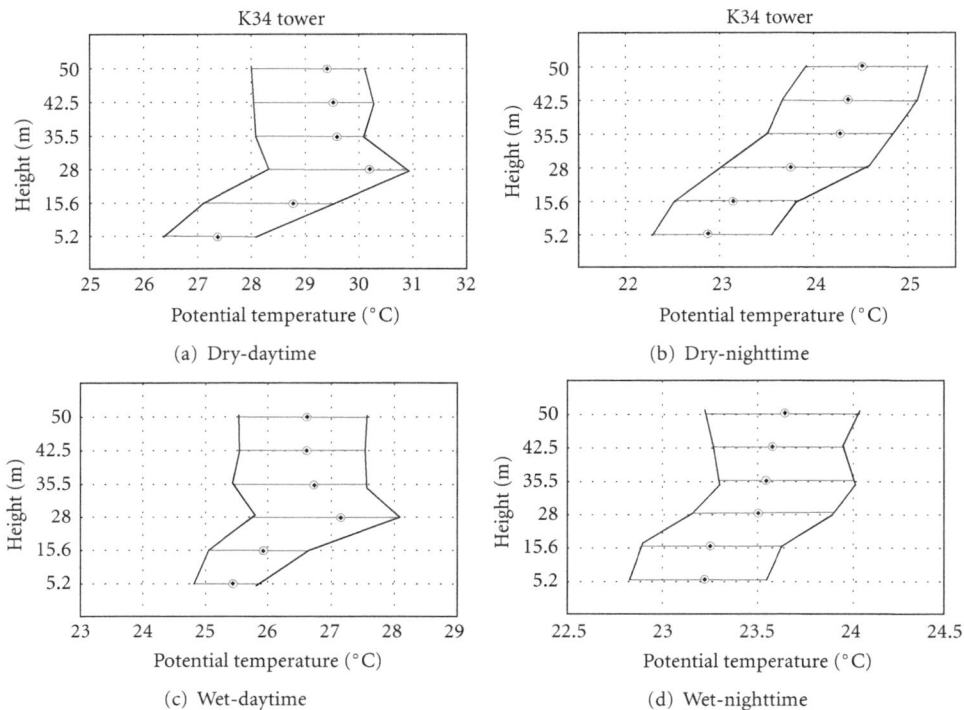

FIGURE 6: Boxplot of the virtual potential temperature vertical profile for dry ((a), (b)) and wet periods ((c), (d)) of the 2006 during night ((b), (d)) and daytime ((a), (c)), on the plateau K34 tower.

3.1. Air Temperature Field

3.1.1. Plateau K34 Tower.
The vertical profiles of air temperature from plateau K34 tower show a very different pattern from that on the slope area, probably due to canopy structure differences (Figure 2(b), [31]). The canopy structure is important for characterizing its thermal regime as it can be seen in Figure 6. The mean canopy layer stores large quantity of heat during the daytime and distributes it downward and upward throughout the nighttime (Figures 6 and 7).

Above the canopy layer, over plateau area, the neutral or unstable conditions were predominant during the daytime for both seasons (Figures 6(a) and 6(c)). During the nighttime, stable conditions dominate during dry period (Figure 6(b)) and neutral-to-stable conditions for the wet period (Figure 6(d)). Similar pattern has been reported elsewhere for plateau forests in the Amazonia [37–40].

The below-canopy layer of ambient air on the plateau area was stable at all times (Figures 6(a), 6(b), 6(c), 6(d)), indicating that this layer is stable where the cold air concentrated in the lower part of the canopy air space as shown in Figure 7.

The Figure 7 presents daily course of the vertical deviation of the virtual potential temperature, for example, $([\theta_v' = \theta_v(z) - \overline{\theta_v(z)}_{5.2}^{55}])$, the temperature differences from each level in relation to the vertical average profile. The subcanopy air space was relatively colder during both dry and wet season, showing a similar feature of strong inversion. The same pattern was reported by Kruijt et al. [39] measured over a tower located 11 km northeast of our site with a similar forest composition.

Note that a very interesting length scale can be extracted from the observation when the deviation is about zero. The

(a) Dry period-2006

(b) Wet period-2006

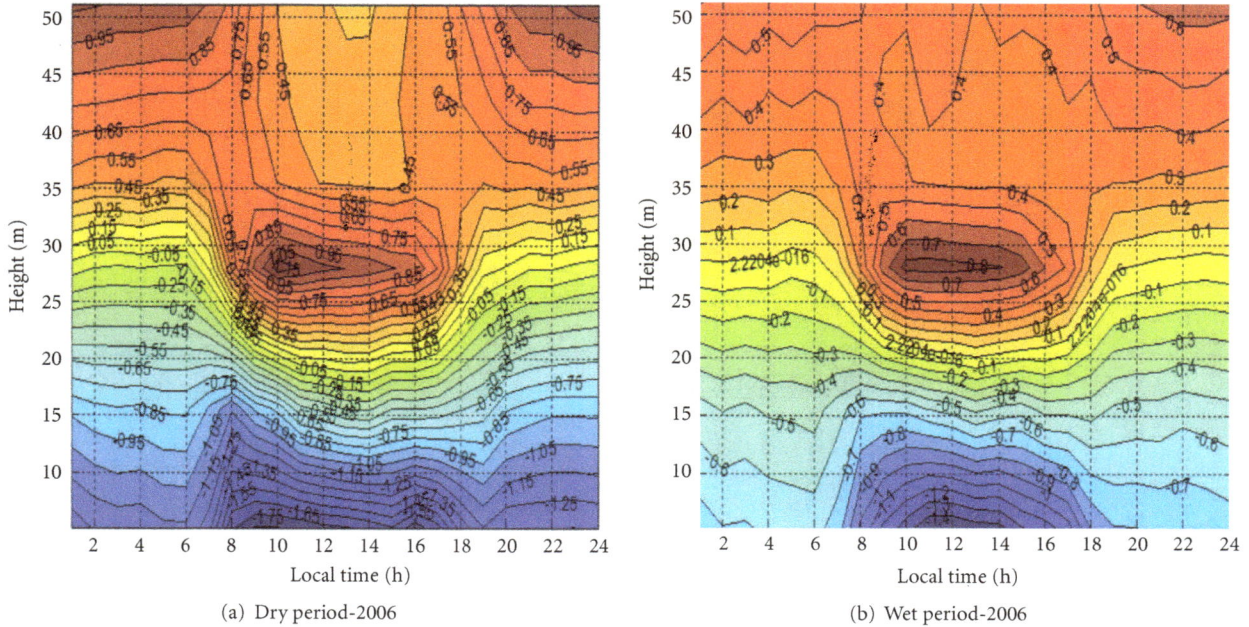

FIGURE 7: Daily course of the vertical deviation of the virtual potential temperature ($[\theta'_v = \theta_v(z) - \overline{\theta_v(z)}^{55}_{5.2}]$), during dry (a) and wet (b) periods of 2006, over plateau K34 tower.

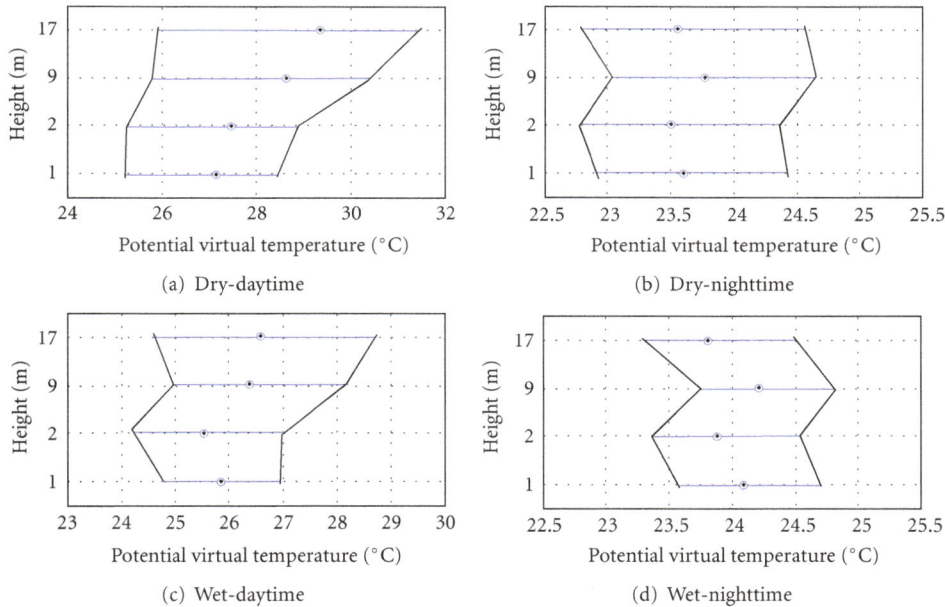

(a) Dry-daytime

(b) Dry-nighttime

(c) Wet-daytime

(d) Wet-nighttime

FIGURE 8: Boxplot of the virtual potential temperature vertical profile for dry ((a), (b)) and wet periods ((c), (d)) of the 2006 during night ((b), (d)) and daytime ((a), (c)), on the slope area of DRAINO System tower (south face; see Figure 2).

vertical length scale has mean value of about 30 m during nighttime and 20 m during daytime (yellow color in the Figures 7(a) and 7(b)). Those values are comparable with above-canopy hydrodynamic instability length scale used in most averaged wind profile models [41–44].

3.1.2. DRAINO System Slope Tower. On the slope area south face (see Figure 2), air temperature at 5 levels underneath

the canopy (heights 17, 10, 3, 2, and 1 m) was measured. The observations of the air temperature profile inside canopy are used to monitor the possible cold or warm air layer that generates drainage flow on the slope area. Figure 8 presents observations of the virtual potential temperature vertical profile for both dry and wet periods, during both day- and nighttime.

The pattern on the slope area is clearly very different when compared with that on the plateau K34 area (Figure 6),

(a) Dry period-2006

(b) Wet period-2006

(c) Vertical Gradient of the Virtual Potential Temperature-DRAINO System south slopE

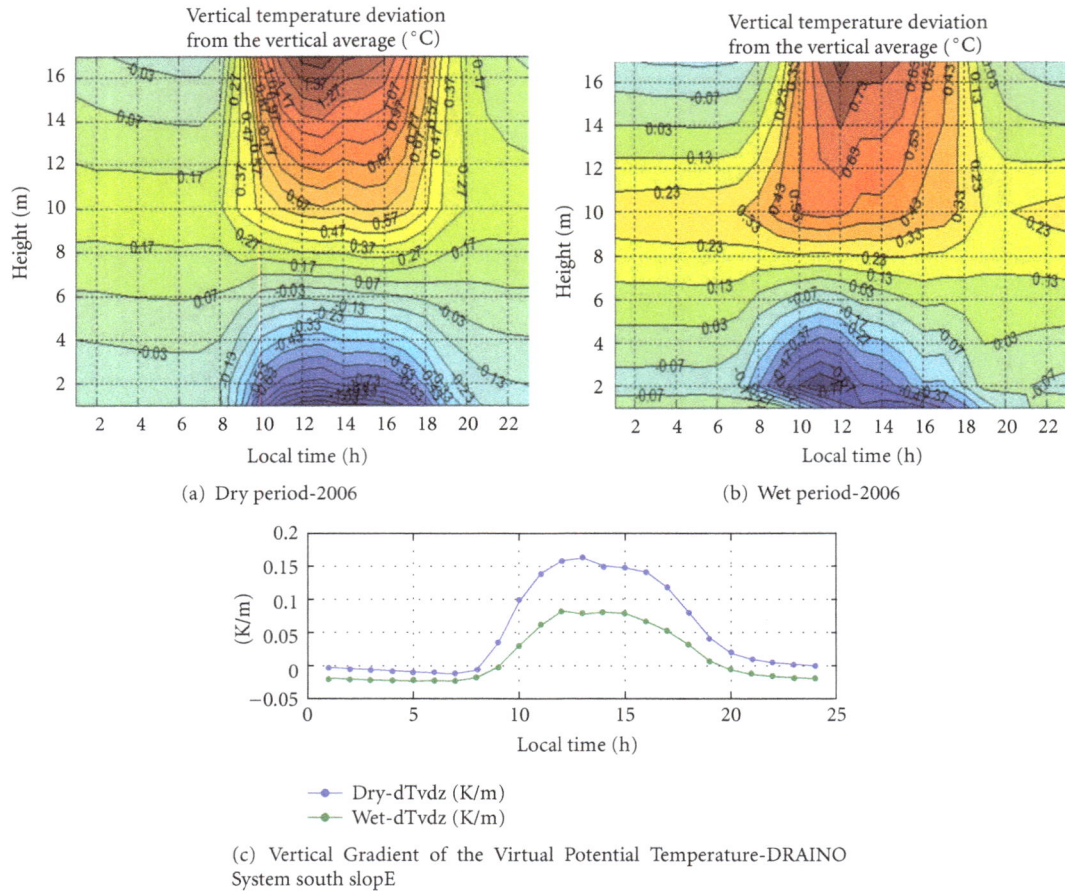

FIGURE 9: Daily course of the vertical deviation of the virtual potential temperature for dry (a) and wet (b) periods of the year 2006, and the virtual potential temperature vertical gradient (c), over slope area DRAINO system tower.

except in dry period during daytime when the air was stable inside the canopy. During nighttime (wet and dry periods) a very stable layer predominates with inversion at about 9 m. These can likely be interpreted as a stable layer between two convective layers is associated with cold air (Figure 8). Yi [25] hypothesized about a similar "*super stable layer*" developing during the night in sloping terrain at the Niwot Ridge AmeriFlux site. This hypothesis suggests that above this layer, vertical exchange is most important (vertical exchange zone) and below it horizontal air flow predominates (longitudinal exchange zone). The relationship between subcanopy thermal structure and the dynamic of the airflow on the slope area will be discussed in the next section.

Figure 9 presents a daily cycle composite of the virtual potential temperature deviation from the vertical average ($[\theta_v(z) - \overline{\theta_v}(z)_1^{18}]$). There is persistent cold air entering during nighttime for both dry and wet periods, a characteristic pattern observed on the slope area. It is a very different vertical thermal structure from that of the plateau area.

The cold air in the subcanopy upper layer is probably associated with top canopy radiative cooling, while the cold air just above floor layer is associated with upslope wind from the valley area (as discussed later in the next section).

The average of the vertical gradients virtual potential temperature was negative during nighttime and positive during daytime for both periods dry and wet (Figure 9). This observation shows that during the daytime a relative cooler subcanopy air layer predominates creating inversion conditions. In contrast, a relative hotter subcanopy air layer generates a lapse conditions during nighttime. In general that is not a classical thermal condition found on the sloping open areas without dense vegetation. This general pattern was present at several specific study cases not shown here due to limited size paper. A similar pattern was reported by Froelich and Schmid [26] during "leaf on" season.

3.2. *Wind Field.* The LBA Manaus site has moderately complex terrain when compared with the Santarem LBA site (Figures 1 and 2). This complexity generates a wind airflow regime much complex to be captured by standard measurement system like a single tower. At the Manaus LBA site, we implemented a complementary measurement system on the slope area to support the plateau K34 tower and better understand how the airflows above and below the canopy interact and also to describe how the valley flow influences the slope airflow regimes. It is important to note that the

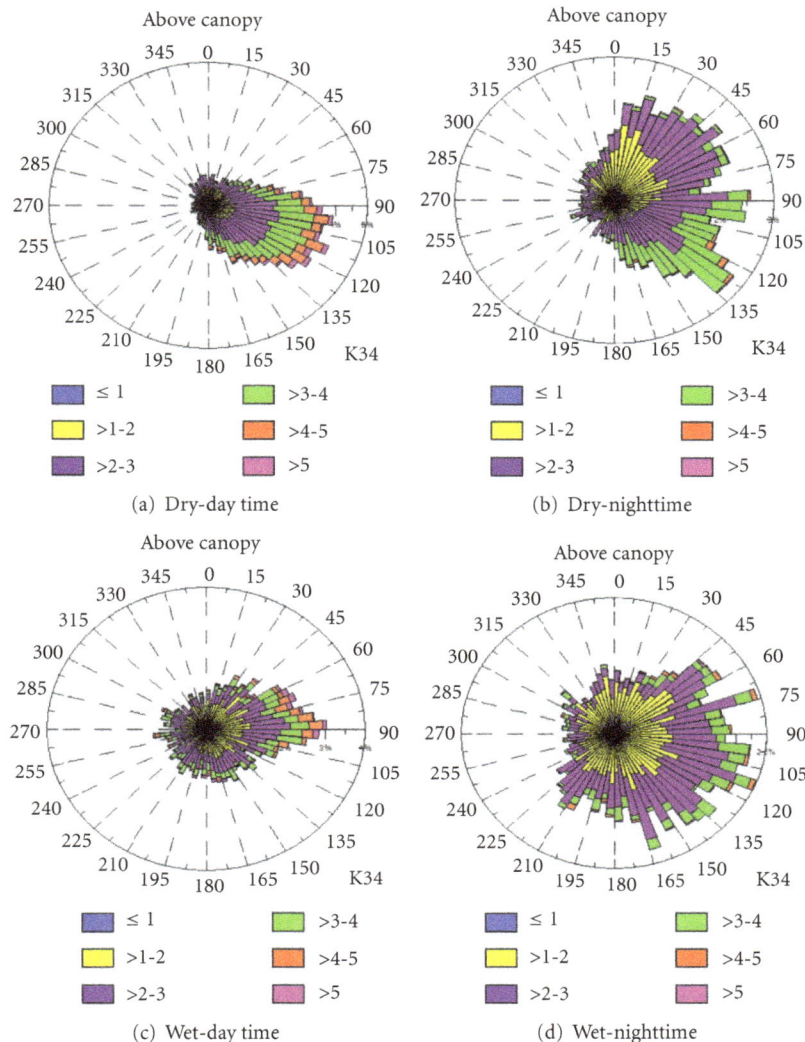

FIGURE 10: Frequency distribution of the wind speed and direction. For dry ((a), (b)) and wet ((c), (d)) periods from the year 2006 during day ((a), (c)) and nighttime ((b), (d)), on the plateau K34 tower.

valley in the microbasin is oriented from east to west (Figures 2 and 4).

3.2.1. Horizontal Wind Regime: Above Canopy

Plateau K34 Tower. Above the canopy (55 m above ground level—a.g.l.) on the plateau area of K34 tower, the wind regime was strongest (most above $2 \, m.s^{-1}$) during daytime for both dry and wet periods of 2006, with direction varying mostly from southeast and northeast for dry and wet period, respectively (Figure 10). During nighttime, the wind regime was slower (most below $3 \, m.s^{-1}$) and with same direction variation from northeast to southeast (Figure 10). As reported by de Araújo [28], the above-canopy valley area's wind speed and direction was different from that of the plateau area, suggesting a decoupling mainly during nighttime. A clear channeling effect on the valley wind regime was observed, which was oriented by microbasin topography during both day- and nighttime, with direction of the flow in the valley area determined by the valley orientation (as also reported by [28]).

DRAINO System Slope Tower. For the above canopy (38 m above ground level—a.g.l.) on the slope area DRAINO system south face (see Figure 4, 3D sonic), the wind regime was very persistent from east quadrant direction during day- and nighttime in both dry and wet periods of the 2006 (Figure 11). The daytime wind speed during the dry season was between 1 and $3 \, m \, s^{-1}$ and much stronger during the wet period with values up to $4 \, m \, s^{-1}$.

During the nighttime the wind speed was slower than $2 \, m \, s^{-1}$, except from northeast during the wet period. The wind direction pattern was similar to that on the plateau K34 tower (Figure 10) prevailing from northeast to southeast. This observation indicates that the airflow above the canopy

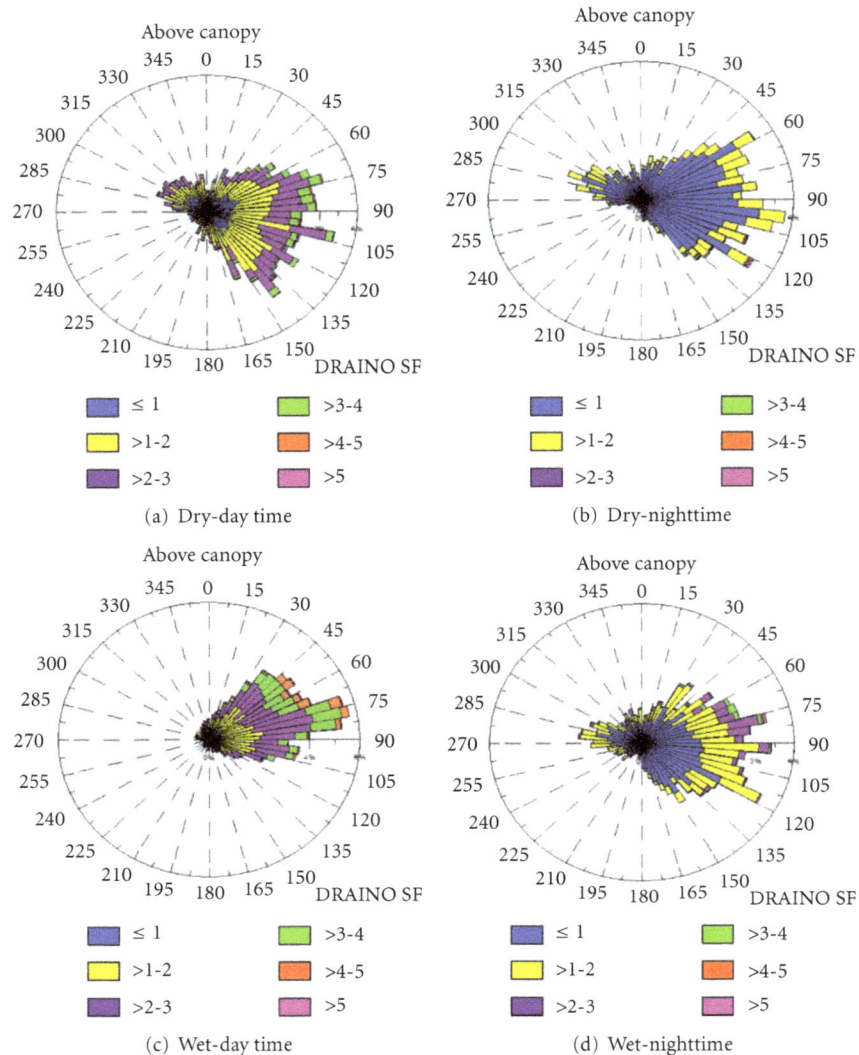

FIGURE 11: Frequency distribution of the wind speed and direction above canopy (38 m above ground level-a.g.l). For dry ((a), (b)) and wet ((c), (d)) periods from the year 2006 during day ((a), (c)) and nighttime ((b), (d)), on the slope area at DRAINO system tower.

on the slope area is related to how the synoptic flow enters in the eastern part of the microbasin (see Figures 2 and 4).

3.2.2. Horizontal Wind Regime: Subcanopy Array Measurements (2 m a.g.l). In Figure 12, the subcanopy array frequency distribution of the wind speed and directions is shown for both dry and wet periods of the year 2006, during both day- and nighttime. The observations show that the airflow in the subcanopy is very persistent and with similar pattern during both dry and wet periods of the year 2006. It is important to observe that the south slope area in the DRAINO System (see Figure 4) is *downslope* from south and *upslope* from north quadrants.

Subcanopy Daytime Wind Regime. During daytime, in both dry (Figures 12(a), 12(b), and 12(c)) and wet periods (Figures 12(g), 12(h), and 12(i)), the wind direction prevailed

from south-southeast (190–150 degrees) on the three slope regions (Figure 12, Top ((a), (d), (g), (j)), Middle ((b), (e), (h), (k)) and Low slope part ((c), (f), (i), (l))). The airflow in the subcanopy was decoupled from the wind regime above the canopy (Figure 11) most of the time. The wind direction in the subcanopy airflow was dominated by a daytime downslope regime during the majority of the period of study, suggesting a systematic daytime katabatic wind pattern.

The wind speed in the subcanopy during the daytime was mostly from 0.1 to 0.4 m/s, and strongest at middle slope region (Figure 12(b), 12(e), 12(h), 12(k)) about 0.3 to 0.5 m/s or above. A similar daytime katabatic wind regime was reported by Froelich and Schmid [26] during "leaf on" season in Morgan-Monroe State Forest (MMSF), Indiana USA.

The daytime downslope wind was also supported by the subcanopy thermal structure (Figure 9), where the air

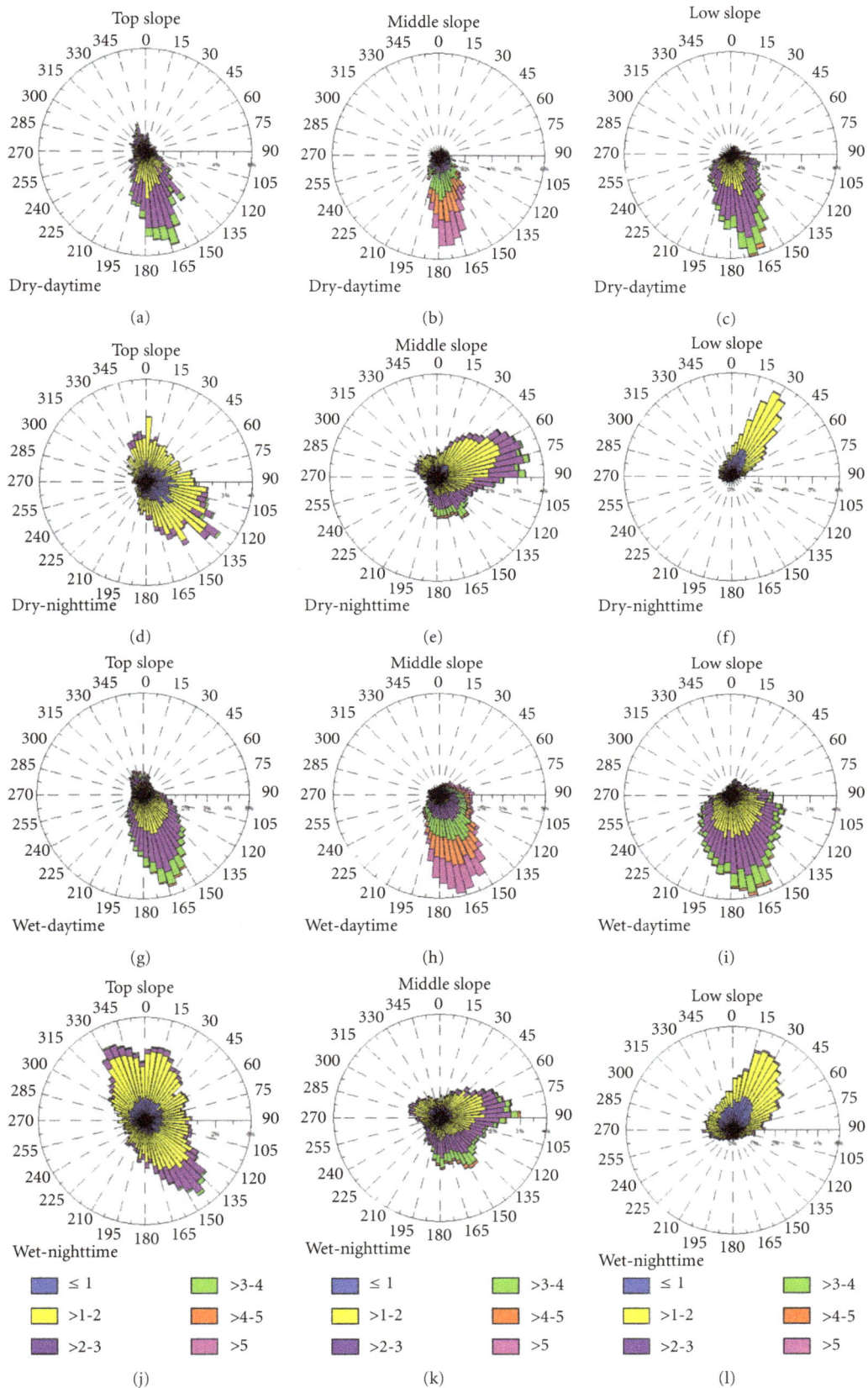

FIGURE 12: Frequency distribution of the wind speed and direction in the subcanopy array (2 m above ground level—a.g.l) on the microbasin south face slope area at DRAINO horizontal array system (see Figure 4). For dry (a–f) and wet (g–l) periods from the year 2006, during day-((a), (b), (c), (g), (h), (i)) and nighttime ((d), (e), (f), (j), (k), (l)).

(a)   (b)

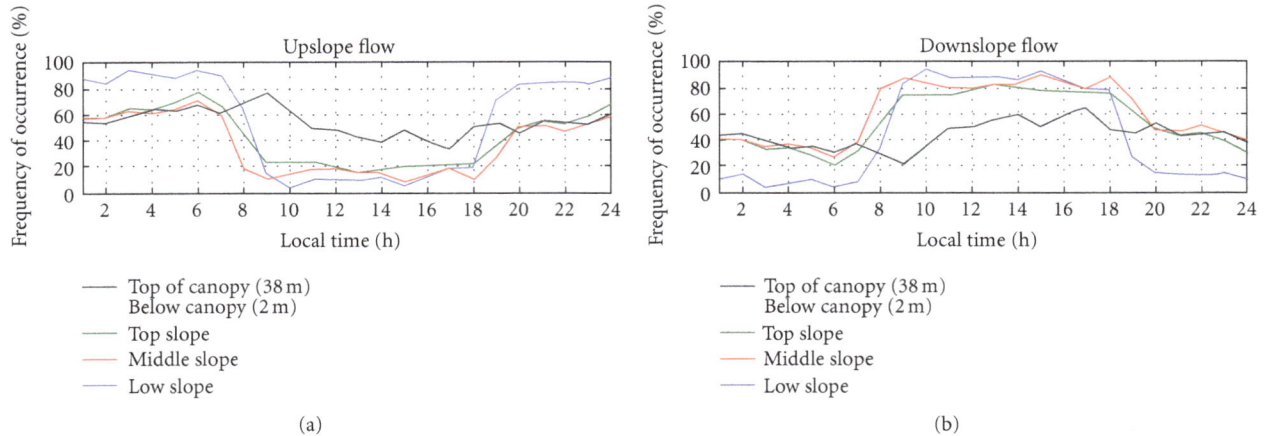

FIGURE 13: Frequency distribution of the subcanopy wind direction (a) upsloping (from north quadrant) and (b) downsloping (from south quadrant) on the south face slope area at the DRAINO horizontal array system (see Figure 4).

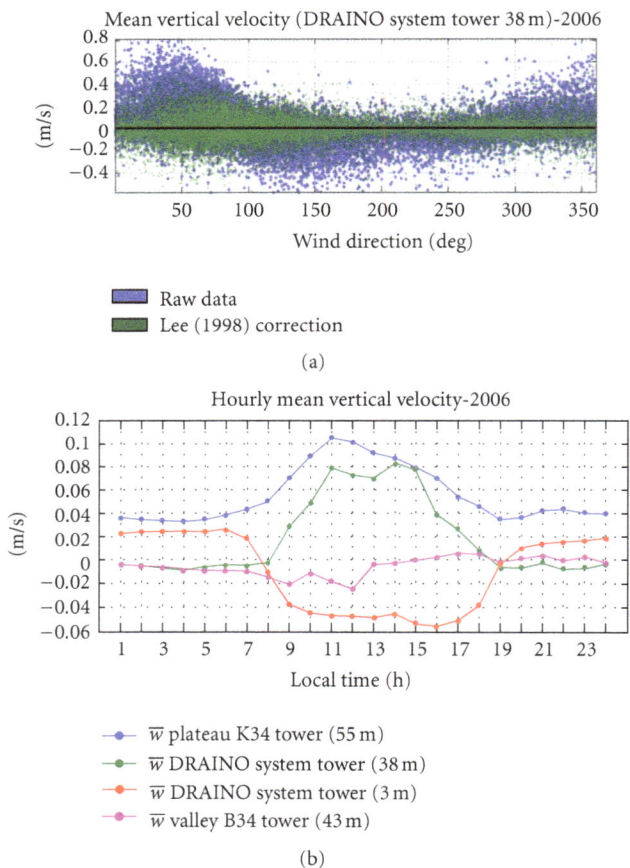

(a)

(b)

FIGURE 14: Mean vertical velocity raw and correct vertical velocity (a) for DRAINO system slope tower (38 m), and hourly mean vertical velocity (b) for: plateau K34 tower (55 m), DRAINO system slope tower (above canopy 38 m and subcanopy 3 m) and for valley B34 (43 m) towers (see Figure 4, for details).

was cooling along the day by inversion of the virtual potential temperature profile with a positive vertical gradient (Figure 9(c)). This results shows that subcanopy flows in a

sloping dense tropical rainforest are opposite to the classical diurnal patterns of slope flows studied elsewhere in the literature (e.g., [45–48]). It is important to note that few studies have been done in forested terrain and it is unclear why similar reversed diurnal patterns have not been observed in studies at other forested sites [15, 17, 49], except by a single point subcanopy measurement observed by Froelich and Schmid [26].

Subcanopy Nighttime Wind Regime. The nighttime subcanopy wind regime on the slope area (see the terrain on Figure 4) was very complex and differentiates from that one above the canopy vegetation. It was observed that, on the upslope part, the nighttime airflow was southeast downsloping direction ($130°-170°$) and northeast-northwest ($45°-340°$) uphill direction (Figures 12(d) and 12(j)). In the middle-part of slope area, the wind moved uphill (from northeast; $30°-90°$) and also downsloping wind direction from southeast (Figures 12(e) and 12(k)), and with lightly higher wind speed. On the lower part of the slope area (Figures 12(f) and 12(l)) the wind direction prevailed from the northeast ($10°-70°$), indicating upsloping pattern (anabatic). It is interesting to note that, on the up-slope area, the wind direction regime (northeast-northwest, $45°-340°$) suggest a reversal lee side airflow (recirculation or separation zone) probably in response to the above canopy wind (see Figures 11(b) and 11(d)). It has been suggested by Staebler [50] and reported by simulations using fluid dynamic models [51, 52].

The upsloping subcanopy flows pattern, on the lower part of the slope area, is supported by the subcanopy's relative heat air layer along the slope during the night, as observed by lapse rate condition of the virtual potential temperature negative vertical gradient (Figure 9(c)). This observation does not follow the classical concept of nighttime slope flow pattern, as commented previously (Section 3.1.2) this is an example of nonclassical microscale slope flow. Froelich and Schmid [26], has reported similar feature where they found anabatic wind regime during nighttime in their

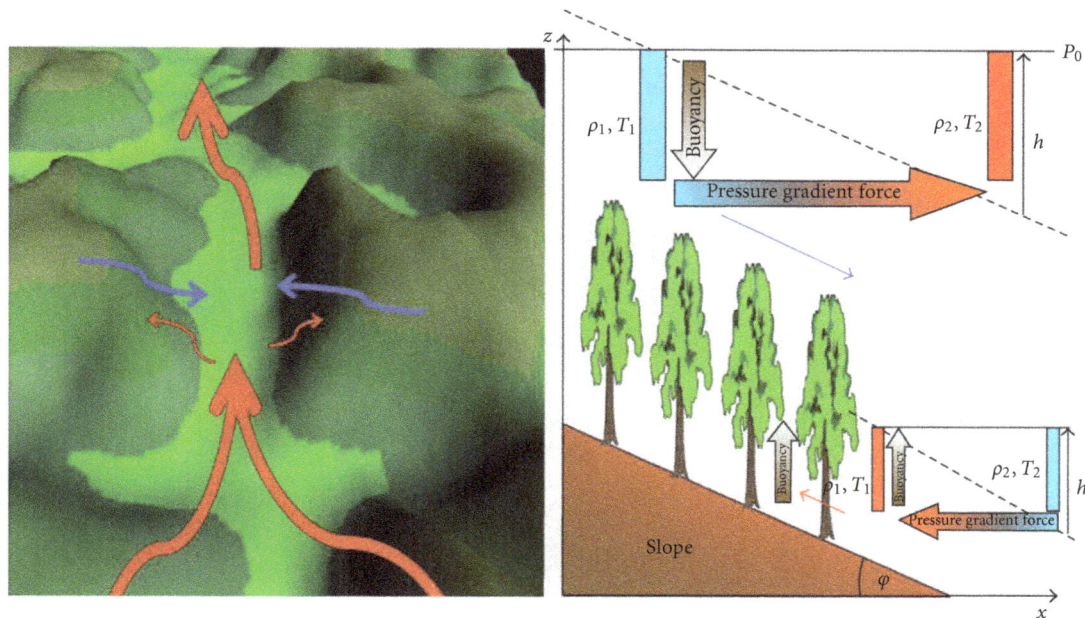

(a)

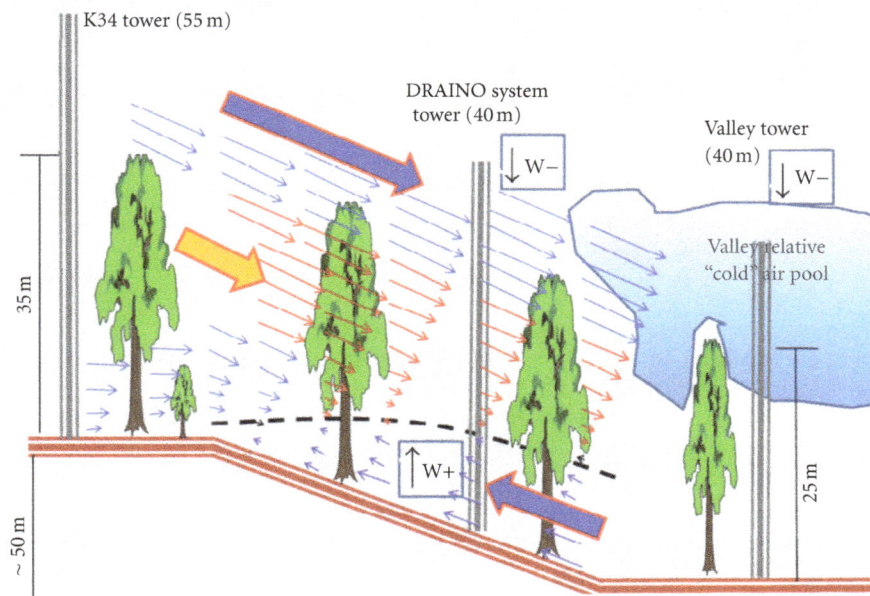

(b)

FIGURE 15: Schematic local circulations in the site studied, valley and slopes flow (a), 2D view from suggested below and above canopy airflow (b).

seasonal forest study area. Figure 13 presents the frequency distribution of the subcanopy wind direction on the south face slope area at the DRAINO horizontal array system during upsloping (from north quadrant) and downsloping (from south quadrant) events.

3.2.3. Mean Vertical Wind Velocity: Subcanopy and Above Canopy. Several correction methods have been proposed to calculate the mean vertical velocity, for example, linear regression method [12], coordinate rotation [53], and the planar fit method [54]. We use the linear regression method by Lee [12] to determine the "true" mean vertical velocity: $\overline{w} = w - a(\alpha_i) - b(\alpha_i)u$, where "$a$" and "$b$" are coefficients to be determine, for each α_i (10° azimuthal wind direction), by a linear regression of measured mean vertical velocity (w) and horizontal velocity (u) in the instrument coordinate system. Figure 14(a) presents the original and the correction results by method application of the mean vertical velocity

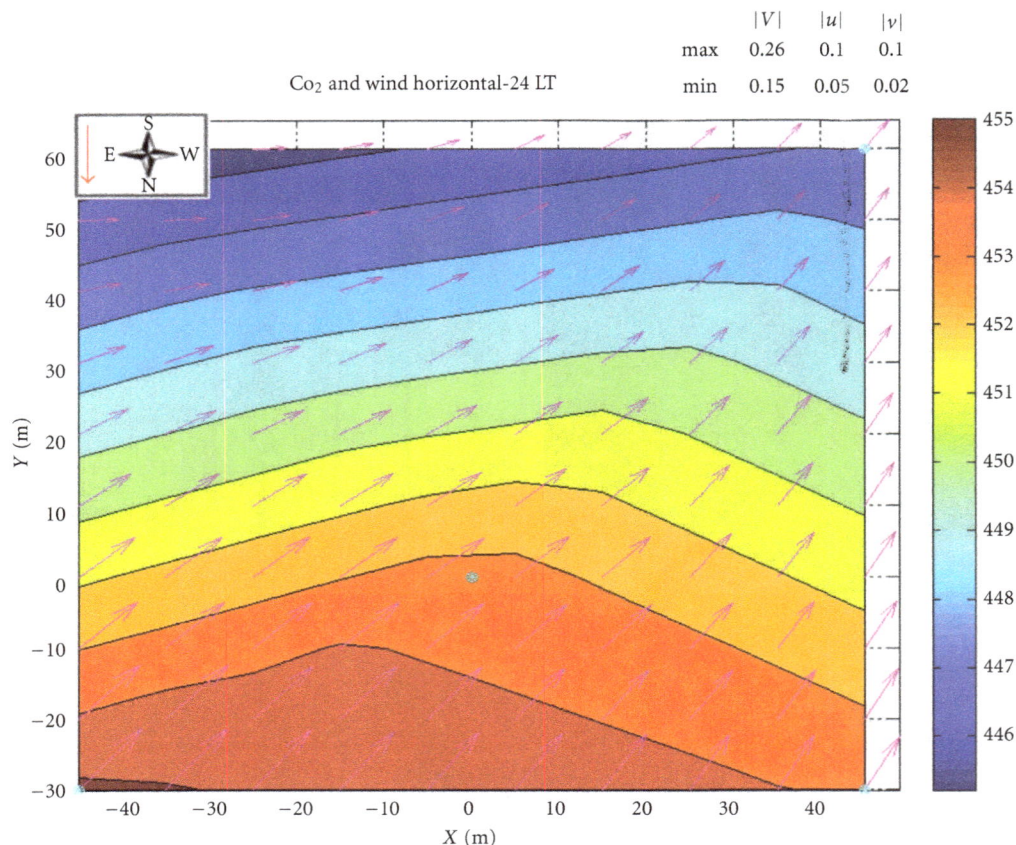

FIGURE 16: Example at midnight (local time) of the horizontal CO_2 concentration (ppmv) over the DRAINO System south face domain including an interpolated horizontal wind field (10 m grid). Note the geographic orientation and the red arrow indicating slope inclination (see Figure 4).

as function of wind direction. In Figure 14(b), the results of the hourly mean vertical velocities for plateau K34, DRAINO system (above and below canopy), and valley B34 towers are shown. As expected, not only low but non-zero values were observed for all points of measurements as well.

On the plateau area, the mean vertical velocity was always positive indicating upward motion or vertical convergence at the top of the hill during night- and daytime. In the valley area during nighttime, negative or zero values were observed, indicating a suppression of vertical motion (mixing) in the valley, as also reported by de Araújo [28].

However, during the daytime, a transition is observed, where beginning in the morning, downward motion is observed, changing after midmorning to upward motion (Figure 14(b)). This suggests that probably the cold air pooled during night moved downslope and started to warm, resulting in a breakdown the inversion over the valley (see [28], for detailed description and references therein for this process). The mechanism of the breakdown, the inversion process over the valley, is consistent with positive vertical velocity observed above canopy at slope area by the DRAINO system tower during daytime (Figure 14(b)).

The subcanopy diurnal pattern of the mean vertical velocity observed shows positive values during nighttime

and negative during daytime, consistent with observed up- and downsloping flow regime, respectively (Figures 13(a) and 13(b)). This is consistent with thermal vertical virtual potential temperature gradient on the slope (see Figure 9(c)), where, during nighttime (daytime), an unstable (inversion) condition is associated with upward (downward) mean vertical velocity (see Figure 9(c)).

3.3. Phenomenology of the Local Circulations: Summary.
Figure 15 shows a schematic cartoon of local flow circulation from the previews sections observations.

In Figure 15(a) is shown the above canopy airflow over valley space (red arrow) and correspondingly (induced) the most probable airflow above canopy over slope areas (blue arrow). In the same figure is shown the main physical mechanisms (pressure gradient force) producing that microscale circulations. The observations result from preview sessions suggesting that the balance of the buoyancy and pressure gradient forces generate the airflow or microcirculation patterns in the site studied.

During nighttime (Figure 15(b)), in the subcanopy, there is an upslope flow reaching about 10 m height above the ground, associated with positive mean vertical velocity (indicating upward movement). Also, above canopy, there

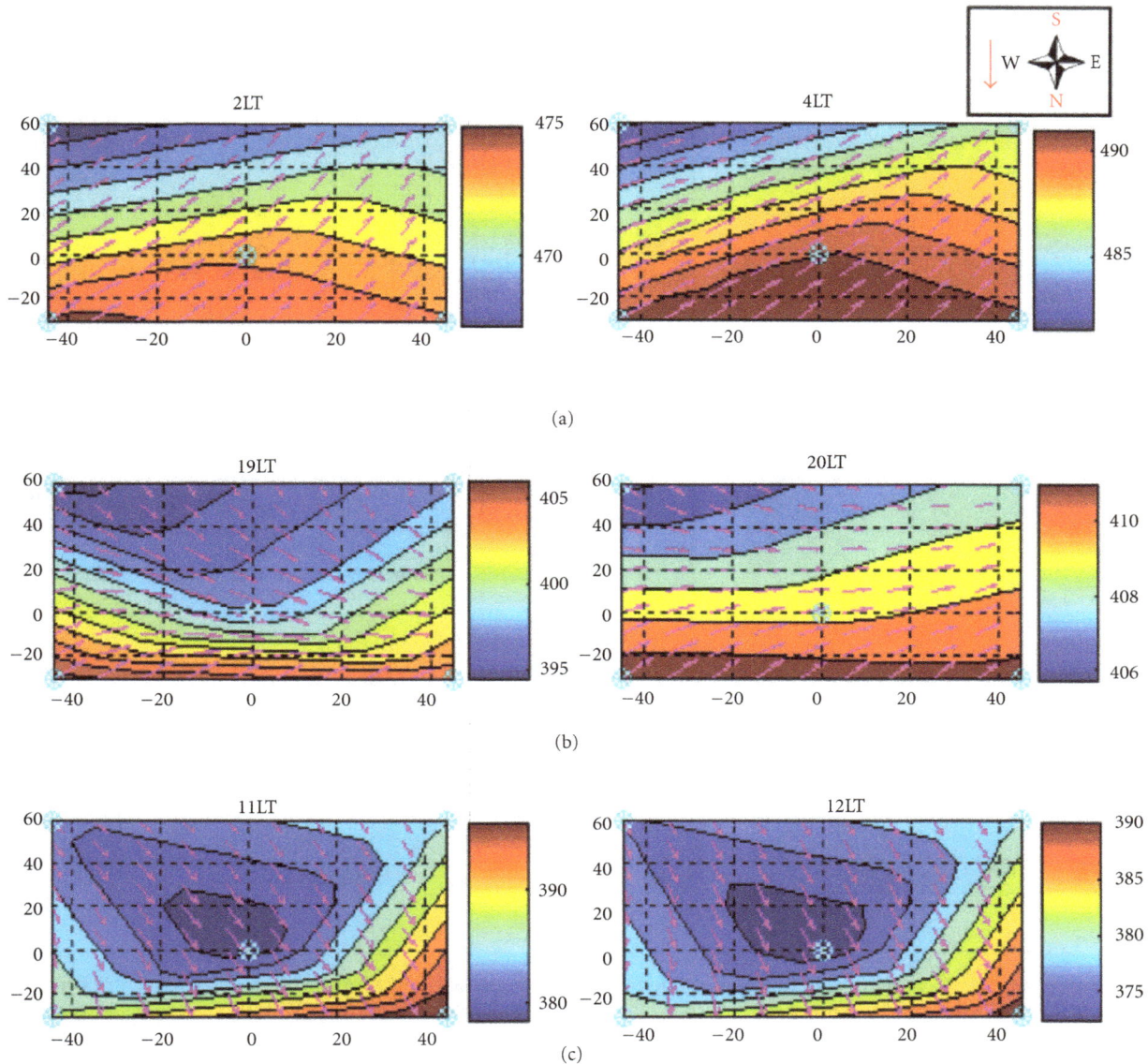

FIGURE 17: Hourly average of the subcanopy (2 m) CO_2 concentration and horizontal wind speed over the DRAINO system south face area during dry period of the year 2006 note the geographic orientation and the red arrow indicating slope inclination (see Figure 4). The axis represents distances from center of the main tower. Daytime (a), transition period—evening (b), established nighttime (c).

is a downslope flow associated with negative mean vertical velocity, with downward convergence above the canopy. The microcirculation along the plateau-slope valley is promoted by a feedback mechanism of accumulation of cold air drainage above canopy into the valley center (Figure 15(b)), creating the forcing needed to sustain nighttime pattern. The air temperature structure above canopy in the valley (see [28]) is a good indication of cold air pool in the center of the valley. Maybe the local pressure gradient force due to the cold air accumulation promotes the upward airflow in both the slopes of the valley. During daytime periods, an inverse pattern is found (not show), indicating that this microcirculation is a systematic pattern in the site.

3.4. CO_2 Concentration and Subcanopy Horizontal Wind Field.
The CO_2 concentration was measured by DRAINO system on the south face slope area for dry and wet periods of the year 2006, and on the north face slope during the dry period (Figure 4). The Figure 16 presents an example, for midnight (local time), of the horizontal wind field and spatial CO_2 concentration over the DRAINO System south face domain.

The wind field was interpolated from the blue points onto a 10 m grid. Similar procedures have been reported in the literature [19, 55]. The horizontal wind regime plays an important role in modulating the horizontal spatial distribution of CO_2 concentration (Figure 16).

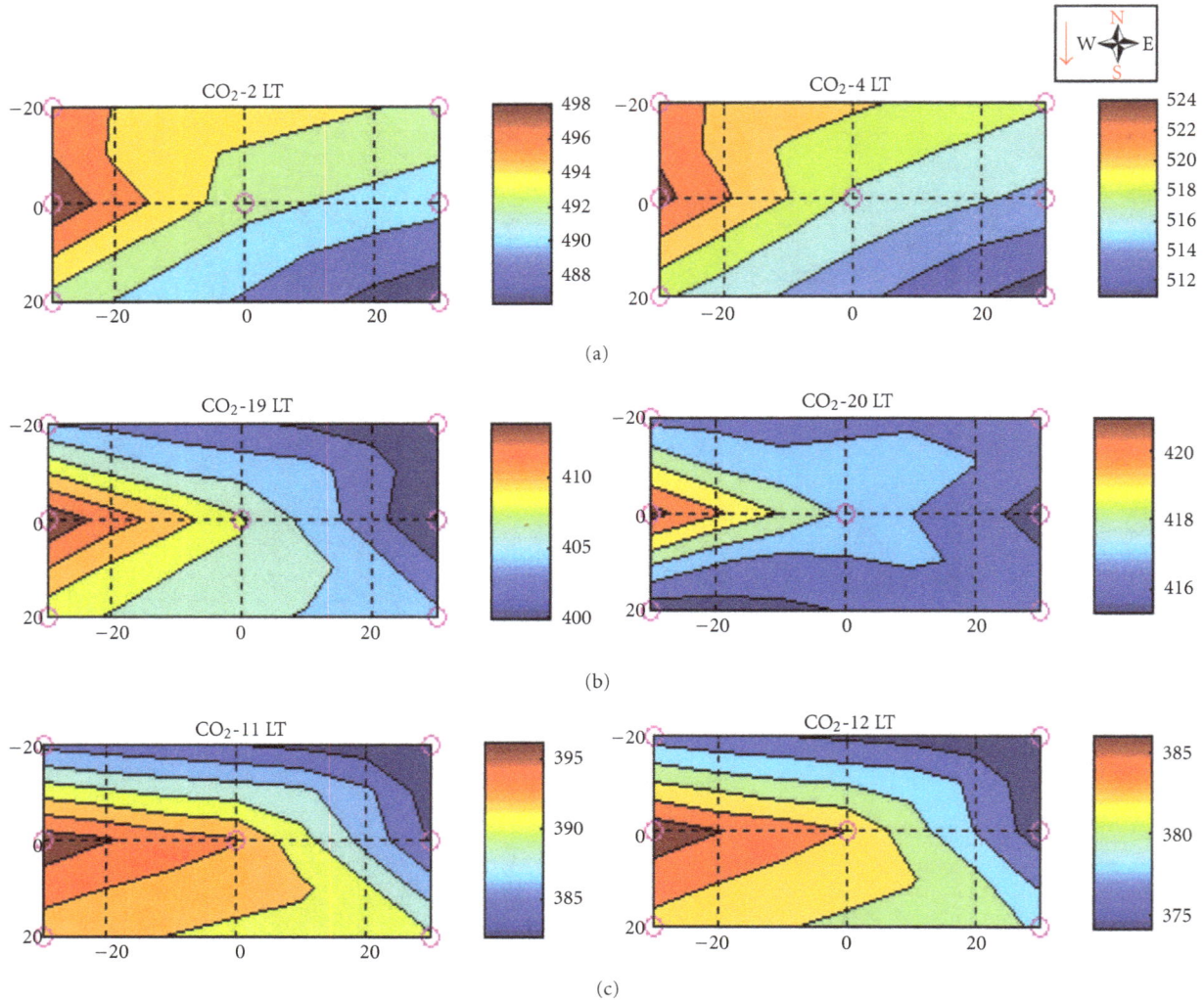

FIGURE 18: Hourly average of the subcanopy (2 m) CO_2 concentration on the DRAINO system north face area during dry period of the year 2006. Note the geographic orientation and the red arrow indicating slope inclination (see Figure 4). The axis represents distances from center of the main tower. Daytime (a), transition period-evening (b), established nighttime (c).

In Figures 17(a), 17(b), and 17(c) the typical pattern observed is shown for both dry and wet periods of the year 2006 measured by the DRAINO system on the south-facing slope area. During the daytime (Figure 17(c)), the wind prevailed downslope inducing a strong horizontal gradient of CO_2 in the slope area (about $0.2 \, \text{ppmv} \, \text{m}^{-1}$). In the evening, periods of changes of the horizontal wind pattern (as described in Section 3.1) show an upsloping regime in the lower part and downsloping in the upper part of the slope areas (Figure 17(b)). The wind regimes produce direct responses in the spatial feature of the horizontal gradient of CO_2 concentration. Later during the night, the upsloping regime is well established and also the horizontal gradient of CO_2 is growing from lower part of the slope to the top (Figure 17(a)).

These observations suggest a subcanopy drainage flow and its influence on the scalar spatial distribution. Therefore, as discussed in the previews' sections, the flow above the

canopy indicates a reverse pattern of downward motion (negative mean vertical velocity; see Section 3.2.3) that suggests vertical convergence and possible horizontally divergent flow during nighttime. The report by Froelich and Schmid [26] and more recently Feigenwinter et al. [23, 24] describe similar features of the airflow interaction between above and below canopy.

The spatial distribution of the horizontal CO_2 concentration, (Figure 18) along the north face, shows a similar pattern than the south face described previously. Despite there is no wind information in that area, if one assumes the same spatial correlation between horizontal wind and CO_2 concentration, it is possible predict that the wind should present an inverse pattern from the south face suggesting that, during daytime, the downslope wind direction should be from the northeast (Figure 18(c), from blue to red color).

During the evening period (Figure 18(b)), it should be indicating downslope (from northeast) in the upper part of the north face slope and upslope (from southeast) in the lower part of the slope, an inverse feature from Figure 17(b). Finally, later in the night, on the north face slope, the wind pattern should present an upslope wind direction regime from southeast, an inverse regime to that one from Figure 17(a) on the south face slope.

One possible explanation to this subcanopy slopes' wind regime and spatial distribution of CO_2 concentration, is the valley wind channeling effect and how it is meandering when it enters in the valley topography (as described by [28]). This valley wind pattern probably causes oscillations as those observed on the CO_2 concentration along the day (Figures 17 and 18), the known "Seiche phenomena" [56].

4. Summary and Conclusions

The main objective of this study was to measure and understand the local circulation over a dense forest site in Manaus with moderately complex terrain and to verify the existence of the drainage flow regimes on slope and valley areas. The main pattern of the airflow above and below the canopy in dense tropical forest in Amazonia was captured by a relative simple measure system, as also has been done by more sophisticated measurement system as those described recently by Feigenwinter et al. [23, 24]. As described and discussed in preview sections, it was identified as drainage flow in both day- and nighttime periods in the site studied. Evidence of the drainage current above canopy was suggested by Goulden et al. [40] similar to the one observed here. The study highlighted that the local microcirculation was complicated and presented tridimensional nature. Where to estimate the advection flux at this site seems uncertain and not possible with the limited measurement system employed. As reported recently by Feigenwinter et al. [23, 24], even using a more sophisticated measurement design, the level of uncertainties is still high and some processes are not yet known and need more exploration perhaps using a more complete spatial observation network or even applying model resources [57, 58].

In summary, the drainage flow exists and is observed at the K34 LBA site. Very large carbon uptake estimates reported previously should be questioned [27, 29], and more research is warranted. The use of nighttime u^* correction to avoid estimating canopy storage is inappropriate. One cannot get by using only the above-canopy turbulence information. The interactions between motions above and below canopy question the foundations of the footprint analysis [59, 60]. The representativeness of the eddy flux tower is most in question for complex terrain, especially on calm nights.

Acknowledgments

This work is a part of the LBA-ECO project, supported by the NASA Terrestrial Ecology Branch under Grants NCC5-283 and NNG-06GE09A to the Atmospheric Sciences Research Center, University at Albany, SUNY, LBA-ECO Team CD-03. The latter grant included a subcontract to the Fundação Djalma Batista, Manaus. J. Tóta was also supported by FAPEAM RH-POSGRAD and by MCT-INPA and Fundação Djalma Batista during 2006-2007, with a fellowship. Thanks to the LBA Manaus Team for support during the field work.

References

[1] D. Baldocchi, "'Breathing' of the terrestrial biosphere: lessons learned from a global network of carbon dioxide flux measurement systems," *Australian Journal of Botany*, vol. 56, no. 1, pp. 1–26, 2008.

[2] M. Aubinet, "Eddy covariance Co_2 flux measurements in nocturnal conditions: an analysis of the problem," *Ecological Applications*, vol. 18, no. 6, pp. 1368–1378, 2008.

[3] M. L. Goulden, J. W. Munger, S. M. Fan, B. C. Daube, and S. C. Wofsy, "Measurements of carbon sequestration by long-term eddy covariance: methods and a critical evaluation of accuracy," *Global Change Biology*, vol. 2, no. 3, pp. 169–182, 1996.

[4] T. A. Black, G. Den Hartog, H. H. Neumann et al., "Annual cycles of water vapour and carbon dioxide fluxes in and above a boreal aspen forest," *Global Change Biology*, vol. 2, no. 3, pp. 219–229, 1996.

[5] D. Baldocchi, E. Falge, L. Gu et al., "FLUXNET: a new tool to study the temporal and spatial variability of ecosystem-scale carbon dioxide, water vapor, and energy flux densities," *Bulletin of the American Meteorological Society*, vol. 82, no. 11, pp. 2415–2434, 2001.

[6] W. J. Massman and X. Lee, "Eddy covariance flux corrections and uncertainties in long-term studies of carbon and energy exchanges," *Agricultural and Forest Meteorology*, vol. 113, no. 1–4, pp. 121–144, 2002.

[7] H. W. Loescher, B. E. Law, L. Mahrt, D. Y. Hollinger, J. Campbell, and S. C. Wofsy, "Uncertainties in, and interpretation of, carbon flux estimates using the eddy covariance technique," *Journal of Geophysical Research D*, vol. 111, no. 21, Article ID D21S90, 2006.

[8] J. Tóta, D. R. Fitzjarrald, R. M. Staebler et al., "Amazon rain forest subcanopy flow and the carbon budget: santarém LBA-ECO site," *Journal of Geophysical Research G*, vol. 114, no. 1, Article ID G00B02, 2009.

[9] L. Gu, E. Falge, T. Boden et al., "Observing threshold determination for nighttime eddy flux filtering," *Agricultural and Forest Meteorologyis*, vol. 128, pp. 179–197, 2005.

[10] S. D. Miller, M. L. Goulden, M. C. Menton et al., "Biometric and micrometeorological measurements of tropical forest carbon balance," *Ecological Applications*, vol. 14, no. 4, pp. S114–S126, 2004.

[11] D. Papale, M. Reichstein, M. Aubinet et al., "Towards a standardized processing of Net Ecosystem Exchange measured with eddy covariance technique: algorithms and uncertainty estimation," *Biogeosciences*, vol. 3, no. 4, pp. 571–583, 2006.

[12] X. Lee, "On micrometeorological observations of surface-air exchange over tall vegetation," *Agricultural and Forest Meteorology*, vol. 91, no. 1-2, pp. 39–49, 1998.

[13] J. Finnigan, "A comment on the paper by Lee (1998): on micrometeorological observations of surface-air exchange over tall vegetation," *Agricultural and Forest Meteorology*, vol. 97, no. 1, pp. 55–67, 1999.

[14] K. T. Paw U, D. D. Baldocchi, T. P. Meyers, and K. B. Wilson, "Correction of eddy-covariance measurements incorporating both advective effects and density fluxes," *Boundary-Layer Meteorology*, vol. 97, no. 3, pp. 487–511, 2000.

[15] M. Aubinet, B. Heinesch, and M. Yernaux, "Horizontal and vertical CO_2 advection in a sloping forest," *Boundary-Layer Meteorology*, vol. 108, no. 3, pp. 397–417, 2003.

[16] C. Feigenwinter, C. Bernhofer, and R. Vogt, "The influence of advection on the short term CO_2-budget in and above a forest canopy," *Boundary-Layer Meteorology*, vol. 113, no. 2, pp. 201–224, 2004.

[17] R. M. Staebler and D. R. Fitzjarrald, "Observing subcanopy CO_2 advection," *Agricultural and Forest Meteorology*, vol. 122, no. 3-4, pp. 139–156, 2004.

[18] R. M. Staebler and D. R. Fitzjarrald, "Measuring canopy structure and the kinematics of subcanopy flows in two forests," *Journal of Applied Meteorology*, vol. 44, no. 8, pp. 1161–1179, 2005.

[19] J. Sun, S. P. Burns, A. C. Delany et al., "CO_2 transport over complex terrain," *Agricultural and Forest Meteorology*, vol. 145, no. 1-2, pp. 1–21, 2007.

[20] M. Aubinet, P. Berbigier, C. Bernhofer et al., "Comparing CO_2 storage and advection conditions at night at different carboeuroflux sites," *Boundary-Layer Meteorology*, vol. 116, no. 1, pp. 63–94, 2005.

[21] B. Marcolla, A. Cescatti, L. Montagnani, G. Manca, G. Kerschbaumer, and S. Minerbi, "Importance of advection in the atmospheric CO_2 exchanges of an alpine forest," *Agricultural and Forest Meteorology*, vol. 130, no. 3-4, pp. 193–206, 2005.

[22] R. Leuning, S. J. Zegelin, K. Jones, H. Keith, and D. Hughes, "Measurement of horizontal and vertical advection of CO_2 within a forest canopy," *Agricultural and Forest Meteorology*, vol. 148, no. 11, pp. 1777–1797, 2008.

[23] C. Feigenwinter, M. Mölder, A. Lindroth, and M. Aubinet, "Spatiotemporal evolution of CO_2 concentration, temperature, and wind field during stable nights at the Norunda forest site," *Agricultural and Forest Meteorology*, vol. 150, no. 5, pp. 692–701, 2010.

[24] C. Feigenwinter, L. Montagnani, and M. Aubinet, "Plot-scale vertical and horizontal transport of CO_2 modified by a persistent slope wind system in and above an alpine forest," *Agricultural and Forest Meteorology*, vol. 150, no. 5, pp. 665–673, 2010.

[25] C. Yi, "Momentum transfer within canopies," *Journal of Applied Meteorology and Climatology*, vol. 47, no. 1, pp. 262–275, 2008.

[26] N. J. Froelich and H. P. Schmid, "Flow divergence and density flows above and below a deciduous forest. Part II. Below-canopy thermotopographic flows," *Agricultural and Forest Meteorology*, vol. 138, no. 1–4, pp. 29–43, 2006.

[27] B. Kruijt, J. A. Elbers, C. Von Randow et al., "The robustness of eddy correlation fluxes for Amazon rain forest conditions," *Ecological Applications*, vol. 14, no. 4, pp. S101–S113, 2004.

[28] A. C. de Araújo, *Spatial variaton of CO_2 fluxes and lateral transport in an area of terra firme forest in central Amazonia*, Ph.D. thesis, Vrije Universiteit Amsterdam, Amsterdam, The Netherlands, 2009.

[29] A. C. Araújo, A. D. Nobre, B. Kruijt, and et tal, "Comparative measurements of carbon dioxide fluxes from two nearby towers in a central Amazonian rainforest: the Manaus LBA site," *Journal of Geophysical Research*, vol. 107, no. 8090, p. D20, 2002.

[30] C. D. Rennó, A. D. Nobre, L. A. Cuartas et al., "HAND, a new terrain descriptor using SRTM-DEM: mapping terra-firme rainforest environments in Amazonia," *Remote Sensing of Environment*, vol. 112, no. 9, pp. 3469–3481, 2008.

[31] G. Parker and D. R. Fitzjarrald, "Canopy structure and radiation environment metrics indicate forest developmental stage,

disturbance, and certain ecosystem functions," in *Proceedings of the 3rd LBA Scientific Conference*, Brazilian Ministry of Science and Technology , Brasilia, Brazil, July 2004.

[32] W. F. Laurance, P. M. Fearnside, S. G. Laurance et al., "Relationship between soils and Amazon forest biomass: a landscape-scale study," *Forest Ecology and Management*, vol. 118, no. 1–3, pp. 127–138, 1999.

[33] C. Castilho, *Variação espacial e temporal da biomassa arbórea viva em64 km2 de floresta de terra-firme na Amazônia Central*, Ph.D. thesis, Instituto Nacional de Pesquisa da Amazônia, Manaus, Amazonas, Brazil, 2004.

[34] R. C. C. Luizao, F. J. Luizao, R. Q. Paiva, T. F. Monteiro, L. S. Sousa, and B. Kruijt, "Variation of carbon and nitrogen cycling processes along a topographic gradient in a central Amazonian forest," *Global Change Biology*, vol. 10, no. 5, pp. 592–600, 2004.

[35] M. J. Waterloo, S. M. Oliveira, D. P. Drucker et al., "Export of organic carbon in run-off from an Amazonian rainforest blackwater catchment," *Hydrological Processes*, vol. 20, no. 12, pp. 2581–2597, 2006.

[36] L. A. Cuartas, J. Tomasella, A. D. Nobre, M. G. Hodnett, M. J. Waterloo, and J. C. Múnera, "Interception water-partitioning dynamics for a pristine rainforest in Central Amazonia: marked differences between normal and dry years," *Agricultural and Forest Meteorology*, vol. 145, no. 1-2, pp. 69–83, 2007.

[37] D. R. Fitzjarrald, K. E. Moore, O. M. R. Cabral, J. Scolar, A. O. Manzi, and L. D. De Abreu Sa, "Daytime turbulent exchange between the Amazon forest and the atmosphere," *Journal of Geophysical Research*, vol. 95, no. 10, pp. 16–838, 1990.

[38] D. R. Fitzjarrald and K. E. Moore, "Mechanisms of nocturnal exchange between the rain forest and the atmosphere," *Journal of Geophysical Research*, vol. 95, no. 10, pp. 16–850, 1990.

[39] B. Kruijt, Y. Malhi, J. Lloyd et al., "Turbulence statistics above and within two Amazon rain forest canopies," *Boundary-Layer Meteorology*, vol. 94, no. 2, pp. 297–331, 2000.

[40] M. L. Goulden, S. D. Miller, and H. R. da Rocha, "Nocturnal cold air drainage and pooling in a tropical forest," *Journal of Geophysical Research D*, vol. 111, no. 8, Article ID D08S04, 2006.

[41] M. R. Raupach, J. J. Finnigan, and Y. Brunet, "Coherent eddies and turbulence in vegetation canopies: the mixing-layer analogy," *Boundary-Layer Meteorology*, vol. 78, no. 3-4, pp. 351–382, 1996.

[42] V. B. Pachêco, *Algumas Características do Acoplamento entre o Escoamento Acima e Abaixo da Copa da Floresta Amazônica em Rondônia*, M.S. thesis, Instituto Nacional de Pesquisas Espaciais, São José dos Campos, Brazil, 2001.

[43] L. D. A. Sá and V. B. Pachêco, "Wind velocity above and inside Amazonian Rain Forest in Rondônia," *Revista Brasileira de Meteorologia*, vol. 21, no. 3a, pp. 50–58, 2006.

[44] I. N. Harman and J. J. Finnigan, "A simple unified theory for flow in the canopy and roughness sublayer," *Boundary-Layer Meteorology*, vol. 123, no. 2, pp. 339–363, 2007.

[45] P. C. Manins and B. L. Sawford, "Katabatic winds: a field case study," *Quarterly Journal, Royal Meteorological Society*, vol. 105, no. 446, pp. 1011–1025, 1979.

[46] A. P. Sturman, "Thermal influences on airflow in mountainous terrain," *Progress in Physical Geography*, vol. 11, no. 2, pp. 183–206, 1987.

[47] K. H. Papadopoulos and C. G. Helmis, "Evening and morning transition of katabatic flows," *Boundary-Layer Meteorology*, vol. 92, no. 2, pp. 195–227, 1999.

[48] M. Kossmann and F. Fiedler, "Diurnal momentum budget analysis of thermally induced slope winds," *Meteorology and Atmospheric Physics*, vol. 75, no. 3-4, pp. 195–215, 2000.

[49] C. Yi, R. K. Monson, Z. Zhai et al., "Modeling and measuring the nocturnal drainage flow in a high-elevation, subalpine forest with complex terrain," *Journal of Geophysical Research D*, vol. 110, no. 22, Article ID D22303, pp. 1–13, 2005.

[50] R. M. Staebler, *Forest subcanopy flows and micro-scale advection of carbon dioxide*, Ph.D. thesis, SUNY Albany, 2003.

[51] G. G. Katul, J. J. Finnigan, D. Poggi, R. Leuning, and S. E. Belcher, "The influence of hilly terrain on canopy-atmosphere carbon dioxide exchange," *Boundary-Layer Meteorology*, vol. 118, no. 1, pp. 189–216, 2006.

[52] D. Poggi, G. G. Katul, J. J. Finnigan, and S. E. Belcher, "Analytical models for the mean flow inside dense canopies on gentle hilly terrain," *Quarterly Journal of the Royal Meteorological Society*, vol. 134, no. 634 A, pp. 1095–1112, 2008.

[53] J. J. Finnigan, R. Clement, Y. Malhi, R. Leuning, and H. A. Cleugh, "Re-evaluation of long-term flux measurement techniques. Part I: averaging and coordinate rotation," *Boundary-Layer Meteorology*, vol. 107, no. 1, pp. 1–48, 2003.

[54] J. M. Wilczak, S. P. Oncley, and S. A. Stage, "Sonic anemometer tilt correction algorithms," *Boundary-Layer Meteorology*, vol. 99, no. 1, pp. 127–150, 2001.

[55] C. Feigenwinter, C. Bernhofer, and U. Eichelmann, "Comparison of horizontal and vertical advective CO_2 fluxes at three forest sites," *Agricultural and Forest Meteorologyis*, vol. 145, pp. 1–21, 2008.

[56] R. H. Spiegel and J. Imberger, "The classification of mixed layer dynamics in lakes of small to medium size," *Journal of Physical Oceanography*, vol. 10, pp. 1104–1121, 1980.

[57] T. Foken, "The energy balance closure problem: an overview," *Ecological Applications*, vol. 18, no. 6, pp. 1351–1367, 2008.

[58] S. E. Belcher, J. J. Finnigan, and I. N. Harman, "Flows through forest canopies in complex terrain," *Ecological Applications*, vol. 18, no. 6, pp. 1436–1453, 2008.

[59] P. H. Schuepp, M. Y. Leclerc, J. I. MacPherson, and R. L. Desjardins, "Footprint prediction of scalar fluxes from analytical solutions of the diffusion equation," *Boundary-Layer Meteorology*, vol. 50, no. 1–4, pp. 355–373, 1990.

[60] H.P. Schmid, "On the "Dos" and "Don't"s of footprint analysis in difficult conditions," in *Proceedings of the iLEAPS Specialist Workshop Flux Measurement in Difficult Conditions*, Boulder, Colo, USA, January 2006.

3

On the Origin of the Air between Multiple Tropopauses at Midlatitudes

Juan Antonio Añel,[1] Laura de la Torre,[2] and Luis Gimeno[2]

[1] Smith School of Enterprise and the Environment, University of Oxford, Oxford OX1 2BQ, UK
[2] Ephyslab, Facultad de Ciencias de Ourense, Universidade de Vigo, 32004 Ourense, Spain

Correspondence should be addressed to Juan Antonio Añel, juan.anel@smithschool.ox.ac.uk

Academic Editors: G.-C. Fang and P. Povinec

Multiple tropopauses are structures that regularly recur in the midlatitudes. Recent studies have relied on the notion of the excursion of tropical air from the upper troposphere into higher latitudes, thereby overlaying the tropopause of the midlatitudes. We herein analyse the origin and characteristics of the air at the Boulder radiosonde station, between the first and second tropopauses combining an analysis of radiosonde data with a Lagrangian approach based on the FlexPart model and ERA-40 analysis data. Our results show that the air between both tropopauses has its origin in midlatitudes.

1. Introduction

Multiple tropopauses (MTs) are structures that regularly recur in the midlatitudes [1–3] and are extremely important for stratosphere-troposphere exchange (STE). Recent studies have relied on the notion of the excursion of tropical air from the upper troposphere towards higher latitudes, which then overlies the tropopause of the midlatitudes and is one of their main causes of MTs. Randel et al. [1] described the coincidence of double tropopause (DT) profiles with reduced amounts of ozone in the lower stratosphere (LS) with regions of increased transport from the tropics to higher latitudes above the subtropical jets. Pan et al. [4] suggested the association of DTs with intrusions of low-latitude air masses with low static stability and low ozone concentrations into the LS at midlatitudes, which they related to Rossby wave breaking events. However, Wang and Polvani [5] have recently suggested using an idealized model that air inside a DT structure comes from high latitudes. Vogel et al. [6] found that, in some cases, the contribution to mixing at levels between 330 K and 350 K is approximately equivalent to stratospheric and tropospheric intrusion.

However, no clear understanding exists of the origin of MT phenomena, which could be associated with the overlapping of the tropical and extratropical tropopause, or to folding of the tropopause linked to the atmospheric circulation phenomena that characterise these latitudes, such as cut-off low (COL) systems [7], the movement of jet-streams, or baroclinic waves [8, 9]. Recent results show a widening of the Earth's tropical belt, which could further complicate this picture [10, 11].

In order to address this gap in our understanding, we herein report on the origin of the air immediately above the first and below the second tropopause at the Boulder radiosonde station. We combine the data obtained from the water vapour soundings launched here with the results of re-analysis, based on Lagrangian analysis and fields of potential vorticity (PV).

Our analysis relies on the fact that, if the origin of the MTs is related to excursions of tropical air over the extratropical tropopause, then we might expect there to be a contribution of air rich in moisture to the extratropical lowermost stratosphere. However, if the origin of the air is associated with the folding of the tropopause and with stratosphere-to-troposphere exchange, we would then expect to find air of stratospheric origin, that is, with a low moisture content and with higher concentrations of ozone.

2. Methods

We herein consider data for the cold season in the Northern Hemisphere (November to March). The period of study was 1980–2009 for the computation of vertical profiles from soundings and 1980–2000 for the Lagrangian analysis, due to the availability of ERA-40 data.

We chose the station of Boulder, Colorado (39.9°N, 105.3°W), for our study, mainly because it is a well-studied station and in a location likely to show the first tropopauses at tropical altitudes during summer and at extratropical altitudes during winter [2]. Sounding data were obtained from http://www.esrl.noaa.gov/gmd/ozwv/.

The first and second tropopauses were computed for 319 different days for the analysis of vertical profiles and for 188 different days for the Lagrangian analysis. When there were days with two soundings at the same time, we chose the sounding that contained more values for water vapour. Tropopauses were defined according to the usual lapse-rate criterion given by the World Meteorological Organization [12] as follows.

(a) The first tropopause is defined as the lowest level at which the lapse rate decreases to 2°C/km or less, provided also the average lapse rate between this level and all higher levels within 2 km does not exceed 2°C/km.

(b) If above the first tropopause the average lapse rate between any level and all higher levels within 1 km exceeds 3°C/km then a second tropopause is defined by the same criterion as under (a). This tropopause may be either within or above the 1 km layer.

Next, we used the Lagrangian particle dispersion model FLEXPART developed by Stohl and James [13, 14], specifically v8.1. Runs were performed having in mind the complete longitudinal extension of the region between 10° and 65° North and with a vertical domain that spanned from sea level up to 22 km. The model was fed with ERA-40 reanalysis data [15]. Following the temporal resolution of this dataset, we used t_0 to represent the time when a double tropopause (DT) was found in a sounding, obtaining results at 6-hour intervals beforehand. The maximum temporal domain for the computation of trajectories was 10 days. Longer computations were not considered to be relevant because 10 days is a typical residence time for water vapour in the atmosphere [16], during which we would expect to find a fingerprint of overlapping of the tropical tropopause. An analysis of the fields of PV was also undertaken, because this can be used to distinguish between tropospheric and stratospheric air masses [10]. The values are the ones given by FLEXPART for each particle.

The density of particles was computed as the sum of the number of particles detected multiplied by the cosine of the latitude in order to weight the different latitudinal contributions. We integrated all the vertical levels in the latitude-longitude representation and all the latitudes in the altitude-longitude representation.

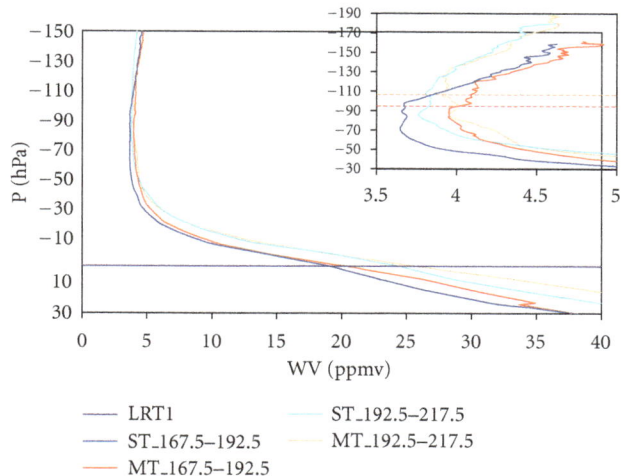

FIGURE 1: Vertical profiles of WV content (in parts per million by volume) relative to the pressure of the first tropopause (LRT1) shown for single (ST) and multiple tropopauses (MT) and over a range of pressures of occurrence (in hPa) of both phenomena. The part of the figure between 3.5 and 5 ppmv is enlarged in the upper right-hand corner in order to obtain a better insight of the changes in the vertical profile. The horizontal lines correspond to the second lapse-rate tropopause (colour denotes the pressure range where it happens).

3. Results

The computed vertical profiles of water vapour are shown in Figure 1 relative to the pressure of the first lapse-rate tropopause (LRT1). They are split into single (ST) and double tropopause (DT) cases and shown for two different vertical layers, namely, 167.5 hPa–192.5 hPa and 192.5 hPa–217.5 hPa. We split them thus because the soundings showed that they were layers in which the incidence of MTs was most common. Furthermore, it allowed us to check whether the MT events were lower or higher than these layers, being more or less representative of the layer between the MTs and the LS, respectively.

It may be seen that, for ST events, the water vapour (WV) content immediately above the tropopause is lower than it is for DT events and that the WV content at LRT1 depends on the pressure of occurrence, independently of being a ST or MT event. It is also clear that the WV contents below the second lapse-rate tropopause (LRT2) are similar and independent of the pressure of occurrence. Moreover, the WV content just above LRT2 is greater for MT events with lower pressure values, suggesting the contribution of air masses rich in moisture, such as those from the tropics.

This result appears to confirm the hypothesis of Pan et al. [4] of a tropical origin of the WV and air masses in MT events in subtropics and therefore support the view that MTs in the subtropics are a consequence of the overlapping of tropical tropopauses with extratropical ones. In order to obtain a better insight of this, we performed a Lagrangian analysis using FlexPart.

The Lagrangian analysis of average specific humidity q is shown in Figure 2. While it is not easy to see much of

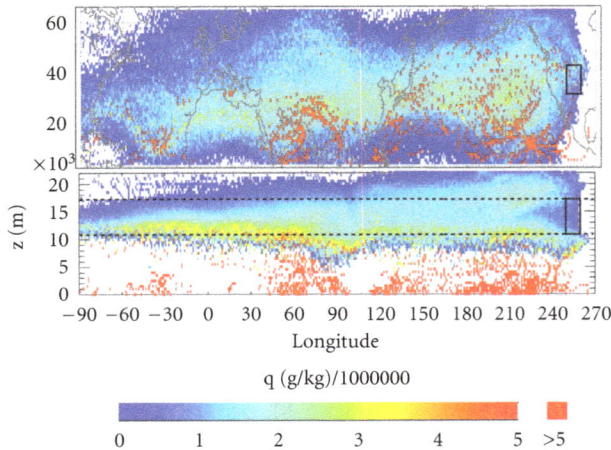

FIGURE 2: Specific humidity averaged for the ten days before time t_0 (color scale). In the upper panel (latitude-longitude representation) the black square on the right-hand side represents the region near Boulder where particles arrive at t_0. In the lower panels (height-longitude representation), the rectangle represents a boundary of 5 degrees of longitude surrounding the station at Boulder. The dotted black lines show the average heights of the first and second tropopauses.

the detail, Figure 2 does show that maximum values occur for tropical and subtropical regions and for tropospheric levels.

Figures 3 and 4 show the density of the particles for the air masses over Boulder, respectively, 10 days and 24 hours before t_0. The plots show the two different perspectives of latitude-longitude and altitude, with lines marking the average position of the first and second tropopauses for all the days when it was possible to compute it.

Figure 3 corresponds to the summed results for the particles between t_{-240} and t_{-6}. Figure 3(a), shows the absolute values. The latitude-longitude plot is less informative because the particles are concentrated around their destination. The use of the same colour scale for each longitude makes it difficult to see where the fastest particles originate. The height-longitude representation makes it possible to see how most of the particles that arrive at Boulder during MT events maintain their altitude for several days.

Figure 3(b) provides more information; this figure is similar to Figure 3(a) but the plot shows the particle density relative to each degree of longitude. It is thus possible to discern the source in terms of latitude and height for the maximum concentration of particles that arrive at Boulder. It is clear that air masses that contribute to the formation of MT events for the sum of the previous ten days are extratropical and that they have their source at similar altitudes to the MT event, above the first extratropical tropopause.

In order to address more accurately the problem of identifying possible contributions of air masses with an extratropical or tropical origin, we performed the same analysis but for air masses whose source was at latitudes to the South of 35°N and to the North of 45°N. The results for the ten days beforehand are similar to those shown in Figure 3. Figure 4 shows the results for t_{-24}; from this figure, it is

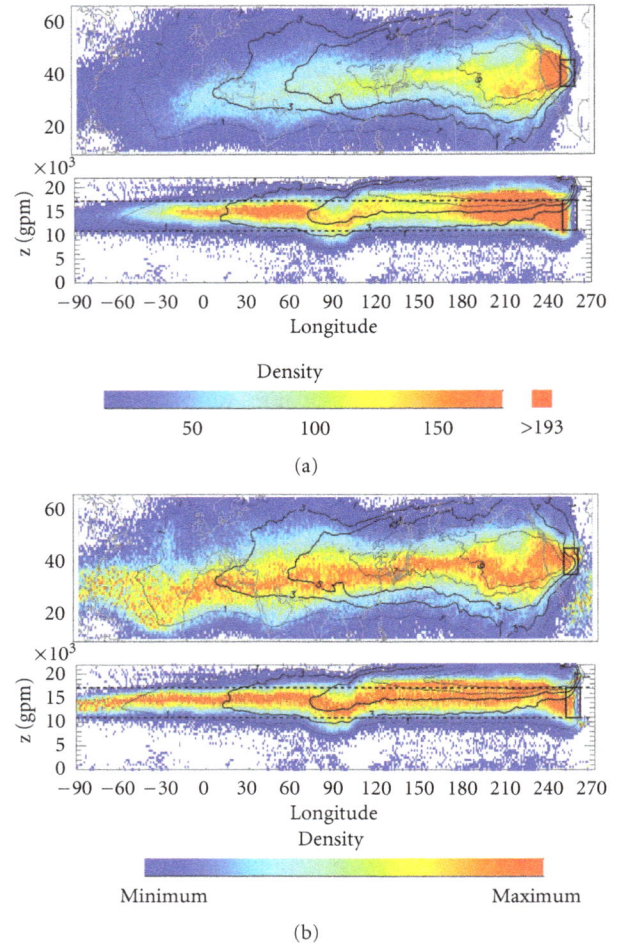

FIGURE 3: Particle density (colour scale) and smoothed contours of averaged PV (isolines of 1, 3, 5, 7, and 9 PVU). In the upper panels (latitude-longitude representation), the black square on the right-hand side represents the region near Boulder where particles arrive at t_0. In the lower panels (height-longitude representation), the rectangle represents a boundary of 5 degrees of longitude surrounding Boulder. The dotted black lines show the average heights of the first and second tropopauses. (a) Sum for the ten days before t_0. (b) Similar to (a) but representing the maximum density for each degree of longitude.

clear that particles arriving at Boulder, which were located below 35°N at any moment during the previous 24 hours, mostly stay at the same latitude and show values of PV greater than 1.5 or 2 PVU. Moreover, these particles usually lose altitude when approaching Boulder. This allows us to conclude that they could be stratospheric in origin and are probably associated with intrusions of stratospheric air in the troposphere, such as those associated with tropopausal folding. However, the effect of the orographic forcing of the Rocky Mountains is clear in the latitude-longitude plot of Figure 3(b). The particles arriving at Boulder that were located above 45°N at any moment in the previous 24 hours maintain the same latitude and altitude and increase their PV, suggesting fast mixing with air masses from upper levels with lower temperatures.

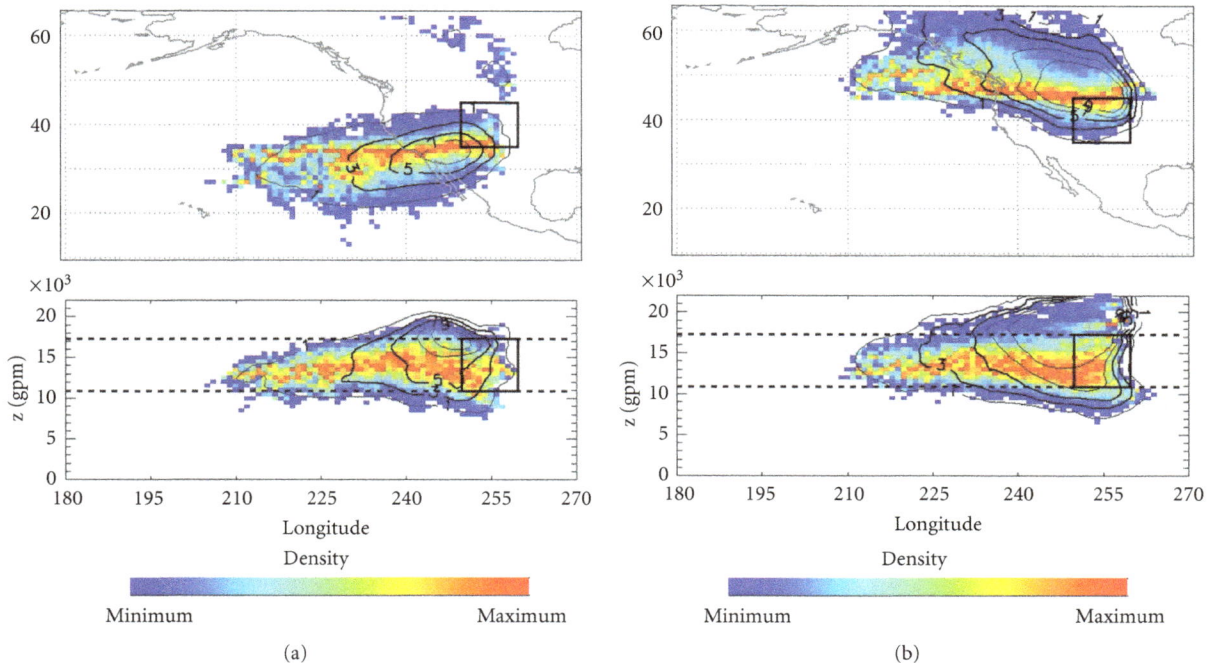

FIGURE 4: Similar to Figure 3(b) but (a) particles between the first and the second tropopauses whose origin 24 hours before t_0 was at a latitude to the South of 35 degrees. (b) Similar to (a) but for particles with their origin at latitudes to the North of 45 degrees.

4. Concluding Remarks

Previous literature on the causes of MTs in extratropical regions is somewhat contradictory, either attributing them to excursions of the tropical over the extratropical tropopause, or to STEs associated with synoptic atmospheric phenomena.

From our results, it is clear that residence times for air masses between the first and second tropopauses are no longer than 24 hours. Moreover, the source of these air masses is mainly stratospheric, as may be concluded from the associated high values of PV (>3 PVU in most cases). It may also be seen that the values of PV associated with these air masses are inversely proportional to their distance from the Boulder station. This could be evidence for a link with a cyclonic circulation at upper levels (local maximum of PV) that is characteristic of COLs. Sometimes the values of PV increase near Boulder and then decrease, which might possibly be an indication of tropopausal folding.

Air masses with tropospheric characteristics are only present in a few cases and when considering periods longer than ten days, which is in agreement with the findings of Vogel et al. [6]. This result should not come as a surprise, however, because meridional circulation patterns increase in importance at longer periods of time. In the same way, given these residence times, it is improbable that the transport of water vapour from lower latitudes towards the extratropical lowermost part of the stratosphere is a main cause of MTs; our findings indicate that it is rather more likely that there is a relationship with the radiative effect of O_3 transported from upper levels to the upper troposphere/lower stratosphere.

In conclusion, our findings provide one additional source of information for the discussion of the origin of MT events at midlatitudes, suggesting that extratropical phenomena associated with exchanges between the stratosphere and the troposphere are in most cases a much more probable cause than overlapping with the tropical tropopause.

Acknowledgments

When this research was being undertaken, J. A. Añel was an Ángeles Alvariño Researcher in the Faculty of Sciences of Ourense, University of Vigo, and was supported by the Xunta de Galicia. L. de la Torre was supported through a travel grant awarded by the Xunta de Galicia. This study was partially funded through the TRODIM Project, which in turn was funded by the Spanish Ministry of Science and Innovation (CGL2007-65891-C05-01) and cofunded through the European Regional Development Fund. The ERA-40 data were obtained from the ECMWF data server.

References

[1] W. J. Randel, D. J. Seidel, and L. L. Pan, "Observational characteristics of double tropopauses," *Journal of Geophysical Research D*, vol. 112, no. 7, Article ID D07309, 2007.

[2] J. A. Añel, J. C. Antuña, L. de la Torre, J. M. Castanheira, and L. Gimeno, "Climatological features of global multiple tropopause events," *Journal of Geophysical Research D*, vol. 114, no. 7, Article ID D00B08, 2009.

[3] T. R. Peevey, J. C. Gille, C. E. Randall, and A. Kunz, "Investigation of double tropopause spatial and temporal global

variability utilizing High Resolution Dynamics Limb Sounder temperature observations," *Journal of Geophysical Research D*, vol. 117, no. 1, Article ID D01105, 2012.

[4] L. L. Pan, W. J. Randel, J. C. Gille et al., "Tropospheric intrusions associated with the secondary tropopause," *Journal of Geophysical Research D*, vol. 114, no. 10, Article ID D10302, 2009.

[5] S. Wang and L. M. Polvani, "Double tropopause formation in idealized baroclinic life cycles: the key role of an initial tropopause inversion layer," *Journal of Geophysical Research D*, vol. 116, no. 5, Article ID D05108, 2011.

[6] B. Vogel, L. L. Pan, P. Konopka et al., "Transport pathways and signatures of mixing in the extratropical tropopause region derived from Lagrangian model simulations," *Journal of Geophysical Research D*, vol. 116, no. 5, Article ID D05306, 2011.

[7] M. Sprenger, M. C. Maspoli, and H. Wernli, "Tropopause folds and cross-tropopause exchange: a global investigation based upon ECMWF analyses for the time period March 2000 to February 2001," *Journal of Geophysical Research D*, vol. 108, article 8518, 11 pages, 2003.

[8] J. M. Castanheira, J. A. Añel, C. A. F. Marques et al., "Increase of upper troposphere/lower stratosphere wave baroclinicity during the second half of the 20th century," *Atmospheric Chemistry and Physics*, vol. 9, pp. 9143–9153, 2009.

[9] J. M. Castanheira and L. Gimeno, "Association of double tropopause events with baroclinic waves," *Journal of Geophysical Research D*, vol. 116, no. 19, Article ID D19113, 2011.

[10] J. A. Añel, A. Gettelman, and J. M. Castanheira, "Tropical widening vs tropopause rising: a PV-theta view of the UTLS," *PLoS ONE*. Under review.

[11] S. M. Davis and K. H. Rosenlof, "A multidiagnostic intercomparison of tropical-width time series using reanalyses and satellite observations," *Journal of Climate*, vol. 25, no. 4, pp. 1061–1078, 2012.

[12] World Meteorological Organization (WMO), "Meterology: a three dimensional science," *WMO Bulletin*, vol. 6, pp. 134–138, 1957.

[13] A. Stohl and P. James, "A Lagrangian analysis of the atmospheric branch of the global water cycle: part 1: method description, validation, and demonstration for the August 2002 flooding in central Europe," *Journal of Hydrometeorology*, vol. 5, no. 4, pp. 656–678, 2004.

[14] A. Stohl and P. James, "A Lagrangian analysis of the atmospheric branch of the global water cycle. Part II: moisture transports between earth's ocean basins and river catchments," *Journal of Hydrometeorology*, vol. 6, no. 6, pp. 961–984, 2005.

[15] S. M. Uppala, P. W. Kållberg, A. J. Simmons et al., "The ERA-40 re-analysis," *Quarterly Journal of the Royal Meteorological Society*, vol. 131, no. 612, pp. 2961–3012, 2005.

[16] A. Numaguti, "Origin and recycling processes of precipitating water over the Eurasian continent: experiments using an atmospheric general circulation model," *Journal of Geophysical Research D*, vol. 104, no. 2, pp. 1957–1972, 1999.

4

Apply Woods Model in the Predictions of Ambient Air Particles and Metallic Elements (Mn, Fe, Zn, Cr, and Cu) at Industrial, Suburban/Coastal, and Residential Sampling Sites

Guor-Cheng Fang and Ci-Song Huang

Department of Safety, Health and Environmental Engineering, Hung Kuang University, Sha-Lu, Taichung 433, Taiwan

Correspondence should be addressed to Guor-Cheng Fang, gcfang@sunrise.hk.edu.tw

Academic Editor: Daniel L. Costa

The main purpose for this study was to monitor ambient air particles and metallic elements (Mn, Fe, Zn, Cr, and Cu) in total suspended particulates (TSPs) concentration, dry deposition at three characteristic sampling sites of central Taiwan. Additionally, the calculated/measured dry deposition flux ratios of ambient air particles and metallic elements were calculated with Woods models at these three characteristic sampling sites during years of 2009-2010. As for ambient air particles, the results indicated that the Woods model generated the most accurate dry deposition prediction results when particle size was $18\,\mu$m in this study. The results also indicated that the Woods model exhibited better dry deposition prediction performance when the particle size was greater than $10\,\mu$m for the ambient air metallic elements in this study. Finally, as for Quan-xing sampling site, the main sources were many industrial factories under process around these regions and were severely polluted areas. In addition, the highest average dry deposition for Mn, Fe, Zn, and Cu species occurred at Bei-shi sampling site, and the main sources were the nearby science park, fossil fuel combustion, and Taichung thermal power plant (TTPP). Additionally, as for He-mei sampling site, the main sources were subjected to traffic mobile emissions.

1. Introduction

Dry deposition is the process by which an atmospheric air pollutant is transferred to the surface of the earth as a result of the turbulent motion of air [1, 2]. The concentrations and size distributions of trace metals are governed by the nature of emissions into the atmosphere as well as by the rates of wet and dry deposition, cloud processing, chemical transformations, and the exchange of air between the boundary layer and free troposphere [3]. It occurs as trace gases and particles are adsorbed or react on objects (plants, soil, water, buildings, etc.) at the earth's surface. Factors that govern dry deposition of gases and particles include atmospheric turbulence intensity, the nature of the gas and particles, and that of the surface [4].

In attrition, fossil fuel and wood combustion, as well as waste incineration and industrial processes, are the main anthropic sources of metals to the atmosphere [5]. Natural emissions (from crustal minerals, forest fires, and oceans) and industrial processes (fuel combustion, waste incineration, automobile exhaust, and mining, and quarrying) are the principal sources of the metals in ambient air. Traffic emissions also represent various potential sources of metals, including combustion products from fuel and oil, road construction materials, road dust, and wear products from tires, brake linings, and bearings [6].

Furthermore, air pollution has become a more and more serious problem in Taiwan, especially in central Taiwan. It includes aerosol, effluvium, secondary pollutant, and organic solvent vapor [7]. In Taiwan, the number of vehicles on the main roads has increased in recent decades. Motor vehicle exhaust is an important source of fine particles [8]. Heavy metals such as Pb, Cu, and Zn mainly are found in the particulate phase [9].

Moreover, considerable research has been conducted to investigate the dry deposition of air pollutants. Among these, heavy metals are of particular interest as most of them are toxic to humans and ecosystems. Dry deposition may be particularly important near urban/industrial areas adjacent to surface waters where particle concentrations and pollutants associated with them are relatively high [10].

Therefore, atmospheric particulates and associated trace metals have been linked with both short- and long-term adverse health effects which mostly include chronic respiratory diseases, heart diseases, lung cancer, and damage to other organs [11–14]. The combined model resolved 97% of PM_{10} mass concentrations and the evaluation analysis showed the results obtained by the combined model were reasonable [15]. The concept of the dry deposition velocity of an air pollutant has been widely utilized/used in modeling air pollution [16, 17].

The goal of this study was (1) monitoring and modeling ambient air particulates and metallic elements Mn, Fe, Zn, Cr, and Cu concentration, dry deposition at three characteristic (suburban/coastal, residential, and industrial) sampling sites; (2) finding the optimum particle size that performs better results with respect to Woods model in the prediction of dry depositions for ambient air particles and metallic elements (Mn, Fe, Zn, Cr, and Cu) in central Taiwan.

2. Dry Deposition Model

Atmospheric particles with aerodynamic diameters <10 μm (PM_{10}) have been under scrutiny as they are easily inhaled and deposited within the respiratory system [18]. Studies show that PM_{10} has a role in the incidence and severity of respiratory diseases [19, 20]. Therefore, 10 μm was used to calculate V_d (Dry deposition Velocity) and modeled dry deposition for comparison with measured dry deposition fluxes at the three characteristic sampling sites.

The descriptions of one model are all described as follows.

2.1. Empirical Equations: Deposition Model by Woods [21, 22]. The deposition across the entire range of particle sizes may be predicted by simple empirical equations and be applied to vertical surfaces across all deposition regimes when configured in the following manner [23]:

$$V_d^+ = k_1 Sc^{-2/3} + k_2 \pi^{+2} \quad \text{if } k_1 Sc^{-2/3} + k_2 \tau^{+2} \leq k_3,$$
$$V_d^+ = k_3 \quad \text{if } k_1 Sc^{-2/3} + k_2 \tau^{+2} > k_3. \tag{1}$$

Table 1 summarizes different values for k_1, k_2, and k_3 found by different investigators [24].

To apply empirical equations of this type to horizontal surfaces, a simple modification to accounts for the effect of gravitational settling on the particle deposition velocity [25]:

$$V_d^+ = k_1 Sc^{-2/3} + k_2 \tau^{+2} + g^+ \tau^+, \tag{2}$$

where g^+ is the dimensionless gravitational acceleration defined by

$$g^+ = \frac{g\nu}{u_\tau^3}, \tag{3}$$

TABLE 1: Recommended values of k_1, k_2 and k_3 for (1), [24].

k_1	k_2	k_3	Investigator
0.084			Cleaver and Yates
0.059			Friedlander
0.045	4.5×10^{-4}	0.13	Wood
0.075		0.3	Davies
0.07	3.5×10^{-4}	0.18	Papavergos and Hedley
	3.79×10^{-4}		Kneen and Strauss
	6×10^{-4}		Liu and Agarwal
		0.14	Fan and Ahmadi

and g is positive for a floor and negative for a ceiling surface. In (2) the first term on the right hand side accounts for Brownian diffusion, the second term accounts for interactions between particle inertia and turbulent eddies and the final term accounts for gravitational settling. However, (2) does not account for surface roughness. In the following (2) is referred to as Woods model.

The previously mentioned equation has been described in detail in previous studies [24].

Average particle sizes of 1.0, 2.5, 10, and 18 μm in TSP were selected in this study to model the particle-bound dry deposition fluxes. Then, the calculated dry deposition velocities are multiplied by the measured ambient air concentrations to obtain calculated dry deposition fluxes for ambient air particles and metallic elements (Mn, Fe, Zn, Cr, and Cu). The calculated dry deposition fluxes are then compared with the measured dry deposition fluxes. The calculated/measured flux ratios for dry deposition were then used to determine whether the over—or under—estimates dry deposition fluxes.

3. Method

3.1. Sampling Program. Figure 1 lists the five characteristic sampling sites. All samples were obtained in 1350–1400 min during the one-day sampling period for each sampling group. They are designated as follows.

The Bei-shi (suburban/coastal) sampling station (24°13′31.82″N, 120°34′09.45″E), which is in Sha-lu, Taichung, Taiwan, is a suburban/coastal station with no nearby obstructions. The immediate vicinity is residential, with an expressway with heavy traffic located approximately 2 km east of the station.

The He-mei (residential) town sampling station (24°06′00.54″N, 120°30′51.34″E) is located in a residential area. The main pollution sources are resident activities and vehicular emissions. The Quan-xing (industrial) sampling station (24°08′37.89″N, 120°29′09.43″E) is located in Shen-kang, a town covering 246.8 hectares, roughly half of which, 126.5 hectares, is occupied by factories and industry. Additionally, the Taichung Thermal Power Plant (TTPP) sits on 281 hectares of the wetland along the coast, on the western side of the sampling site. This plant burns coal to supply central Taiwan with 4,400 MW of electricity daily.

Apply Woods Model in the Predictions of Ambient Air Particles and Metallic Elements (Mn, Fe, Zn, Cr, and Cu) at
Industrial, Suburban/Coastal, and Residential Sampling Sites

37

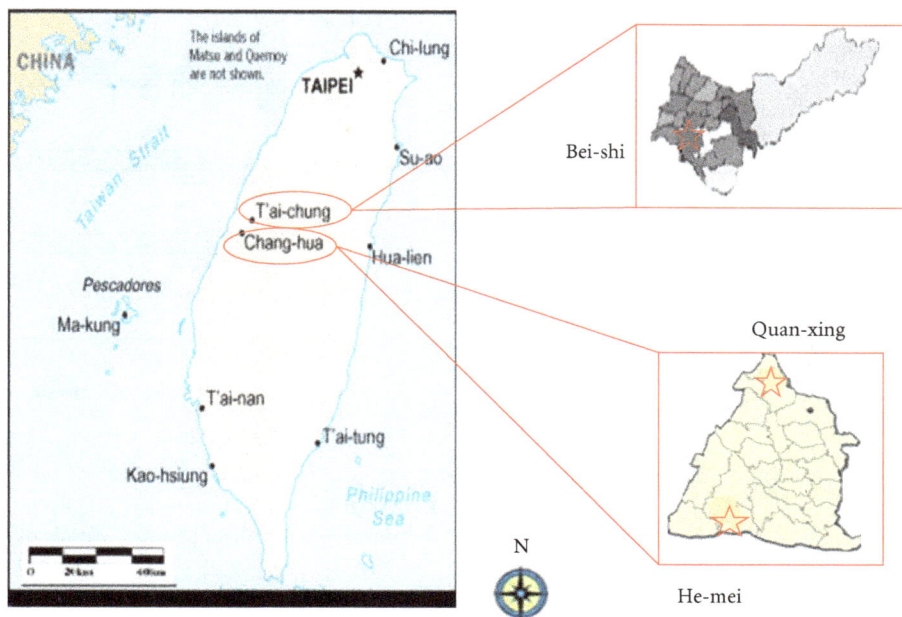

FIGURE 1: Geographical location for three characteristic sampling sites in central Taiwan.

3.2. Sampling Program

3.2.1. PS-1 (Total Suspended Particulate) Sampler.

The PS-1 is a complete air sampling system designed to collect suspended airborne particles (GMW High-Volume Air Sampler; Graseby-Andersen, USA). Maximum particle size in this study was roughly $100\,\mu m$. Sampler flow rate was 200 L/min. A quartz filter 10.2 cm in diameter filtered suspended particles. Filters were first conditioned for 24 h under an electric chamber under a humidity $35 \pm 5\%$ and temperature $25 \pm 5°C$ prior to both on and off weighing. Filters were in a sealed CD box when transported and stored. The sampling device and procedures are similar to those in previous study [26].

3.2.2. Dry Deposition Plate.

The dry deposition plate (DDP) had a smooth horizontal surrogate surface, providing a lower-bound estimate of dry deposition flux. The polyvinyl chloride (PVC) DDP measured 21.5 cm long, 8.0 cm wide by 0.8 cm thick. The DDP also had a sharp leading edge that was pointed into the prevailing wind. All filters were maintained under 50% relative humidity at 25°C for over 48 h. Prior to sampling, all filters were weighed to 0.0001 Gram-significant digits [26].

3.3. Chemical Analysis.

Samples were placed in an oven for 12 h before being weighed. One quarter of each filter was cut and selected before the digestion process. The filters were then cut into thin pieces and placed in a Teflon cup. Then, 3 mL hydrochloric acid (HCl) and 9 mL nitrate (HNO_3) were mixed together and then poured into this cup. Samples were then heated at 50°C on a hotplate for 2 h. Samples after digestion on the hotplate were then filtered. Following filtration, the sample solution was added to 0.2% HNO_3 to create 100 mL solution. The samples were maintained at 4°C in the refrigerator for ICP-AES analysis. Concentrations of metallic elements Mn, Fe, Zn, Cr, and Cu were measured in this study. ICP-AES was conducted using a Perkin Elmer Optima 2100 Plasma Emission Spectrometer to analyze the metallic elements. A time and an argon gas plasma flow rate of 15 L/min were applied. The nebulizer flow rate was set to 0.65 L/min and the sample flow rate was set to 1.5 mL/min.

3.4. Quality Control.

Blank test background contamination was assessed using operational blanks (unexposed projection film and a quartz filter) that were processed simultaneously with field samples. The field blanks were exposed in the field when the field sampling box was opened to remove and replace filters. This study accounted for background contamination of metallic elements by subtracting field blank values from concentrations. Field blank values were extremely low, generally below or around the method detection limits. In this study, the background contamination is insignificant and can be ignored. Blank test results were 0.21, 0.35, 0.19, 0.22, and 0.25 μg for Mn, Fe, Zn, Cr, and Cu. Respectively.

4. Results and Discussion

The meteorological parameters were measured by using weather station Model 525 (Spectrum Technologies, Inc., Taichung County, Taiwan).

Table 1 shows the meteorological conditions and average metallic elements (Mn, Fe, Zn, Cr, and Cu) in total suspended particulates (TSPs) and dry deposition at three characteristic sampling sites during 2009-2010.

The results indicated that the average temperature, relative humidity, and wind speed were 24.1°C, 76.9%, and

1.9 m/sec at Bei-shi sampling site, respectively. The results also showed that the average temperature, relative humidity, and wind speed were 24.5°C, 75.2%, and 2.2 m/sec at He-mei sampling site, respectively. Finally, as for Quan-xing sampling site, the results indicated that the average temperature, relative humidity, and wind speed were 24.4°C, 78.3%, and 2.6 m/sec, respectively.

As for metallic element Mn, the results indicated that the average concentrations order in TSP for location variations was Quan-xing (industrial) > He-mei (residential) > Bei-shi (suburban/coastal) and the average dry deposition order was Bei-shi (suburban/coastal) > Quan-xing (industrial) > He-mei (residential) (Table 2). Furthermore, for metallic element Fe, the results also indicated that the average concentration order in TSP for location variations was Quan-xing (industrial) > Bei-shi (suburban/coastal) > He-mei (residential) and the average dry deposition order was Bei-shi (suburban/coastal) > Quan-xing (industrial) > He-mei (residential). Moreover, metallic element Zn, the results indicated that the average concentration order in TSP for location variations was Quan-xing (industrial) > Bei-shi (suburban/coastal) > He-mei (residential) and the average dry deposition order was Bei-shi (suburban/coastal) > Quan-xing (industrial) > He-mei (residential). Besides, for metallic element Cr, the results also indicated that the average concentration order in TSP for location variations was He-mei (residential) > Quan-xing (industrial) > Bei-shi (suburban/coastal) and the average dry deposition order was He-mei (residential) > Bei-shi (suburban/coastal) > Quan-xing (industrial). Finally, metallic element Cu, the results indicated that the average concentration order in TSP for location variations was Quan-xing (industrial) > Bei-shi (suburban/coastal) > He-mei (residential) and the average dry deposition order was Bei-shi (suburban/coastal) > He-mei (residential) > Quan-xing (industrial).

Figure 2 displayed the average calculated/modeled ratios results for ambient air particles by using Woods models for various particles sizes (1.0 μm, 2.5 μm 10 μm, and 18 μm) at three sampling sites. The results indicated that the average calculated/measured flux ratios for particle sizes by using Woods model for (1.0 μm, 2.5 μm 10 μm, and 18 μm) particle size were 0.0014, 0.0079, 0.1210, and 0.4381 at Bei-shi sample site, respectively. And the average calculated/measured flux ratios for particle sizes by using Woods model for (1.0 μm, 2.5 μm, 10 μm, and 18 μm) particle size were 0.0016, 0.0089, 0.1330, and 0.4792 at He-mei sample site, respectively. Moreover, the results indicated that the average calculated/measured flux ratios for particle sizes by using Woods model for (1.0 μm, 2.5 μm, 10 μm, and 18 μm) particle size were 0.0017, 0.0093, 0.1411, and 0.5027 at Quan-xing sample site, respectively.

Figure 3 displayed the average calculated/measured flux ratios by using Woods models in the prediction of ambient air metallic elements dry deposition for various particle sizes (1 μm, 2.5 μm, 10 μm, and 18 μm) at Bei-shi sample site. The results indicated that the average calculated/measured flux ratios for metallic elements Mn, Fe, Zn, Cr, and Cu by using Woods model for 1 μm particle size were 0.01, 0.02, 0.01, 0, and 0.01, respectively. And the average calculated/measured

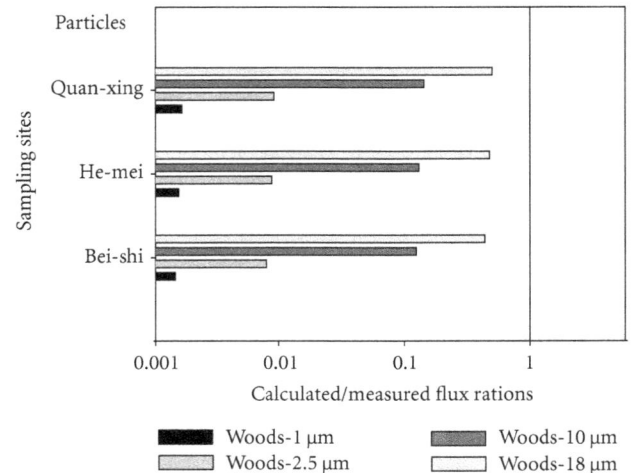

FIGURE 2: It displays the average calculated/modeled ratios results for ambient air particles by using Woods models for various particles sizes (1.0 μm, 2.5 μm, 10 μm, and 18 μm) at three sampling sites.

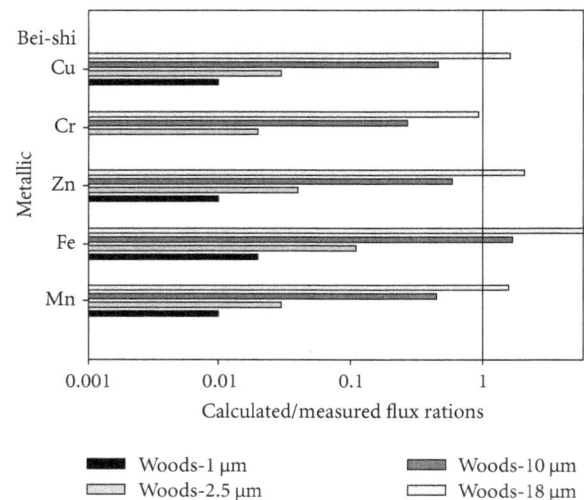

FIGURE 3: It displays the average calculated/measured flux ratios by using Woods models in the prediction of ambient air metallic elements dry deposition for various particle sizes (1 μm, 2.5 μm, 10 μm, and 18 μm) at Bei-shi sample site.

flux ratios for metallic elements Fe, Zn, Cr, and Cu by using Woods model for 2.5 μm particle size were 0.03, 0.11, 0.04, 0.02, and 0.03, respectively. In addition, the results indicated that the average calculated/measured flux ratios for metallic elements Mn, Fe, Zn, Cr, and Cu by using Woods model for 10 μm particle sizes were 0.45, 1.68, 0.59, 0.27, and 0.46, respectively. Finally, the results also indicated that the average calculated/measured flux ratios for metallic elements Mn, Fe, Zn, Cr, and Cu by using Woods model for 18 μm particle size were 1.57, 5.87, 2.06, 0.93, and 1.61 at Bei-shi sample site, respectively.

Figure 4 displayed the average calculated/measured flux ratios by using Woods models in the prediction of ambient air metallic elements dry deposition for various particle sizes

Apply Woods Model in the Predictions of Ambient Air Particles and Metallic Elements (Mn, Fe, Zn, Cr, and Cu) at
Industrial, Suburban/Coastal, and Residential Sampling Sites

39

TABLE 2: Meteorological conditions and average metallic elements (Mn, Fe, Zn, Cr, and Cu) in total suspended particulates (TSPs) and dry deposition at three characteristic sampling sites during year 2009-2010.

	Bei-shi (suburban/coastal) Average ($N = 60$)	He-mei (residential) Average ($N = 60$)	Quan-xing (industrial) Average ($N = 60$)
Temp (°C)	24.1	24.5	24.4
RH (%)	76.9	75.2	78.3
WS (m/s)	1.9	2.2	2.6
PWD	SE	SE	SE
	TSP (ng/m³)	TSP (ng/m³)	TSP (ng/m³)
Mn	16.0	60.1	80.8
Fe	290.5	93.7	2698.6
Zn	34.3	23.5	110.1
Cr	17.5	78.9	20.1
Cu	39.6	31.9	68.4
	Dry deposition (ng/m²/min)	Dry deposition (ng/m²/min)	Dry deposition (ng/m²/min)
Mn	36.3	14.1	17.6
Fe	2547.8	34.9	311.6
Zn	103.6	17.2	44.2
Cr	23.5	36.4	10.4
Cu	89.9	35.1	26.5

Temp: Temperature, RH: relative humidity, WS: wind speed, PWD: prevailing wind.

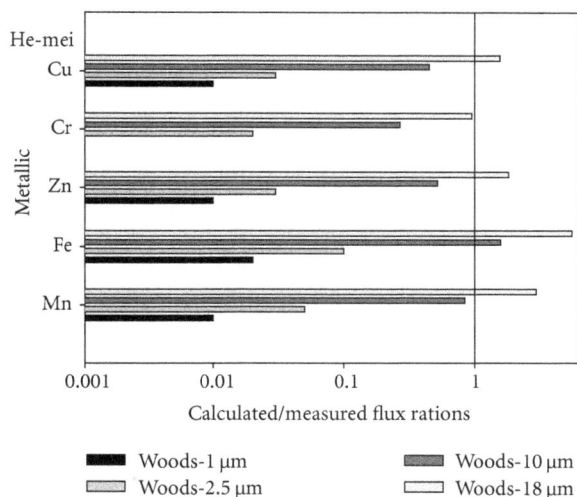

FIGURE 4: It displays the average calculated/measured flux ratios by using Woods models in the prediction of ambient air metallic elements dry deposition for various particle sizes (1 μm, 2.5 μm, 10 μm, and 18 μm) at He-mei sample site.

(1 μm, 2.5 μm, 10 μm, and 18 μm) at He-mei sample site. The results indicated that the average calculated/measured flux ratios for metallic elements Mn, Fe, Zn, Cr, and Cu by using Woods model for 1 μm particle size were 0.01, 0.02, 0.01, 0, and 0.01, respectively. And the average calculated/measured flux ratios for metallic elements Fe, Zn, Cr, and Cu by using Woods model for 2.5 μm particle size were 0.05, 0.10, 0.03, 0.02, and 0.03, respectively. Furthermore, the results

indicated that the average calculated/measured flux ratios for metallic elements Mn, Fe, Zn, Cr, and Cu by using Woods model for 10 μm particle sizes were 0.84, 1.59, 0.52, 0.27, and 0.45, respectively. Finally, the results also indicated that the average calculated/measured flux ratios for metallic elements Mn, Fe, Zn, Cr, and Cu by using Woods model for 18 μm particle size were 2.96, 5.57, 1.83, 0.95, and 1.57 at He-mei sample site, respectively.

Figure 5 displayed the average calculated/measured flux ratios by using Woods models in the prediction of ambient air metallic elements dry deposition for various particle sizes (1 μm, 2.5 μm, 10 μm, and 18 μm) at Quan-xing sample site. The results indicated that the average calculated/measured flux ratios for metallic elements Mn, Fe, Zn, Cr, and Cu by using Woods model for 1 μm particle size were 0.01, 0.02, 0.01, 0, and 0.01, respectively. And the average calculated/measured flux ratios for metallic elements Fe, Zn, Cr, and Cu by using Woods model for 2.5 μm particle size were 0.06, 0.12, 0.04, 0.02, and 0.03, respectively. Moreover, the results indicated that the average calculated/measured flux ratios for metallic elements Mn, Fe, Zn, Cr, and Cu by using Woods model for 10 μm particle sizes were 0.88, 1.91, 0.63, 0.34, and 0.49, respectively. Finally, the results also indicated that the average calculated/measured flux ratios for metallic elements Mn, Fe, Zn, Cr, and Cu by using Woods model for 18 μm particle size were 3.09, 6.69, 2.19, 1.19, and 1.70 at Quan-xing sample site, respectively.

From the above analysis, the results indicate that the Woods model generated the most accurate dry deposition prediction results for ambient air particles when particle size was at 18 μm in this study.

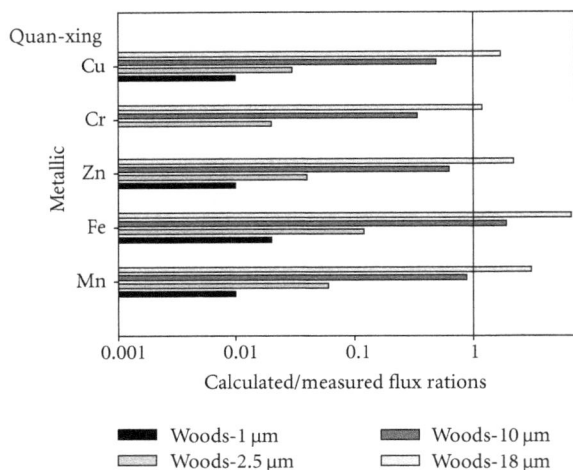

FIGURE 5: It displays the average calculated/measured flux ratios by using Woods models in the prediction of ambient air metallic elements dry deposition for various particle sizes (1 μm, 2.5 μm, 10 μm, and 18 μm) at Quan-xing sample site.

The results also indicated that the Woods model exhibited better dry deposition prediction performance for the ambient air metallic elements (i.e., Mn, Fe, Zn, Cr, and Cu) when the particle size was at 10 μm in this study. As for the meteorological conditions, average temperature, humidity, wind direction, and wind speed did not display any significant variations. Finally, the summarized important issue is that the highest average concentrations of Mn, Fe, Zn, and Cu species in TSP occurred at Quan-xing (industrial) sampling site, with many industrial factories under process around these regions that were severely polluted areas. In addition, the summarized important issue is that the highest average dry deposition for Mn, Fe, Zn, and Cu species occurred at Bei-shi (suburban/coastal) sampling site, and the main sources were the nearby science park, fossil fuel combustion, and Taichung thermal power plant (TTPP). Additionally, the summarized important issue is that the highest average concentrations of Cr species in TSP and dry deposition occurred at He-mei (residential) sampling site, and the main sources were subjected to traffic mobile emissions.

5. Conclusions

The main conclusions for this study are listed as follows.

(1) As for ambient air particles, the modeling results showed that the Woods model exhibits better average calculated/modeled ratios for 18 μm particle size at all sampling sites of this study.

(2) As for metallic elements Mn, Fe, Zn, and Cu, the results showed that the Woods model exhibits better dry deposition flux results for 10 μm particle size at Bei-shi, Quan-xing, and He-mei sampling sites. However, as for metallic element Cr, the results showed that the Woods model exhibits better dry deposition flux results for 18 μm particle size at all sampling sites.

(3) As for the these three sampling sites, the results indicated that the average highest metallic elements Mn, Fe, Zn, Cr, and Cu concentrations in TSP and dry deposition were occurred at Quan-xing (industrial) and Bei-shi (suburban/coastal) areas, respectively, and that average lowest metallic elements (Mn, Fe, Zn, Cr, and Cu) concentrations in TSP and dry deposition occurred at Bei-shi (suburban/coastal) and Quan-xing (industrial) area, respectively.

(4) From the point of view of metallic elements concentrations, metallic element Fe has the average highest metallic elements concentrations for any of sampling site in this study. And the average lowest metallic elements concentrations were Mn, Zn, and Cr at Bei-Shi, He-mei, and Quan-xing sampling sites, respectively. From the point of view of dry deposition, the average lowest dry deposition Cr at both Bei-shi and Quan-xing sampling sites while the lowest metallic element dry deposition at Her-Mei sampling site was Zn.

Acknowledgment

The authors gratefully acknowledge the National Science Council of the ROC (Taiwan) for financial support under Project no. NSC 99-2221-E-241-006-MY3.

References

[1] C. H. Lin, C. H. Lai, Y. L. Wu, and M. J. Chen, "Simple model for estimating dry deposition velocity of ozone and its destruction in a polluted nocturnal boundary layer," *Atmospheric Environment*, vol. 44, no. 35, pp. 4364–4371, 2010.

[2] M. L. Wesely and B. B. Hicks, "A review of the current status of knowledge on dry deposition," *Atmospheric Environment*, vol. 34, no. 12-14, pp. 2261–2282, 2000.

[3] C. Samara and D. Voutsa, "Size distribution of airborne particulate matter and associated heavy metals in the roadside environment," *Chemosphere*, vol. 59, no. 8, pp. 1197–1206, 2005.

[4] D. M. Moreira, T. Tirabassi, M. T. Vilhena, and A. G. Goulart, "A multi-layer model for pollutant dispersion with dry deposition to the ground," *Atmospheric Environment*, vol. 44, no. 15, pp. 1859–1865, 2010.

[5] A. P. Pereira, A. L. Wilson, S. C. Luiz et al., "Atmospheric concentrations and dry deposition fluxes of particulate trace metals in Salvador, Bahia, Brazil Pedro," *Atmospheric Environment*, vol. 41, pp. 7837–7850, 2007.

[6] M. El-Fadel and Z. Hashisho, "Vehicular emissions in roadway tunnels: a critical review," *Critical Reviews in Environmental Science and Technology*, vol. 31, no. 2, pp. 125–174, 2001.

[7] G. C. Fang, C. N. Chang, Y. S. Wu, P. P. C. Fu, K. F. Chang, and D. G. Yang, "The characteristic study of TSP, $PM_{2.5--10}$ and $PM_{2.5}$ in the rural site of central Taiwan," *Science of the Total Environment*, vol. 232, no. 3, pp. 177–184, 1999.

[8] C. G. Nolte, J. J. Schauer, G. R. Cass, and B. R. T. Simoneit, "Trimethylsilyl derivatives of organic compounds in source samples and in atmospheric fine particulate matter," *Environmental Science and Technology*, vol. 36, no. 20, pp. 4273–4281, 2002.

Apply Woods Model in the Predictions of Ambient Air Particles and Metallic Elements (Mn, Fe, Zn, Cr, and Cu) at
Industrial, Suburban/Coastal, and Residential Sampling Sites

41

[9] P. Buat-Menard, "Global change in atmospheric metal cycles," in *Global Atmospheric Chemical Change*, pp. 271–311, Elsevier, Amsterdam, The Netherlands, 1993.

[10] U. Shahin, S. M. Yi, R. D. Paode, and T. M. Holsen, "Long-term elemental dry deposition fluxes measured around Lake Michigan with an automated dry deposition sampler," *Environmental Science and Technology*, vol. 34, no. 10, pp. 1887–1892, 2000.

[11] D. L. Costa and K. L. Dreher, "Bioavailable transition metals in particulate matter mediate cardiopulmonary injury in healthy and compromised animal models," *Environmental Health Perspectives*, vol. 105, pp. 1053–1060, 1997.

[12] O. K. Magas, J. T. Gunter, and J. L. Regens, "Ambient air pollution and daily pediatric hospitalizations for asthma," *Environmental Science and Pollution Research*, vol. 14, no. 1, pp. 19–23, 2007.

[13] H. Prieditis and I. Y. R. Adamson, "Comparative pulmonary toxicity of various soluble metals found in urban particulate dusts," *Experimental Lung Research*, vol. 28, no. 7, pp. 563–576, 2002.

[14] P. Wild, E. Bourgkard, and C. Paris, "Lung cancer and exposure to metals: the epidemiological evidence," *Methods in Molecular Biology*, vol. 472, pp. 139–167, 2009.

[15] F. Zeng, G. L. Shi, X. Li et al., "Application of a combined model to study the source apportionment of PM10 in Taiyuan, China," *Aerosol and Air Quality Research*, vol. 10, no. 2, pp. 177–184, 2010.

[16] J. S. Chang, R. A. Brost, I. S. A. Isaksen et al., "A three-dimensional Eulerian acid deposition model: physical concepts and formulation," *Journal of Geophysical Research*, vol. 92, no. 12, pp. 14681–14700, 1987.

[17] L. Zhang, J. R. Brook, and R. Vet, "A revised parameterization for gaseous dry deposition in air-quality models," *Atmospheric Chemistry and Physics*, vol. 3, no. 6, pp. 2067–2082, 2003.

[18] C. A. I. Pope, D. W. Dockery, and J. Schwartz, "Review of epidemiological evidence of health effects of particulate air pollution," *Inhalation Toxicology*, vol. 7, pp. 1–18, 1995.

[19] C. A. Pope and D. W. Dockery, "Epidemiology of particle effects," in *Air Pollution and Health*, S. T. Holgate, J. M. Samet, H. S. Koren, and R. L. Maynard, Eds., pp. 673–706, Academic Press, 1999.

[20] B. Brunekreef and S. T. Holgate, "Air pollution and health," *The Lancet*, vol. 360, no. 9341, pp. 1233–1242, 2002.

[21] N. B. Wood, "A simple method for the calculation of turbulent deposition to smooth and rough surfaces," *Journal of Aerosol Science*, vol. 12, no. 3, pp. 275–290, 1981.

[22] N. B. Wood, "Mass transfer of particles and acid vapour to cooled surfaces," *Journal of the Institute of Energy*, vol. 54, no. 419, pp. 76–93, 1981.

[23] M. R. Sippola and W. W. Nazaroff, "Particle deposition from turbulent flow: review of published research and its applicability to ventilation ducts in commercial buildings," Report LBNL 51432, Lawrence Berkeley National Laboratory, 2002.

[24] L. Jonsson, E. Karlsson, and P. Jönsson, "Aspects of particulate dry deposition in the urban environment," *Journal of Hazardous Materials*, vol. 153, no. 1-2, pp. 229–243, 2008.

[25] W. Kvasnak and G. Ahmadi, "Deposition of ellipsoidal particles in turbulent duct flows," *Chemical Engineering Science*, vol. 51, no. 23, pp. 5137–5148, 1996.

[26] G. C. Fang, W. J. Huang, J. C. Chen, J. H. Huang, and Y. L. Huang, "Study of ambient air particle-bound As(p) and Hg(p) in dry deposition, total suspended particulates (TSP) and seasonal variations in Central Taiwan," *Environmental Forensics*, vol. 12, no. 1, pp. 7–13, 2011.

5

Comparison of BTX Profiles and Their Mutagenicity Assessment at Two Sites of Agra, India

Vyoma Singla, Tripti Pachauri, Aparna Satsangi, K. Maharaj Kumari, and Anita Lakhani

Department of Chemistry, Faculty of Science, Dayalbagh Educational Institute, Dayalbagh, Agra 282110, India

Correspondence should be addressed to Anita Lakhani, anitasaran2003@yahoo.co.in

Academic Editors: G.-C. Fang and A. Gabric

In the present study, the concentrations of three volatile organic compounds (VOCs), namely, acronym for benzene, toluene, and xylenes (BTX) were assessed because of their role in the tropospheric chemistry. Two representative sites, a roadside and a petrol pump, were chosen for sample collection. VOCs were collected using SKC-activated charcoal tubes and SKC personal sampler and characterized by gas chromatograph using flame ionization detector. Among BTX, benzene had the highest concentration. At the roadside, mean concentration of benzene, toluene, o-,m-xylene, and p-xylene were $14.7 \pm 2.4\,\mu\mathrm{gm}^{-3}$, $8.1 \pm 1.2\,\mu\mathrm{gm}^{-3}$, $2.1 \pm 0.8\,\mu\mathrm{gm}^{-3}$, and $5.1 \pm 1.2\,\mu\mathrm{gm}^{-3}$, respectively. At the petrol pump, the mean concentrations of benzene, toluene, o-,m-xylene and p-xylene were $19.5 \pm 3.7\,\mu\mathrm{gm}^{-3}$, $12.9 \pm 1.1\,\mu\mathrm{gm}^{-3}$, $3.6 \pm 0.5\,\mu\mathrm{gm}^{-3}$ and $11.1 \pm 1.5\,\mu\mathrm{gm}^{-3}$, respectively, and were numerically higher by a fraction of 2. Monthly variation of BTX showed maximum concentration in winter. Inter-species ratios and inter-species correlation indicated traffic as the major source of BTX. Extracts of samples were positive in both Salmonella typhimurium tester strains TA98 and TA100 without metabolic activation suggesting the presence of direct mutagens in ambient air that can cause both frame-shift and base-pair mutation. The mutagenic response was greater for TA100 than TA98 suggesting greater activity for base-pair mutagenicity than frame-shift mutagenicity and was found to be statistically significant.

1. Introduction

Volatile organic compounds (VOCs) comprise a wide range of compounds including aliphatic and aromatic hydrocarbons, alcohols, aldehydes, ketones, esters, and halogenated compounds. The main emission sources of VOCs are evaporative emissions from solvent utilization especially paints and protective coatings and combustion emissions from automobile exhaust [1]. VOCs play an important role in the formation of photochemical smog and tropospheric ozone. They modify the oxidizing capacity of the atmosphere by reacting with hydroxyl radicals (OH) in the presence of NOx to form ozone [2].

Aromatic VOCs in the atmosphere have a high photochemical ozone creation potential, which depends on their structure and reactivity. This has led to the development of scales of "reactivity" or "ozone formation potential" for VOC, of which the most widely used are the "maximum incremental reactivity" scale (MIR), developed by Carter and coworkers [3–5], and the "photochemical ozone creation potential" (POCP) scale, developed by Derwent et al. [6].

VOCs have direct toxic effects from carcinogenesis to neurotoxicity on human beings, and long-term exposure to VOCs is detrimental to human health, for example, by causing sick building syndrome (SBS) [7]. The carcinogenic effects of VOCs are becoming increasingly clear through genotoxicity studies using bacteria [8], mammalian cells [9], animals [10], and epidemiological investigations of large groups of people [11]. Of these assay methods, the salmonella/microsome (Ames) assay is the most widely used, convenient, and effective method to evaluate mutagenicity of airborne particles and source emissions [12].

In urban areas, a group of aromatic VOCs (benzene, toluene, ethylbenzene, and xylene) collectively called as BTEX are of significant concern and constitute up to 60% of nonmethane VOCs [13]. Among BTEX, benzene is an important representative of aromatic hydrocarbons and has been a high-priority urban air pollutant for assessment [14, 15]. It has been reported near heavy road traffic, mobile sources account for 75–85% of the benzene emissions of which 70% is from exhaust. The highest emissions are related to the use of gasoline in noncatalytic cars [16]. Benzene is a well-known human carcinogen for all routes of exposure. According to toxicological studies, Environmental Protection Agency (EPA) has classified benzene as a Group A, human carcinogen, and International Agency for Research on Cancer (IARC) considers benzene as confirmed and probable carcinogen [17, 18]. In addition to being carcinogenic, benzene is also known to be mutagenic while other VOCs are known to have effects on the central nervous system. Toluene is less toxic and causes drowsiness, and impaired coordination. Xylene exhibits neurological effects. High levels from exposure for acute (14 days or less) or chronic periods (more than 1 year) can cause headaches, lack of muscle coordination, dizziness, confusion, and alterations in body balance. Exposure of people to high levels of xylene for short periods can also cause irritation of the skin, eyes, nose, and throat, difficulty in breathing, and other problems with the lungs, delayed reaction time, memory difficulties, stomach discomfort, and possibly adverse effects on the liver and kidneys. Also the reaction of BTEX with hydroxyl radicals and/or nitrate (NO_3) radicals serves as the dominant degradation processes for aromatic VOCs in the atmosphere [14].

VOCs also contribute to stratospheric ozone depletion and enhancement of the global greenhouse effect [19]. Moreover, the acute and chronic health effects related to VOCs require their monitoring in risk areas.

Knowledge of ambient levels of VOCs is necessary to evolve a proper strategy to control tropospheric ozone build-up and maintain healthy air quality. Inspite of the well-known toxic effects of VOCs, information on VOC levels for Indian cities is limited [15, 20–24]. Aromatic hydrocarbons represent a significant fraction of the volatile organic compounds emitted in urban environment by road traffic [16] and the use of unleaded gasoline which is rich in aromatic hydrocarbons have increased worldwide. Therefore, monitoring of these hydrocarbons in urban area has become an important issue. With this view, the present study aims to report the ambient levels of BTX measured along a roadside carrying heavy vehicular traffic and at a petrol pump from April 2010 to March 2011.

2. Methodology

2.1. Study Area.
Agra is situated at latitude of 27°10′ N and longitude of 78°05′ E with an altitude of 169 m above the sea level in the semiarid zone of India. It is positioned with the Thar Desert of Rajasthan to the west, central hot plains to the south, Gangetic plains to the east, and cooler hilly regions to the north. Agra has a continental type of climate characterized by extreme dryness in summer and cold winters with calm periods. The summer in Agra is hot with intense solar radiation and is dominated by strong southeastern winds (wind speed ranges from 10–16 km h^{-1}). Intense solar radiation varies from 19–23 W m^{-2}. The winters are associated with calm periods with wind speeds reduced to about 40% (wind speed ranges from 0.5–5 km h^{-1}).

On the basis of rainfall, the study period was divided into four seasons, namely, summer (March–June), monsoon (July-August), post-monsoon (September-October),

and winter season (November–February). During the sampling period, meteorological profile measured by automatic weather station (AWS 321) was observed to vary significantly in summer and winter months. In summer months, the average temperature ranged from 35–40°C and average humidity varied from 25–50%. Summers were also associated with strong dust storms (wind speed varied from 10–16 km h^{-1}). On the other hand, winters were observed to be associated with calm periods. The average temperature varied from 22–28°C and relative humidity varied from 70–75% in winter months. During monsoon (July–September), relative humidity reached up to 80–85% and average rainfall was observed to be 160 mm. Table 1 gives the prevailing meteorological conditions during the study period and Figure 1 shows the observed wind direction pattern.

Two sites were selected based on the traffic flow as combustion of fuel specifically gasoline serves as the main source of BTX emissions. Site 1 (roadside) lies in the north of the city and is located along a busy road which caters a mean traffic volume of 10^6 per day. The site is also surrounded by a stretch of shops of paints and varnishes. Site 2 (A petrol pump) lies about 1 km away from Site 1. These sites were selected to study the variation in emissions of BTX in ambient air and at a point source (petrol pump) as fuel storage and delivery activities serve as a supplementary source to vehicular traffic at petrol pump.

2.2. Sampling Period and Sampling Procedure.
Ambient air samples were collected during the period of April 2010 to March 2011 along a roadside and a petrol pump. Sampling was carried out by active grab sampling method using a battery-operated pump. A battery, operated portable sampling pump (SKC 224-XR) was used to draw air at the rate of 2 l min^{-1} through SKC adsorption cartridge containing activated charcoal (60–80 mesh). The sampling pump was calibrated by a flow meter SKC making electronic digital flow calibrator. The monitoring schedule followed 4 hourly samples during peak hours, that is, 08:00–12:00 and 16:00–20:00 h in a day. On an average, 6 samples were collected every month.

2.3. Protection and Storage of Samples.
Protection of the exposed charcoal tubes is a very important part of the VOC determination method. Therefore, as soon as the pump was turned off, the tubes were removed from the sampling train and the two open sides were tightly closed using special caps to avoid any desorption and contamination. The sample tubes were put into polythene bags that were tightly closed and were kept in a box in a freezer until processed.

2.4. Analysis.
Before analysis, all sample tubes were taken from the freezer, contents of the adsorber tubes were emptied into 5 mL tubes containing 2 mL of HPLC-grade carbon disulphide (CS_2). The tubes were then placed in ultrasonic bath for at least 1 h to obtain the final sample solution. Immediately after this, the microsyringe was washed five times with the sample, washings were discarded successively, and finally 1 μL of aliquot was withdrawn from the samples and injected into a Shimadzu gas chromatograph (GC-17 A)

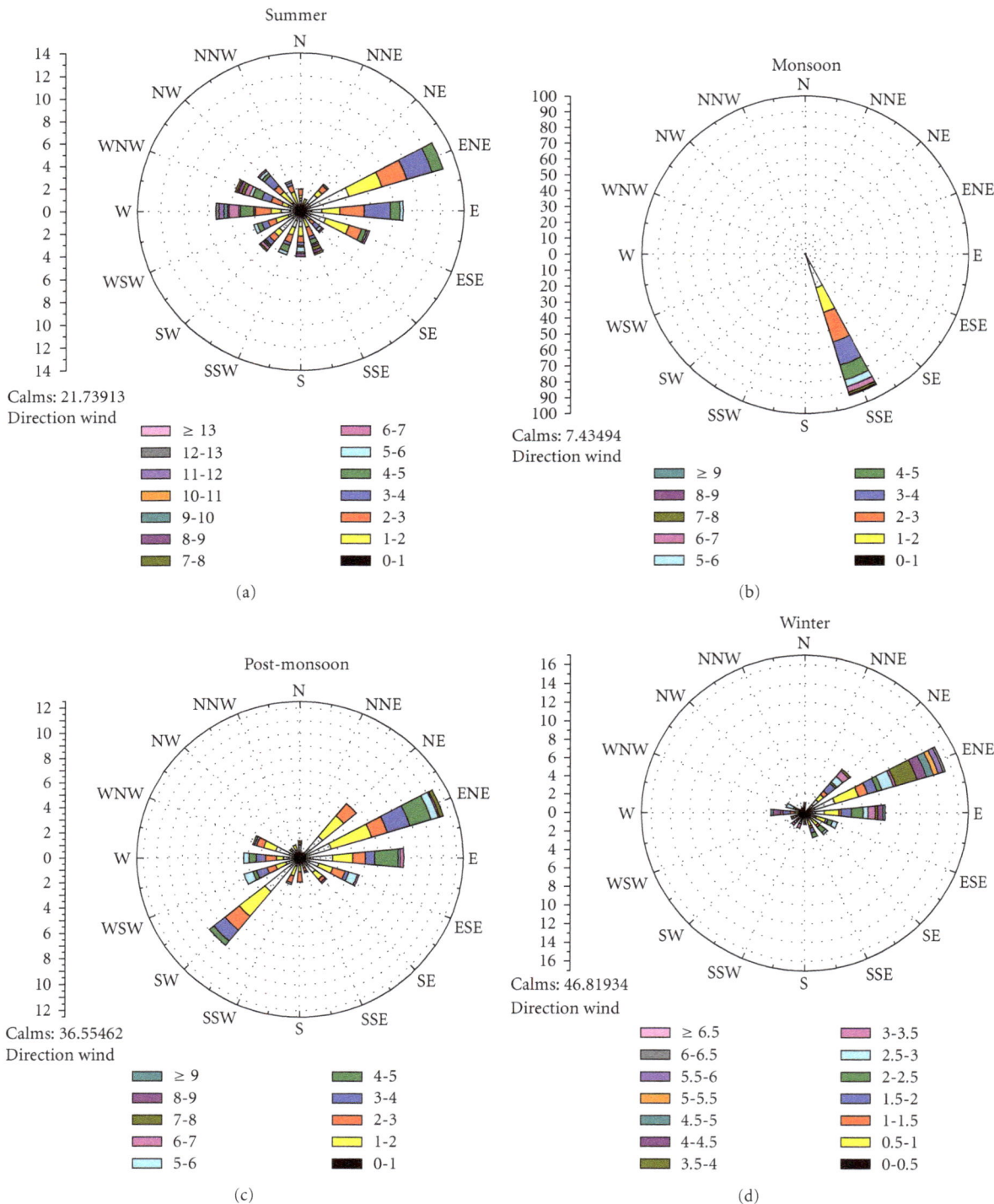

Figure 1: Local prevailing winds at sampling site.

equipped with a flame ionization detector (FID), BP1 capillary column (25 m length and 0.3 mm internal diameter), and GC Solution software. GC oven was programmed for 50°C hold for 4 min and ramped to 250°C at a rate of 10°C/min with 10 min hold at 250°C. Injector and detector temperature was maintained at 250°C. Nitrogen was used as a carrier gas with flow rate of 1 mL min^{-1} and split ratio

1:10. The external calibration standard mixture (MISA non-halogenated volatiles group 17 mix) containing benzene, toluene, ethylbenzene, o-xylene, m-xylene, p-xylene, and styrene, procured from Sigma-Aldrich, was used for the calibration. The samples were quantified against a five-point calibration curve prepared using standards between 2 and 20 μg/mL. Duplicate measurements were done for more than

TABLE 1: Range of temperature, relative humidity, and wind speed observed during the study period.

Sampling period	Temperature (°C)	Humidity (%)	Wind speed (km h^{-1})
April'10	17–45	4–27	5–12.5
May'10	23–46	4–45	9–24
June'10	25–48	12–63	4–19
July'10	24–40	41–94	3–17
August'10	22–38	45–100	3–12
September'10	19–39	56–100	1–14
October'10	12–36	55–100	1–17
November'10	7–31	48–100	1–13
December'10	3–25	84–100	1–3
January'11	2–24	53–100	1–4
February'11	5–29	57–100	1–7
March'11	8–37	11–66	2–12

10% of the samples so as to control the quality of the samples. A blank sample (unexposed tubes) was also run to check for any contamination of the tubes.

2.5. Mutagenicity Assay. Mutagenicity assay in the extracts was conducted by the Ames test [25]. In the Ames test, bacteria are mixed with the test substance and then incubated. The number of revertant colonies, called revertants, formed after incubation, serves as an index of activity. Most mutagens and carcinogens identified in air such as certain unsubstituted VOCs require the addition of liver homogenates (referred as S9) to the incubation mixture before they become mutagenic in the Ames test. The liver contains enzymes that metabolically activate mutagens to reactive intermediates. Such compounds are called as indirect-acting mutagens. In this study, two tester strains of Salmonella typhimurium (his$^-$) TA100 and TA98 suggested for base-pair mutation and frame shift mutation, respectively, were used. These strains are shown to be most sensitive to mutagens of extract of VOCs [26]. These strains were obtained as lyophilized cultures from Microbial Type Culture Collection and Gene Bank (MTCC), Institute of Microbial Technology (IMTECH), Chandigarh, India.

For performing the Ames test, the extracts of sample were dissolved in 4 mL DMSO, immediately before use. The bacterial strains were initially cultured in culture tubes containing beef extract agar medium, 15–18 h before the experiment. The test was performed in plates containing two layers of agar, the bottom agar (1.5%) providing a suitable support media and the top agar (0.07%) for applying the test chemical and the desired strain. All reagents were prepared as per Maron and Ames [25]. Vogel-Bonner minimal media E (50X) supplemented with agar and glucose was used as bottom agar (25 mL). 2 mL of the top agar consisted of agar, NaCl, and histidine/biotin. The extracts were applied to the plate at dose levels of 1, 2, 4, 8, 16, and 32 μL plate^{-1}. In addition to the test plates, a negative control consisting of DMSO solvent blank was used to establish the number of colonies that arise spontaneously for each of the tester strains. Positive control containing known mutagens, dibenzanthracene (DbA) and chrysene (Chy), was also included to confirm

the reversion properties and specificities of the strains. DbA and Chy were applied at five dose levels 10, 20, 30, 45, and 60 μg mL^{-1}. The plates were incubated at 37°C for 63 hours in dark in inverted position and subsequently the revertant colonies formed in each plate were counted. Counting was performed using a handheld digital colony counter (LA 663, Himedia). Mutagenic response was classified as positive if a reproducible dose-dependent increase of the number of revertant colonies over and above those in the negative control was observed [27]. Mutagenic activity was calculated from the positive slope of the dose-response curve using the regression coefficient b, where $y = a + bx$. Mutagenicity was expressed as number of revertant colonies μL^{-1} of extract. The bacterial background lawn was regularly checked by microscopy, as high doses of the extracts may prove toxic to the tester strains, resulting in a thinning out of background. If massive cell death occurs, the background lawn on the test plates would be sparse compared to control plates.

3. Results and Discussion

3.1. BTX Concentrations in Ambient Air. The minimum, maximum, and mean concentration of BTX measured at the roadside and petrol pump site is shown in Table 2. Vehicular emissions are reported to be responsible for more than 60% of total VOC emissions [28, 29] and, therefore, traffic seems to be a major source of ambient VOCs in many urban areas. The concentration of BTX at any site would depend upon traffic density and fleet composition. Benzene and toluene with an average value of $14.7 \pm 2.4 \mu$g m^{-3} and $8.1 \pm 1.2 \mu$g m^{-3}, respectively, were observed to be most abundant at roadside. The mean concentration of p-xylene and o-,m-xylene was found to be $5.1 \pm 1.2 \mu$g m^{-3} and $2.1 \pm 0.8 \mu$g m^{-3}, respectively. Similar variation was also observed at petrol pump but the levels were elevated.

The higher concentration of benzene and toluene might be probably due to their emission from gasoline (major fuel of vehicles) which is known to contain high proportion of these compounds, their longer lifetime, and lesser reactivity with OH radical as compared to xylenes (shorter lifetime and

TABLE 2: Concentrations of BTX ($\mu g\,m^{-3}$) at roadside and petrol pump.

Hydrocarbon	Roadside ($N = 74$)			Petrol pump ($N = 50$)		
	Minimum	Maximum	Mean ± standard deviation	Minimum	Maximum	Mean ± standard deviation
Benzene	5.1	20.4	14.7 ± 2.4	9.3	25.8	19.5 ± 3.7
Toluene	2.7	9.9	8.1 ± 1.2	2.1	25.2	12.9 ± 1.1
o-,m-xylene	1.5	3.6	2.1 ± 0.8	4.8	9.9	3.6 ± 0.5
p-xylene	3.9	6.9	5.1 ± 1.2	4.8	13.2	11.1 ± 1.5

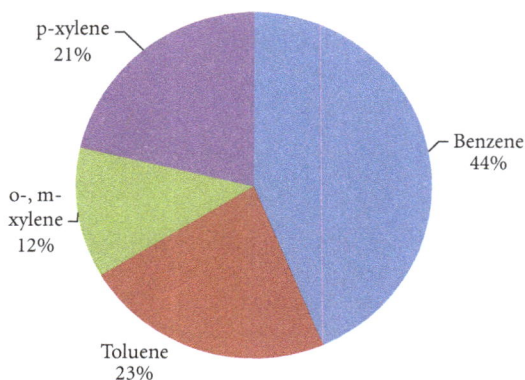

FIGURE 2: Percent abundance of VOCs at roadside.

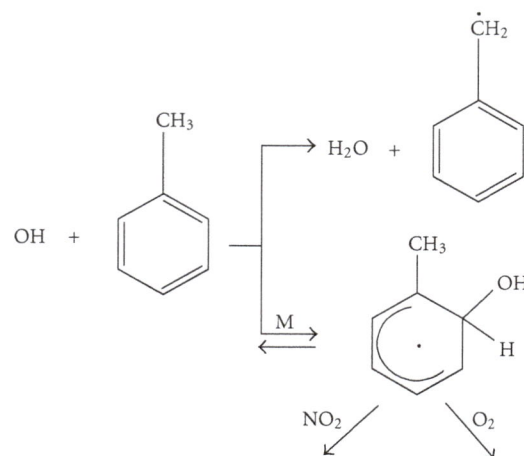

SCHEME 1

higher reactivity). Figure 2 shows the percent abundance of observed BTX. It is observed that benzene (44%) and toluene (23%) contribute to more than 50% of total VOCs (sum of benzene, toluene, o-,m-xylene, and p-xylene) emitted.

Benzene and toluene with atmospheric lifetimes of 12.5 and 2 days, respectively, are known to be relatively stable and do not dissipate into the environment immediately after release. Xylene, however, has a lifetime of 7.8 h only and usually does not remain long in the atmosphere [15]. Further, benzene and toluene being monoaromatic hydrocarbons react very slowly with O_3 and NO_3 radicals, their rate constants being in the order of $<10^{-20}$ and 10^{-16} $cm^3\,mole^{-1}\,s^{-1}$ [30]. Thus their depletion in the atmosphere via these reactions is negligible. The only significant process for their atmospheric loss is their photochemical reaction with OH radicals, having rate constants 1.23×10^{-16} and 5.96×10^{-12} $cm^3\,mol^{-1}\,s^{-1}$, respectively [30].

As evident from temperature-dependent kinetic studies, the OH radical reactions proceed by two pathways; H-atom abstraction from the C–H bonds of the alkyl substituent groups, and addition of the OH radical to the aromatic ring as shown in Scheme 1.

The photochemical reactions with OH radicals are known to have temperature dependence. During the study period (April 2010 to March 2011), the temperature varied from approximately 300–320°K in summer months and 285–300°K in winter months. Based on temperature conditions, it is expected that the removal of BTX would have probably taken place through abstraction of H-atom mechanism in summer months which proceeds under higher temperature conditions. Therefore, comparatively lower levels of BTX

were observed in summer months than winter months. The annual variation of VOCs is better reflected if the concentration is plotted as a function of temperature observing a significant negative correlation between benzene concentration and temperature as observed in previous studies [31, 32]. As expected, considerable negative correlation ($R^2 = 0.73$) was observed between benzene concentration and temperature as shown in Figure 3. Toluene and xylenes are largely used as solvents and, therefore, their existence may come from sources other than traffic too. In addition, photochemical degradation of toluene and xylenes is faster than that of benzene [31], implying that it is easier to see the relationship between benzene and traffic density.

Figure 4 shows the comparison of VOCs observed at two sites. It is observed that comparatively higher concentrations of benzene and toluene were observed at petrol pump than roadside measurements. The concentration of Benzene varied from $9.3\,\mu g\,m^{-3}$ to $25.8\,\mu g\,m^{-3}$, Toluene varied from $2.1\,\mu g\,m^{-3}$ to $25.2\,\mu g\,m^{-3}$, o-,m-xylenes varied from $4.8\,\mu g\,m^{-3}$ to $9.9\,\mu g\,m^{-3}$, and p-xylene varied from $4.8\,\mu g\,m^{-3}$ to $13.2\,\mu g\,m^{-3}$ at petrol pump. The concentration levels were found to be comparable with the levels reported at filling stations in France [33] and Greece [34]. It can be concluded that the air at the petrol pump does not have the same proportion of VOCs as the general city air. Presumably, the VOC concentrations at petrol pump are influenced by both moving vehicular traffic and the petrol pump activities. These activities include fuel storage and delivery, filling, and emptying activities of tanks containing petrol causing

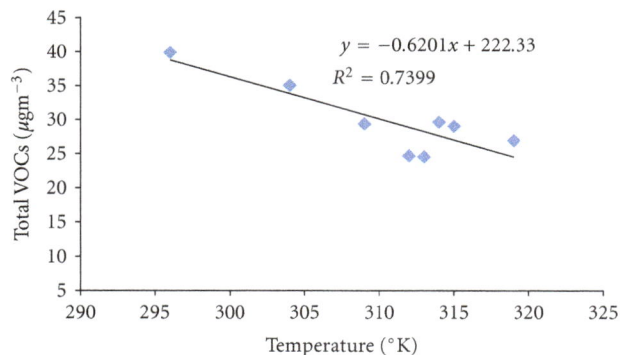

FIGURE 3: Benzene concentration (monthly mean) plotted as a function of temperature.

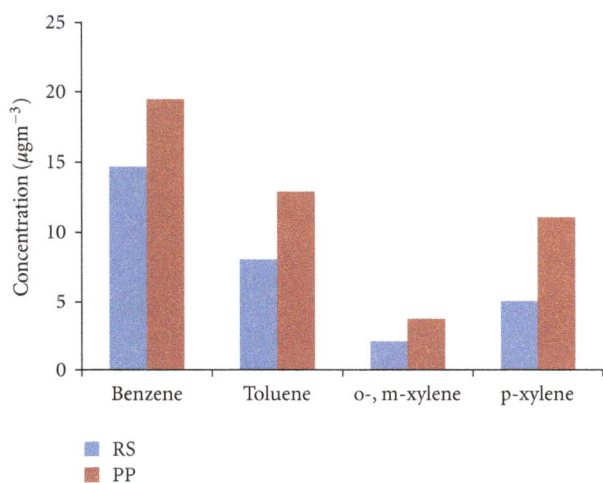

FIGURE 4: Comparison of BTX at roadside and petrol pump.

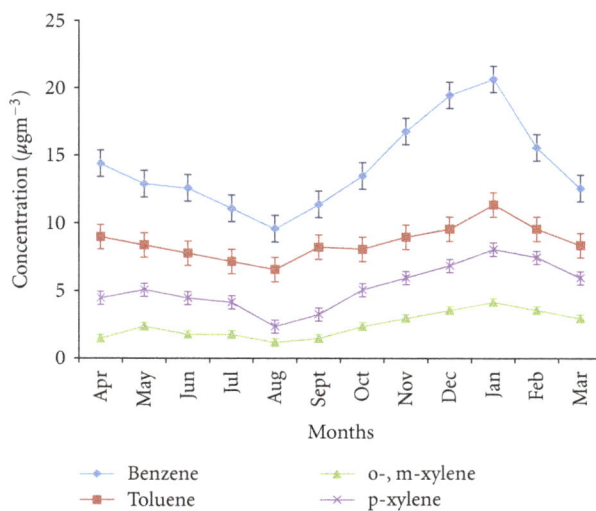

FIGURE 5: Monthly variation of BTX.

displacement losses. As we go further from petrol pump, the VOC concentrations are closer to that found in the general air.

The independent samples t-test was applied to further confirm whether the concentrations of benzene, toluene, o-,m-xylene, and p-xylene along the roadside differ significantly from the concentrations observed at petrol pump site. The null hypothesis indicates that concentrations of benzene, toluene, o-,m-xylene, and p-xylene along the roadside and petrol pump are similar. On applying t-test, t-value and P-value were found to be 2.63 and 0.018 for benzene, 3.42 and 0.004 for toluene, 1.55 and 0.146 for o-,m-xylene, and 2.34 and 0.063 for p-xylene. On comparing calculated t-value and P-value with the tabulated value, it was found that null hypothesis is rejected at 5% significance level for benzene and toluene and accepted for o-,m-xylene and p-xylene. Hence, t-test is found to be statistically significant with the 95% significance level for benzene and toluene, that is, the concentration of benzene and toluene along the roadside differs significantly from the concentration at petrol pump. This might be probably because benzene and toluene are constituents of gasoline and are emitted into the atmosphere by car exhausts [20]. Although the concentrations are nume-

rically higher at petrol pump site, t-test indicates variation at two sites is not statistically significant for o-,m-xylene and p-xylene.

3.2. Comparison with Other Sites. Table 3 presents the comparison of Benzene and Toluene levels at Agra with other sites. The levels of Benzene and Toluene were observed to be comparable to Northern Germany [31], Mexico [35], Naples [36], and Barelona City [32] and much lower than other megacities like Mumbai [24], Delhi [21], China [37], Taiwan [38], Hyderabad [39], Zabrze [40], and Kolkata [41]. Comparatively higher concentrations of Benzene and Toluene might be due to more traffic volume and hence greater exhaust emissions.

3.3. Monthly Variation of BTX. Figure 5 shows the monthly variations of the concentrations of the BTX measured at this site. Benzene was found to be the most abundant component followed by Toluene. BTX shows the expected summer minimum and winter maximum concentrations. Benzene, toluene, o-,m-xylene, and p-xylene varied from its maximum concentration $20.7\,\mu g\,m^{-3}$, $11.4\,\mu g\,m^{-3}$, $4.2\,\mu g\,m^{-3}$, $8.1\,\mu g\,m^{-3}$ in winter months, respectively, to as low as $9.6\,\mu g\,m^{-3}$, $6.6\,\mu g\,m^{-3}$, $1.2\,\mu g\,m^{-3}$, $2.4\,\mu g\,m^{-3}$ in summer months, respectively. Monthly variations are primarily dependent on emission patterns, meteorology, variations in the source strength, and most essentially photochemistry between VOCs and OH that results in the removal of VOC species from the atmosphere [42].

Agra has a continental type of climate with dry summers and winters associated with calm periods occurring for almost 40% of the time. Although traffic emissions are expected to remain reasonably constant throughout the year, the higher winter values might be probably due to calm conditions and stability of the atmosphere wherein dispersion and dilution of pollutants is restricted due to temperature inversion phenomenon and low mixing heights. Higher levels of BTX have also been reported by Hoque et al. [20]

TABLE 3: Comparison of benzene and toluene with other sites.

	Benzene (μg m^{-3})	Toluene (μg m^{-3})	Reference
Mumbai	13.4–38.6	10.9–33.5	[24]
Hannover (Northern Germany)	9.6	25.7	[31]
China	15.4–67.3	28.6–106.9	[37]
Mexico City Metropolitan Zone	5.29	28.22	[35]
Kaohsing City Site 1	10.97	43.36	[38]
Taiwan Site 2	13.28	54.49	
Delhi	174.7–369.4	—	[21]
Zabrze Aug-Sep 2001	0.3–145.4	0.4–100.7	[40]
Aug-Sep 2005	0.3–113.7	0.4–200.6	[32]
Barcelona City Rural	0.2–8.3	0.5–95.4	
Urban	0.5–12.4	4.3–121.3	
Naples Metropolitan Area	4.4–17.2	15.8–57.7	
Near Suburban Area	3.6–11.8	8.1–27.3	[36]
Far Suburban Area	2.3–12.8	5.6–30.3	
Hyderabad	120–173	110–126	[39]
Kolkata	13.8–72.0	21.0–83.2	[41]
Agra	14.7 ± 2.4	8.1 ± 1.2	Present study

TABLE 4: Interspecies ratio.

	Ratio
Toluene/benzene	0.74
Xylene/benzene	0.26

at Delhi. On the other hand, summer experiences unstable atmosphere associated with high solar radiation, high temperature, and frequent occasions of dust storms locally known as *andhi*. The high solar radiation acts as a catalyst for the easy dissociation of species like ozone, aldehydes, and so forth, leading to formation of OH radical. The higher concentration of OH radical and frequent dust storms in summer plays a key role in enhanced mixing and rapid dissipation of pollutants resulting in their lower levels in the atmosphere through atmospheric clean up and degradation of VOCs. Hakola et al. [43] have reported higher OH radical concentration in summer months as compared to winter months in Central Finland.

3.4. Interspecies Ratios. Interspecies ratios of the aromatic VOCs were used as an indicator to compare the BTX emission sources. The ratios of VOCs with varying reaction rates against OH provide information about the characteristics of air at the sampling site. Interspecies ratio is listed in Table 4.

Toluene-to-Benzene (T/B) ratio has been commonly used as an indicator of traffic emissions. Benzene and Toluene are chief constituents of gasoline emitted into the environment by vehicular exhausts. Lee et al. [13] suggested that T/B ratio increases with increasing traffic volume, industrial emissions, and other urban sources. In the present study, value of T/B ratio was observed to be 0.74. The (T/B) ratios found in the present study were comparable with those

observed in Thane (0.61; [24]) and Southern Taiwan (0.2–4.6; [44]) and lower than those found in Rome (2.8), Izmir (1.87–2), and Copenhagen (2.2) [14, 45, 46]. The difference of (T/B) ratio among these cities is due to differences in vehicle types, fuel composition, and industrial activities.

The lower values of (X/B) ratio (0.26) indicate higher Benzene concentration and suggest that more reactive species have been exposed to photochemical degradation. This lower (X/B) ratio also indicates aging of air mass. The ratio at this site is observed to be higher than ratio observed at Thane (0.06, [24]) and lower than China (0.71, [47]).

3.5. Interspecies Correlation. Several work [14, 36, 48] have used correlation analysis to elucidate the possible sources of the aromatic VOCs. A good mutual correlation among the species indicates that they might primarily originate from the same source.

In the present study, Pearson's correlation (2-tailed) of concentration of BTX was calculated (Table 5). A good mutual correlation between o-,m-xylene and benzene ($r = 0.9$) indicates that they might possibly originate from gasoline sources [15, 49]. Since benzene mainly comes from traffic source, it can be used as an indicator of other aromatic compounds in heavy traffic areas; but benzene shows lower "r" values with toluene and p-xylene thereby indicating spiking of these VOCs from some additional source apart from vehicular source. The additional VOC source might be a chain of shops of paints and varnishes at this site.

3.6. Ranking of BTX with respect to Ozone Formation Potential. The ranking of airborne pollutants on a mass concentration (μg m^{-3}) basis is of interest in order to assess human exposure to toxic compounds like benzene, it is also of interest to examine the relative importance of these pollutants for

TABLE 5: Interspecies correlation.

	Benzene	Toluene	o-,m-xylene	p-xylene
Benzene	1			
Toluene	0.7	1		
o-,m-xylene	0.9	0.6	1	
p-xylene	0.5	0.5	0.4	1

their role in photochemical smog formation [3] including production of ozone [4].

Generally, the maximum incremental reactivity (MIR) is popular in the assessment of ozone formation potential of various VOC compounds [50]. Carter's MIR is the amount (in grams) of ozone formed per gram of VOC added to an initial VOC-NOx mixture, indicating how much a compound may contribute to ozone formation in the air mass [4]. These unitless MIR coefficients are intended for use in relatively high NOx conditions, which may be used as an important tool in ozone control programs. The reactivity of VOC with OH radical depicts the ability of the hydrocarbon to form higher oxidized products like aldehydes, ketones, acids, organic peroxy radicals, and so forth.

The ranking of the BTX species according to mass concentration, ozone formation potential, and reaction with OH is given in Table 6. The rate constants of VOC-OH reactions and MIR coefficients were taken from the literature [3, 4, 51]. The concentration of benzene was found to be maximum ($14.7\,\mu g\,m^{-3}$) followed by toluene ($8.1\,\mu g\,m^{-3}$), p-xylene ($5.1\,\mu g\,m^{-3}$), and o-,m-xylene ($2.1\,\mu g\,m^{-3}$). Although the concentration of xylenes were observed to be minimum among the BTX, they were found to be the most dominant contributor to ozone formation based on the MIR scale and reaction of VOCs with OH. The results obtained are found to be in agreement with the observations of McCann and Ames [52] at Seoul and Grosjean et al. [53] at Porto Alegre. Benzene being the most abundant and hazardous species among BTX showed minimum potential to ozone formation. Toluene was found to have the second largest contribution to ozone formation.

3.7. Mutagenicity of VOCs. A number of environmental chemicals have been detected as mutagens in *in vitro* tests and have subsequently been identified as animal carcinogens. It is likely that environmental factors initiate most human cancer, and it is becoming increasingly apparent that the causative agents in these environmental factors are likely to act by damaging DNA, for example, cigarette smoke, asbestos, and known human chemical carcinogens [54]. With this view, Salmonella typhimurium test strains TA98 and TA100 were exposed to VOC samples without the addition of metabolizing enzymes (S9). Therefore, only direct acting mutagens were detected. Simultaneously, negative control of DMSO and positive control of known mutagens (DbA and Chy) were also tested. According to the criteria given by Maron and Ames [25], results were considered positive if the number of revertants on the plates containing the test concentrations gets double with the spontaneous reversion rate (with respect to negative control) and a reproducible dose-res-

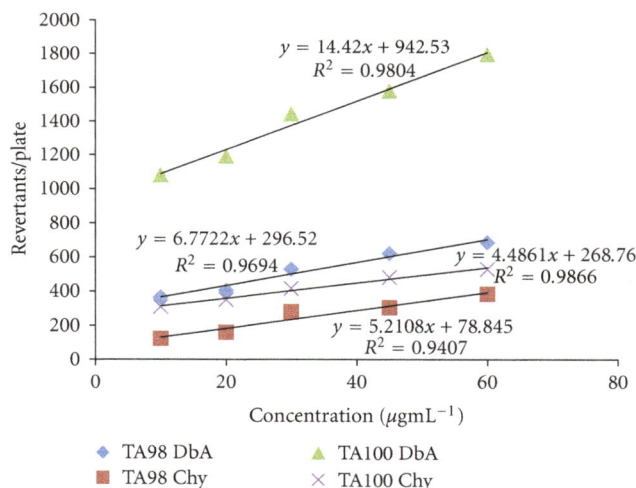

FIGURE 6: Dose-response curves for standard mutagens (DbA and Chy).

ponse relationship is obtained. The dose-response curves were obtained by regression plot of dose versus number of revertant colonies (Figure 6). The mutagenic activity of the extracts, expressed as the number of revertants μL^{-1} of extract was calculated from the slope of the linear dose-response curve: $y = a + bx$, where y is the number of revertants plate^{-1}, x is the concentration of the extract (μL plate^{-1}), a is the intercept, and b is the slope of the regression line (the number of revertants μL^{-1}).

Positive results were found for both the standard compounds (DbA and Chy) while the negative controls showed insignificant growth of colonies. From the slope of the regression line (Figure 6), it is evident that the mutagenic activity of DbA is 2-3 folds greater than Chy in both the strains, suggesting its greater carcinogenic potency as compared to Chy. This finding is also consistent with the order of potential carcinogencity and bioactivity of PAH [55].

Figure 7 shows the linear dose-response curves of VOC samples exposed to the two bacterial strains TA98 and TA100 within the range of the selected concentrations, that is, 1–$32\,\mu L^{-1}$. The number of revertants plate^{-1} increased linearly for TA98 strain TA100 strain. Hence, significant increase in the number of revertants plate^{-1} was observed. From the slope of regression line, it is observed that extracts of these samples induced and increased number of revertant colonies linearly with the increase of dose level. This linear increase is an indication of presence of both frame shift mutagens as well as mutagens capable of causing base pair substitutions. Further, the higher slope value of TA100 ($m = 4.248$) than TA98 ($m = 1.525$) indicated greater mutagenic activity of extract toward TA100 strain and, therefore, it was concluded that the extracts had a higher potential for base pair mutagenicity as compared to frame shift mutagenicity.

Several other researchers have also reported higher revertants plate^{-1} for TA100 in contrast to TA98 for organic extracts from diesel exhaust particles [56], biofuel combustion [57], and photocatalytic degradation of toluene [58]. Similar studies on mutagenicity assays have been conducted on

TABLE 6: MIR coefficient, VOC-OH rate constant, and ranking of BTX according to mass concentration, ozone formation potential, and reaction with OH leading to formation of oxidants.

Hydrocarbon	MIR coefficient	OH*	VOC ($\mu g\,m^{-3}$)	O_3 formation potential[a]	Reaction with OH[b]
Benzene	0.42	1.23	14.7	6.2	1.8
Toluene	2.7	5.96	8.1	21.9	12.6
o-,m-xylene	8.2	23.6	2.1	16.9	11.1
p-xylene	6.5	13.7	5.1	33.2	15.8

[a] VOC × MIR.
[b] VOC in ppb × VOC-OH* rate const. ($10-12\,cm^3$/molecule/s) multiplied by 10^{12}.

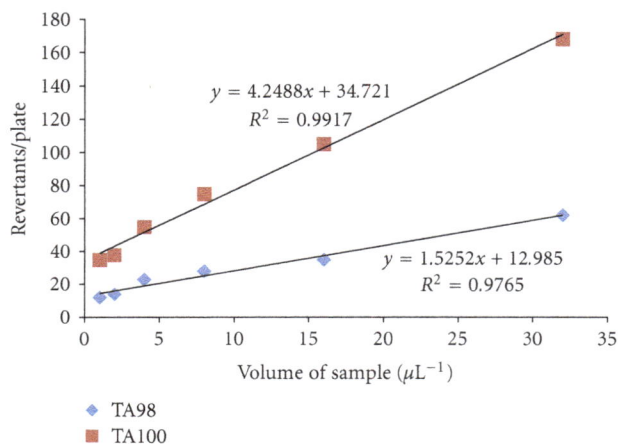

FIGURE 7: Dose-response curves for VOC extracts.

exhaust from mobile sources like natural gas fueled truck [59], diesel passenger cars [60], and motorcycles exhaust [61]. The higher mutagenicity of TA100 strain than TA98 strain was further validated by applying independent t-test. The null hypothesis indicates that the numbers of revertant colonies obtained using TA98 and TA100 strain are similar. The t-value and P-value were found to be 2.291 and 0.035, respectively. On comparing calculated t-value and P-value with the tabulated values, t-test is found to be significant with 95% significance level, and the null hypothesis is rejected at 5% significant level. Here, it is further concluded that the numbers of revertant colonies obtained using TA98 and TA100 strains are different.

4. Conclusion

Analysis of ambient air samples at two sites, roadside and petrol pump indicated that BTX concentration levels are mainly, influenced by the road traffic. The mean concentrations of benzene, toluene, o-,m-xylene, and p-xylene were $4.9\pm0.4\,\mu g\,m^{-3}$, $2.7\pm0.8\,\mu g\,m^{-3}$, $0.69\pm0.2\,\mu g\,m^{-3}$, and $1.7\pm0.2\,\mu g\,m^{-3}$, respectively, at roadside and the mean concentrations of benzene, toluene, o-,m-xylene, and p-xylene mean concentrations were $6.5\pm1.7\,\mu g\,m^{-3}$, $4.3\pm1.1\,\mu g\,m^{-3}$, $1.2\pm0.5\,\mu g\,m^{-3}$, and $3.7\pm0.5\,\mu g\,m^{-3}$, respectively, at petrol pump. The higher BTX concentrations at petrol pump were found to be statistically significant at 95% significance level and are presumably due to contribution from both moving vehicular traffic and petrol pump activities. BTX levels were higher

in winter than summer. The toluene/benzene ratio (0.74) and xylenes/benzene (0.26) ratios reflected the emission of road traffic and ageing of air mass, respectively. Interspecies correlation indicates the possible origin of BTX from gasoline sources. The extracts of samples were mutagenic. TA100 strains showed a greater mutagenic response indicating higher potential for base pair mutagenicity compared to frame shift mutagenicity and independent t-test was significant at 95% significance level.

Acknowledgments

The authors are thankful to the Director of Dayalbagh Educational Institute, Agra, and the Head of Department of Chemistry for necessary help. The authors gratefully acknowledge the financial support for this work which is provided by ISRO-GBP under AT-CTM Project. V. Singla is grateful to the above project for providing JRF.

References

[1] L. Zou, Y. Luo, M. Hooper, and E. Hu, "Removal of VOCs by photocatalysis process using adsorption enhanced TiO₂-SiO₂ catalyst," *Chemical Engineering and Processing: Process Intensification*, vol. 45, no. 11, pp. 959–964, 2006.

[2] R. Atkinson, "Atmospheric chemistry of VOCs and NO(x)," *Atmospheric Environment*, vol. 34, no. 12–14, pp. 2063–2101, 2000.

[3] W. P. L. Carter, "A detailed mechanism for the gas-phase atmospheric reactions of organic compounds," *Atmospheric Environment*, vol. 24, no. 3, pp. 481–518, 1990.

[4] W. P. L. Carter, "Development of ozone reactivity scales for volatile organic compounds," *Journal of the Air and Waste Management Association*, vol. 44, no. 7, pp. 881–899, 1994.

[5] W. P. L. Carter, J. A. Pierce, D. Luo, and I. L. Malkina, "Environmental chamber study of maximum incremental reactivities of volatile organic compounds," *Atmospheric Environment*, vol. 29, no. 18, pp. 2499–2511, 1995.

[6] R. G. Derwent, M. E. Jenkin, S. M. Saunders, and M. J. Pilling, "Photochemical ozone creation potentials for organic compounds in northwest Europe calculated with a master chemical mechanism," *Atmospheric Environment*, vol. 32, no. 14-15, pp. 2429–2441, 1998.

[7] J. Jeong, K. Sekiguchi, and K. Sakamoto, "Photochemical and photocatalytic degradation of gaseous toluene using short-wavelength UV irradiation with TiO₂ catalyst: comparison of three UV sources," *Chemosphere*, vol. 57, no. 7, pp. 663–671, 2004.

[8] H. B. Mansour, D. Barillier, D. Corroler, K. Ghedira, C. G. Leila, and R. Mosrati, "In vitro study of dna damage induced by acid orange 52 and its biodegradation derivatives," *Environmental Toxicology and Chemistry*, vol. 28, no. 3, pp. 489–495, 2009.

[9] H. Y. Choi, Y. J. Kim, H. K. Jeon, and J. C. Ryu, "Study on genotoxicity of crocin, a component of gardenia fruit, in bacterial and mammalian cell systems," *Molecular Cellular Toxicology*, vol. 4, pp. 285–292, 2008.

[10] E. Dybing, J. O'Brien, A. G. Renwick, and T. Sanner, "Risk assessment of dietary exposures to compounds that are genotoxic and carcinogenic-An overview," *Toxicology Letters*, vol. 180, no. 2, pp. 110–117, 2008.

[11] L. D. Claxton and G. M. Woodall, "A review of the mutagenicity and rodent carcinogenicity of ambient air," *Mutation Research*, vol. 636, no. 1–3, pp. 36–94, 2007.

[12] J. Krahl, G. Knothe, A. Munack et al., "Comparison of exhaust emissions and their mutagenicity from the combustion of biodiesel, vegetable oil, gas-to-liquid and petrodiesel fuels," *Fuel*, vol. 88, no. 6, pp. 1064–1069, 2009.

[13] S. C. Lee, M. Y. Chiu, K. F. Ho, S. C. Zou, and X. Wang, "Volatile organic compounds (VOCs) in urban atmosphere of Hong Kong," *Chemosphere*, vol. 48, no. 3, pp. 375–382, 2002.

[14] D. Brocco, R. Fratarcangeli, L. Lepore, M. Petricca, and I. Ventrone, "Determination of aromatic hydrocarbons in urban air of Rome," *Atmospheric Environment*, vol. 31, no. 4, pp. 557–566, 1997.

[15] V. Tiwari, Y. Hanai, and S. Masunaga, "Ambient levels of volatile organic compounds in the vicinity of petrochemical industrial area of Yokohama, Japan," *Air Quality, Atmosphere and Health*, vol. 3, no. 2, pp. 65–75, 2010.

[16] R. Kerbachi, M. Boughedaoui, L. Bounoua, and M. Keddam, "Ambient air pollution by aromatic hydrocarbons in Algiers," *Atmospheric Environment*, vol. 40, no. 21, pp. 3995–4003, 2006.

[17] ASTDR, *Toxicological Profile for Toluene. Agency for Toxic Substances and Disease Registry*, U.S. Public Health Service, U.S. Department of Health and Human Services, Atlanta, Ga, USA, 1992.

[18] T. Sorahan, L. J. Kinlen, and R. Doll, "Cancer risks in a historical UK cohort of benzene exposed workers," *Occupational and Environmental Medicine*, vol. 62, no. 4, pp. 231–236, 2005.

[19] M. Thammakhet, T. Villeneuve, V. Munisawang, P. Thavarungkul, and P. Kanatharana, "Monitoring of BTX by passive sampling in Hat Yai," *Songklanakarin Journal of Science and Technology*, vol. 26, supplement 1, pp. 151–160, 2004.

[20] R. R. Hoque, P. S. Khillare, T. Agarwal, V. Shridhar, and S. Balachandran, "Spatial and temporal variation of BTEX in the urban atmosphere of Delhi, India," *Science of the Total Environment*, vol. 392, no. 1, pp. 30–40, 2008.

[21] A. Srivastava, A. E. Joseph, S. Patil, A. More, R. C. Dixit, and M. Prakash, "Air toxics in ambient air of Delhi," *Atmospheric Environment*, vol. 39, no. 1, pp. 59–71, 2005.

[22] P. K. Srivastava, G. G. Pandit, S. Sharma, and A. M. Mohan Rao, "Volatile organic compounds in indoor environments in Mumbai, India," *Science of the Total Environment*, vol. 255, no. 1–3, pp. 161–168, 2000.

[23] G. Chattopadhyay, G. Samanta, S. Chatterjee, and D. Chakraborti, "Determination of benzene, toluene and xylene in ambient air of Calcutta for three years during winter," *Environmental Technology*, vol. 18, no. 2, pp. 211–218, 1997.

[24] A. M. Mohan Rao, G. G. Pandit, P. Sain, S. Sharma, T. M. Krishnamoorthy, and K. S. V. Nambi, "Non-methane hydrocarbons in industrial locations of Bombay," *Atmospheric Environment*, vol. 31, no. 7, pp. 1077–1085, 1997.

[25] D. M. Maron and B. N. Ames, "Revised methods for the Salmonella mutagenicity test," *Mutation Research*, vol. 113, no. 3-4, pp. 173–215, 1983.

[26] L. D. Claxton, "Characterization of automotive emissions by bacterial mutagenesis bioassay: a review," *Environmental Mutagenesis*, vol. 5, no. 4, pp. 609–631, 1983.

[27] J. McDonald and M. W. Spear, "Biodiesel: effects on exhaust constitutents," in *Plant Oils as Fuel: Present State of Science and Future Developments*, N. Martini and J. Schell, Eds., pp. 92–103, Springer, Berlin, Germany, 1998.

[28] J. Duan, J. Tan, L. Yang, S. Wu, and J. Hao, "Concentration, sources and ozone formation potential of volatile organic compounds (VOCs) during ozone episode in Beijing," *Atmospheric Research*, vol. 88, no. 1, pp. 25–35, 2008.

[29] A. Kumar and S. K. Tyagi, "Benzene and toluene profiles in ambient air of Delhi as determined by active sampling and GC analysis," *Journal of Scientific and Industrial Research*, vol. 65, no. 3, pp. 252–257, 2006.

[30] B. J. Finlayson-Pitts and J. N. J. Pitts, *Chemistry of the Upper and Lower Atmosphere*, Academic Press, San Diego, Calif, USA, 2000.

[31] E. Ilgen, N. Karfich, K. Levsen et al., "Aromatic hydrocarbons in the atmospheric environment: part I. Indoor versus outdoor sources, the influence of traffic," *Atmospheric Environment*, vol. 35, no. 7, pp. 1235–1252, 2001.

[32] E. Gallego, F. X. Roca, X. Guardino, and M. G. Rosell, "Indoor and outdoor BTX levels in Barcelona City metropolitan area and Catalan rural areas," *Journal of Environmental Sciences*, vol. 20, no. 9, pp. 1063–1069, 2008.

[33] N. Gonzalez-Flesca, S. Vardoulakis, and A. Cicolella, "BTX concentrations near a Stage II implemented petrol station," *Environmental Science and Pollution Research*, vol. 9, no. 3, pp. 169–174, 2002.

[34] S. P. Karakitsios, C. L. Papaloukas, P. A. Kassomenos, and G. A. Pilidis, "Assessment and prediction of exposure to benzene of filling station employees," *Atmospheric Environment*, vol. 41, no. 40, pp. 9555–9569, 2007.

[35] H. Bravo, R. Sosa, P. Sánchez, E. Bueno, and L. González, "Concentrations of benzene and toluene in the atmosphere of the southwestern area at the Mexico City Metropolitan Zone," *Atmospheric Environment*, vol. 36, no. 23, pp. 3843–3849, 2002.

[36] P. Iovino, R. Polverino, S. Salvestrini, and S. Capasso, "Temporal and spatial distribution of BTEX pollutants in the atmosphere of metropolitan areas and neighbouring towns," *Environmental Monitoring and Assessment*, vol. 150, no. 1–4, pp. 437–444, 2009.

[37] X. M. Wang, G. Y. Sheng, J. M. Fu et al., "Urban roadside aromatic hydrocarbons in three cities of the Pearl River Delta, People's Republic of China," *Atmospheric Environment*, vol. 36, no. 33, pp. 5141–5148, 2002.

[38] C. H. Lai, K. S. Chen, Y. T. Ho, and M. S. Chou, "Characteristics of C2-C15 hydrocarbons in the air of urban Kaohsiung, Taiwan," *Atmospheric Environment*, vol. 38, no. 13, pp. 1997–2011, 2004.

[39] P. V. Rekhadevi, M. F. Rahman, M. Mahboob, and P. Grover, "Genotoxicity in filling station attendants exposed to petroleum hydrocarbons," *Annals of Occupational Hygiene*, vol. 54, no. 8, pp. 944–954, 2010.

[40] H. Pyta, "BTX air pollution in Zabrze, Poland," *Polish Journal of Environmental Studies*, vol. 15, no. 5, pp. 785–791, 2006.

[41] D. Makumdar, A. K. Mukherjeea, and S. Sen, "BTEX in ambient air of a Metropolitan City," *Journal of Environmental Protection*, vol. 2, pp. 11–20, 2011.

[42] H. B. Singh and P. B. Zimmerman, "Atmospheric distributions and sources of nonmethane hydrocarbons," in *Gaseous Pollutants: Characterization and Cycling*, J. O. Nriagu, Ed., pp. 177–225, John Wiley & Sons, New York, NY, USA, 1992.

[43] H. Hakola, V. Tarvainen, T. Laurila, V. Hiltunen, H. Hellén, and P. Keronen, "Seasonal variation of VOC concentrations above a boreal coniferous forest," *Atmospheric Environment*, vol. 37, no. 12, pp. 1623–1634, 2003.

[44] L. T. Hsieh, H. H. Yang, and H. W. Chen, "Ambient BTEX and MTBE in the neighborhoods of different industrial parks in Southern Taiwan," *Journal of Hazardous Materials*, vol. 128, no. 2-3, pp. 106–115, 2006.

[45] R. Hartmann, U. Voght, G. Baumbach, R. Seyfioglu, and A. Muezzinoglu, "Results of emission and ambient air measurements of VOC in Izmir," *Environment Research Funders' Forum*, vol. 7-8, pp. 107–112, 1997.

[46] A. B. Hansen and F. Palmgren, "VOC air pollutants in Copenhagen," *Science of the Total Environment*, vol. 189-190, pp. 451–457, 1996.

[47] C. Liu, Z. Xu, Y. Du, and H. Guo, "Analyses of volatile organic compounds concentrations and variation trends in the air of Changchun, the northeast of China," *Atmospheric Environment*, vol. 34, no. 26, pp. 4459–4466, 2000.

[48] K. Na and Y. P. Kim, "Seasonal characteristics of ambient volatile organic compounds in Seoul, Korea," *Atmospheric Environment*, vol. 35, no. 15, pp. 2603–2614, 2001.

[49] J. M. Baldasano, R. Delgado, and J. Calbó, "Applying receptor models to analyze urban/suburban VOCS air quality in Martorell (Spain)," *Environmental Science and Technology*, vol. 32, no. 3, pp. 405–412, 1998.

[50] C. Hung-Lung, T. Jiun-Horng, C. Shih-Yu, L. Kuo-Hsiung, and M. Sen-Yi, "VOC concentration profiles in an ozone non-attainment area: a case study in an urban and industrial complex metroplex in southern Taiwan," *Atmospheric Environment*, vol. 41, no. 9, pp. 1848–1860, 2007.

[51] R. Atkinson, "Gas-phase tropospheric chemistry of volatile organic compounds: 1. Alkanes and alkenes," *Journal of Physical and Chemical Reference Data*, vol. 26, no. 2, pp. 215–290, 1997.

[52] J. McCann and B. N. Ames, "A simple method for detecting environmental carcinogens as mutagens," *Annals of the New York Academy of Sciences*, vol. 271, pp. 5–13, 1976.

[53] E. Grosjean, R. A. Rasmussen, and D. Grosjean, "Ambient levels of gas phase pollutants in Porto Alegre, Brazil," *Atmospheric Environment*, vol. 32, no. 20, pp. 3371–3379, 1998.

[54] K. Na, K. C. Moon, and P. K. Yong, "Source contribution to aromatic VOC concentration and ozone formation potential in the atmosphere of Seoul," *Atmospheric Environment*, vol. 39, no. 30, pp. 5517–5524, 2005.

[55] D. Hoffman, S. S. Herht, and E. L. Wynder, *In Modifying the Risk for the Smoker*, United States Department of Health and Human Services, 1976.

[56] E. Rivedal, O. Myhre, T. Sanner, and I. Eide, "Supplemental role of the Ames mutation assay and gap junction intercellular communication in studies of possible carcinogenic compounds from diesel exhaust particles," *Archives of Toxicology*, vol. 77, no. 9, pp. 533–542, 2003.

[57] S. Mukherji, A. K. Swain, and C. Venkataraman, "Comparative mutagenicity assessment of aerosols in emissions from biofuel combustion," *Atmospheric Environment*, vol. 36, no. 36-37, pp. 5627–5635, 2002.

[58] L. Sun, G. Li, S. Wan, and T. An, "Mechanistic study and mutagenicity assessment of intermediates in photocatalytic degradation of gaseous toluene," *Chemosphere*, vol. 78, no. 3, pp. 313–318, 2010.

[59] C. A. Lapin, M. Gautam, B. Zielinska, V. O. Wagner, and R. O. McClellan, "Mutagenicity of emissions from a natural gas fueled truck," *Mutation Research*, vol. 519, no. 1-2, pp. 205–209, 2002.

[60] S. Isotalo, T. Kuljukka-Rabb, L. Rantanen, S. Mikkonen, and K. Savela, "The effect of diesel fuel reformulation on the exhaust measured by ames mutagenicity and DNA adducts," *International Journal of Environmental Analytical Chemistry*, vol. 82, no. 2, pp. 87–95, 2002.

[61] H. H. Yang, S. A. Lee, D. P. H. Hsieh, M. R. Chao, and C. Y. Tung, "PM2.5 and associated polycyclic aromatic hydrocarbon and mutagenicity emissions from motorcycles," *Bulletin of Environmental Contamination and Toxicology*, vol. 81, no. 4, pp. 412–415, 2008.

6

Wind Tunnel Measurement of Turbulent and Advective Scalar Fluxes: A Case Study on Intersection Ventilation

Libor Kukačka,[1, 2] **Štěpán Nosek,**[2] **Radka Kellnerová,**[1, 2]
Klára Jurčáková,[2] **and Zbyněk Jaňour**[2]

[1] *Department of Meteorology and Environment Protection, Faculty of Mathematics and Physics, Charles University in Prague, V Holešovičkách 2, 180 00 Prague, Czech Republic*
[2] *Institute of Thermomechanics AS CR, v.v.i., Dolejškova 1402/5, 182 00 Prague, Czech Republic*

Correspondence should be addressed to Libor Kukačka, kukacka@it.cas.cz

Academic Editor: Alan W. Gertler

The objective of this study is to determine processes of pollution ventilation in the X-shaped street intersection in an idealized symmetric urban area for the changing approach flow direction. A unique experimental setup for simultaneous wind tunnel measurement of the flow velocity and the tracer gas concentration in a high temporal resolution is assembled. Advective horizontal and vertical scalar fluxes are computed from averaged measured velocity and concentration data within the street intersection. Vertical advective and turbulent scalar fluxes are computed from synchronized velocity and concentration signals measured in the plane above the intersection. All the results are obtained for five approach flow directions. The influence of the approach flow on the advective and turbulent fluxes is determined. The contribution of the advective and turbulent flux to the ventilation is discussed. Wind direction with the best dispersive conditions in the area is found. The quadrant analysis is applied to the synchronized signals of velocity and concentration fluctuation to determine events with the dominant contribution to the momentum flux and turbulent scalar flux.

1. Introduction

Dispersion of air pollution within urban areas is an important aspect of the environment quality for a significant part of the population. Vehicle emissions represent the main source of pollutants in large cities, Fenger [1], Colvile et al. [2]. The dispersion in street canyons determines a spatial distribution of pollutants and their dangerous impact. Short-time average concentrations measured especially in lower parts of the street canyons often reach threshold values. Improvement of air quality in urban areas is necessary to avoid risk for human health, Hoek et al. [3], Nyberg et al. [4].

We can define ventilation of an urban area as a process of changing polluted and fresh air within street canyons, which improves the air quality. Ventilation is directly caused by horizontal and vertical transport of pollution out from the area.

Wind tunnel investigations provide an environment where flow and dispersion can be explored in relatively stationary conditions and allow facile changes of model geometry. Several wind tunnel studies focused on concentrations within canyon for a tracer emitted at street level and flow perpendicular to the street, Kastner-Klein and Plate [5], Pavageau and Schatzmann [6]. The transport of pollution to the air above the roof level was estimated from measurements of concentrations in these works.

Wind tunnel and field studies for relatively symmetrical and regular street canyons arrangements express influence of geometry of streets and intersections in pollutant dispersion and hence ventilation in urban areas, see Brown et al. [7].

Mixing and transport processes in a simple street and its ventilation were elaborated by Belcher [8]. In this work ventilation fluxes were determined for estimation of the mean scalar transport within the urban street network. Barlow and Belcher [9] focused on studying the ventilation characteristics of a street canyon for the simple case of wind perpendicular to the street. Wind tunnel experiments published by Robins [10] show that the mass exchange between street canyons may be significantly changed due to

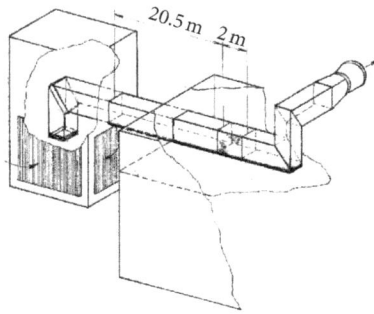

FIGURE 1: The scheme of the open low-speed wind tunnel.

a small variations of the building geometry. These results were obtained from computing scalar fluxes determining pollution transport. Results from numerical simulation published by Scaperdas and Colvile [11] show a very complex behaviour of the flow in an urban area. This work shows configuration of the street canyon and the wind direction when air exchange between alongwind and crosswind streets is reversed. Numerical and wind tunnel simulations of the flow and dispersion near regular and irregular street intersections were studied by Wang and McNamara [12].

Presented papers demonstrate high sensitivity of flow and dispersion processes to the intersection geometry and wind direction that are naturally connected with ventilation of an urban area.

Several publications have been focused on the air quality near street intersection in detail, for example, Dabberdt et al. [13]. Significantly higher pollution concentrations have been observed near intersections than along the streets with a continuous traffic, see Claggett et al. [14]. The reason of this observed phenomenon is that vehicles spend longer period of time near junctions, in driving modes that generate more pollutants (decelerating or accelerating), than in relatively steady movement in streets. The review of the traffic pollution modeling was published by Sharma and Khare [15].

The objective of this study is to determine processes of the traffic pollution transport within the X-shaped intersection in an idealized symmetric urban area for several approach flow directions. Pollutant is emitted into the urban area from the point source simulating "pollution hotspot"—the place with higher emission of traffic pollution situated near a junction, Soulhac et al. [16], Tomlin et al. [17].

2. Experimental Setup

2.1. Wind Tunnel. The experiment was conducted in the open low-speed wind tunnel of Institute of Thermomechanics Academy of Sciences of the Czech Republic in Nový Knín. The crossdimension of the tunnel test section was 1.5×1.5 m, the length of the test section was 2 m. The scheme of the tunnel is depicted in Figure 1.

Fully turbulent boundary layer was developed by the 20.5 m long development section of the tunnel. This section was equipped by turbulent generators at the beginning and covered by 50 mm and 100 mm high roughness elements on the floor, see the photo in Figure 2.

2.2. Urban Area Model. The model of idealized symmetric urban area with apartment houses was designed according to the common central European inner-city area. Regular blocks of apartment houses with pitched roofs formed a perpendicular arrangement of the street canyons and X-shaped intersections, see Figure 2.

The model was scaled down to $1:200$. The model buildings were formed by the body of height 100 mm and width 50 mm with pitched roof of height 20 mm. We setup the characteristic building height $H = 120$ mm (24 m in full scale) as the height of building body with the roof.

The width of street canyons was $L = 100$ mm. The aspect ratio of the street canyons given by the building height H and the street width S was $H/L = 1.2$.

A point pollution source simulating a "pollution hotspot" (the place with higher emission of traffic pollution situated near a junction) was placed at the bottom of the street canyon in front of the studied intersection, see Figure 2.

2.3. Measurement Techniques. The flow characteristics were measured using a two-dimensional optical fibre Laser Doppler Anemometry (LDA), based on DANTEC BSA F-60 burst processor. Tracing particles (glycerine droplets with approximately 1 μm diameter) were produced by a commercial haze generator placed at the beginning of the tunnel generating section, in front of turbulent generators. We got the air flow in the test section equally filled by seeding particles after running the haze generator inside the tunnel for several minutes. Data rate reached about 100 Hz at the bottom levels of street canyons $z \lesssim 0.5H$ and up to 1000 Hz at the roof top level $z \approx H$. The time of recording was 180 s in all the cases.

Point concentration measurements of tracer gas were realised by Slow-Response Flame Ionisation Detector (SFID) and Fast-Response Flame Ionisation Detector (FFID). We used SFID (type ROSEMONT NGA 2000) for mean concentration measurement within the studied intersection. Simultaneous vertical velocity and concentration measurement at the roof top level above the intersection were realised using LDA and FFID (type Cambustion Ltd. HFR400 Atmospheric Fast FID). The FFID was set to acquire data at a data rate of 1 KHz. The sampling time was 180 s in all of the cases.

We used ethane as the tracer gas simulating passive pollutants. Ethane is passive and nonreactive gas with its own density $\rho_{Ethane} = 1.24$ kg $\cdot$ m^{-3} close to density of the air $\rho_{Air} = 1.28$ kg $\cdot$ m^{-3}.

SFID and FFID were calibrated approximately every four hours of measurement. The differences in output voltage reached up to 5% through the measuring campaign. All the concentration values were computed from measured voltage signal using linear interpolated values from two calibrations realised before and after the recorded data set.

We applied a standard three-point calibration for the SFID measurement using clean air (air sucked into the wind tunnel from the atmosphere) and two span gases of known hydrocarbons concentrations.

For simultaneous velocity and concentration measurement, the four-point FFID calibration using clean air (air sucked into the wind tunnel from the atmosphere), air

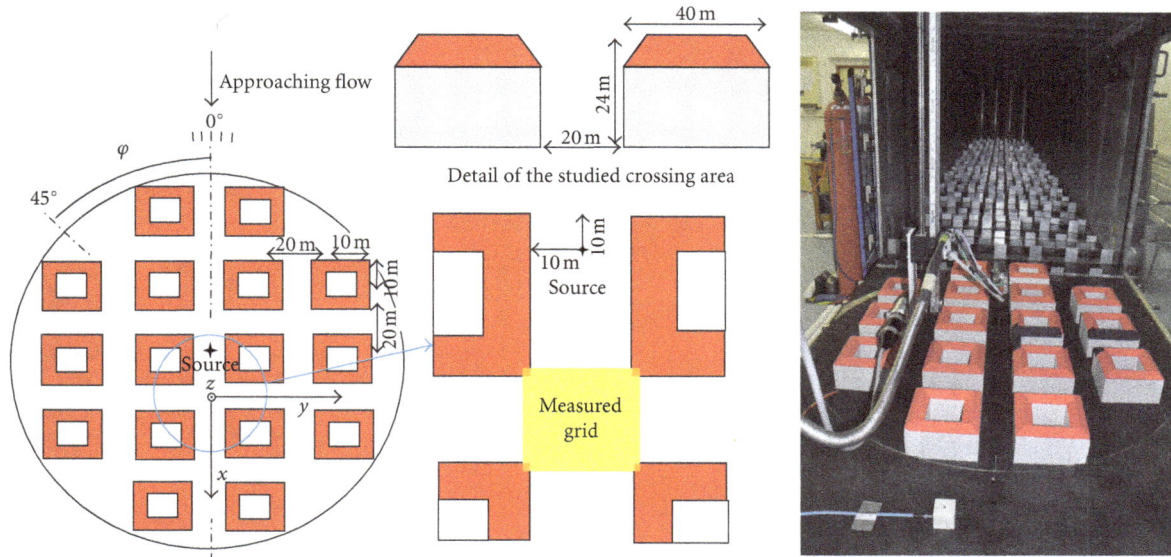

FIGURE 2: Scheme of the idealized symmetric urban area model (left), the studied X-shaped intersection (middle), and the photograph of the model placed in the wind tunnel (right).

equally filled by seeding particles, and two span gases of known hydrocarbons concentrations was obtained.

As expected, the presence of the seeding particles in the air during simultaneous LDA and FFID measurement influenced FFID output signal. At first we got isolated spikes in the recorded concentration signal probably due to suction of combustible aerosol particles from air into the FFID probe. The problem was mentioned by Hall and Emmott [18], Contini et al. [19]. Unlike these published results, we got similar count of spikes in time series obtained from measurements in clean air and in air contained seeding particles in most cases. We neglected the influence of spikes on the results because the frequency of isolated spikes was about 0.006% of used sampling data rate.

The second influence of seeding particles on the measured concentration data was an almost constant shift of recorded concentration values caused obviously by sucking seeding particles by FFID probe. This shift reached about 0.5% of the FFID measuring range. The shift was corrected by the calibration sequence mentioned above.

For simultaneous velocity and concentration measurement, LDA and FFID probes were mounted on the traverse system in a way that the measuring volume of the LDA was close to the intake of the FFID sampling tube. The sampling tube intake was placed 1.5 mm above, 1 mm behind, and 1 mm beside the centre of the LDA measuring volume. Several test measurements with different positions of both probes demonstrated a negligible influence of FFID sampling tube placed close to the LDA measuring volume on the flow. The configuration of probes is captured in Figure 3.

2.4. Boundary Layer Characteristics. Fully turbulent boundary layer was developed by spires and roughness elements placed it the tunnel. The characteristics of the boundary

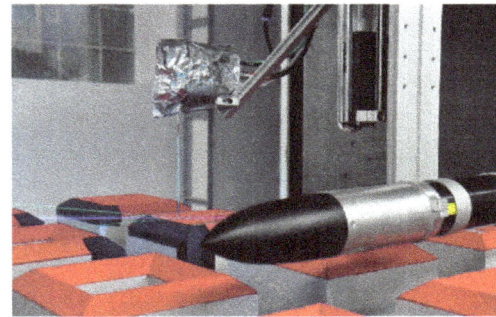

FIGURE 3: The configuration of the FFID (left) and LDA (right) probes mounted on the traverse system in the wind tunnel.

layer above the urban area model were measured with a two-dimensional LDA system in four vertical profiles placed above, upstream, and downstream from the studied intersection, see Figure 4.

The vertical profile of mean longitudinal velocity is depicted in Figure 5(a), the momentum flux profile can be found in Figure 5(b). The vertical profiles of longitudinal and vertical turbulent intensity are plotted in Figures 5(c) and 5(d). The high above the surface is expressed in full scale.

Vertical profiles of measured turbulent approach flow characteristics were fitted by the logarithmic and the power law. Mean roughness length z_0, displacement d_0, and friction velocity u_* (alias square-root of constant value of Reynolds stress within the inertial sublayer) were obtained from the logarithmic fit. Power exponent α was obtained from the power fit. The parameters are listed in Table 1.

Categories of boundary layer are defined according to classification in VDI [20]. Measured parameters corresponded to a neutrally stratified boundary layer flow above

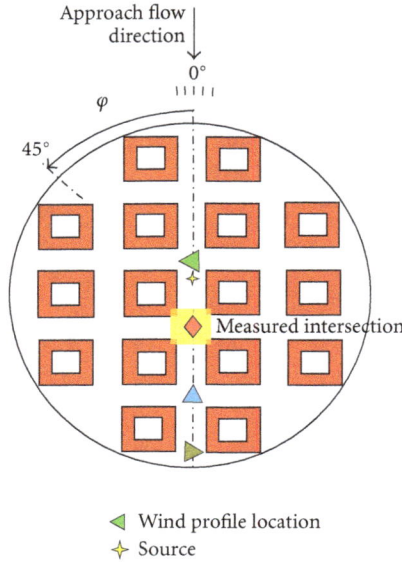

FIGURE 4: Wind profile measurement locations.

TABLE 1: Parameters of modelled boundary layer above the measured area (in full scale).

z_0 [m]	d_0 [m]	α [−]	u_*/U_{2H} [−]
0.83	13.40	0.24	0.096

a densely built-up area without much obstacle height variation.

To verify requirements for the Townsend hypothesis, see Townsend [21], the critical Reynolds building number Re_B was found. This criterion was used by Meroney et al. [22] and Pavageau and Schatzmann [6] for the flow within street canyons to be independent of viscous effects. The Reynolds building number modified for our experiment was given by

$$Re_B = \frac{U_{2H}H}{\nu}, \qquad (1)$$

where ν is kinematic viscosity. The experiment was carried out by $Re_B \approx 21000$ that lies on the lower edge of determined interval for valid Townsend hypothesis. Free stream velocity was approximately $4\,\mathrm{ms}^{-1}$.

3. Results

Horizontal velocity of the flow and concentration of the tracer gas was measured in vertical cuts (cross-sections) labelled A, B, C, and D. These cuts were placed in the exit planes of the street canyons connected to the studied intersection, see Figure 2. Cuts were placed 5 mm inward to the canyons because of the high gradients of measured quantities and the strongly unstable flow at the exact exit planes of the street canyons.

Furthermore, the vertical velocity and tracer gas concentration were simultaneously measured in a horizontal plane at the roof level $z = H$ above the studied intersection. We used a reference velocity U_{2H} measured at the reference height $z = 2H$. Results were obtained from five different values of the approach flow angle $\varphi = 0°$, $5°$, $15°$, $30°$, and $45°$.

In order to get an understandable image of the results, we used a transformation of the measured three-dimensional grid to a horizontal plane, see Figure 6. Vertical cuts of the measured grid were tipped out to the horizontal plane given by the roof level of the intersection. An orientation of horizontal velocity vectors in the vertical cuts was maintained in the transformed horizontal plain image.

3.1. Mean Velocity Fields. The flow inside the canopy was strongly three-dimensional and vortices of various scales are formed within and above the canyons and intersections. Measured components of velocity vector are expressed by the dimensionless form given by

$$\frac{U}{U_{2H}}, \qquad \frac{V}{U_{2H}}, \qquad \frac{W}{U_{2H}}, \qquad (2)$$

where U and V are the horizontal velocity components measured in vertical cuts placed in the exit planes of the street canyons connected to the studied intersection, W is vertical velocity of the flow measured in the horizontal plane at the roof level $z = H$ above the intersection. U_{2H} means a reference velocity measured at the reference height $z = 2H$.

A contour plots of velocity magnitude were added to the images of the velocity field. The orientation of horizontal velocity components is given by plotted vectors. The orientation of vertical velocity is given by a sign of the scalar values: the positive sign means an upward direction of vertical velocity and the negative sign means a downward direction.

A roughly symmetrical velocity field was formed by $\varphi = 0°$ (Figure 7(a)). The main stream was situated to an alongwind street parallel with the approach flow (Cuts A and C). A vortex with vertical axis was formed within the crosswind streets (Cuts B and D). The horizontal velocity decreased in levels towards the bottom of the street canyons (further form the middle of the picture). The vertical velocity on the top of the intersection was negligible in this case.

We observed an obvious change in the velocity field by $\varphi = 5°$ (Figure 7(b)). The main stream was still situated to an alongwind street, but the horizontal velocity increased in the left transverse street and decreased in the right transverse street. There was a small increase of upwards vertical velocity on the right side. A region with upward vertical velocity was formed near the right leeward corner.

As for the angle $\varphi = 15°$ as well as $\varphi = 30°$, a significant stream was formed within crosswind streets (Figure 7(c)). The increase in the upwards vertical velocity continued on the right side; however, it was not so important as in comparison with the changes of the vertical velocity.

An almost symmetrical velocity field was formed by $\varphi = 45°$ (Figure 7(d)). The main stream was divided into the alongwind and left crosswind streets (Cuts B and C). Asymmetry of flow was probably caused by minor geometrical deviations of the model in case of approach flow angle $\varphi \approx 0°$.

(a) The vertical profiles of mean longitudinal velocity

(b) The vertical profiles of mean momentum flux

(c) The vertical profiles of longitudinal turbulent intensity

(d) The vertical profiles of vertical turbulent intensity

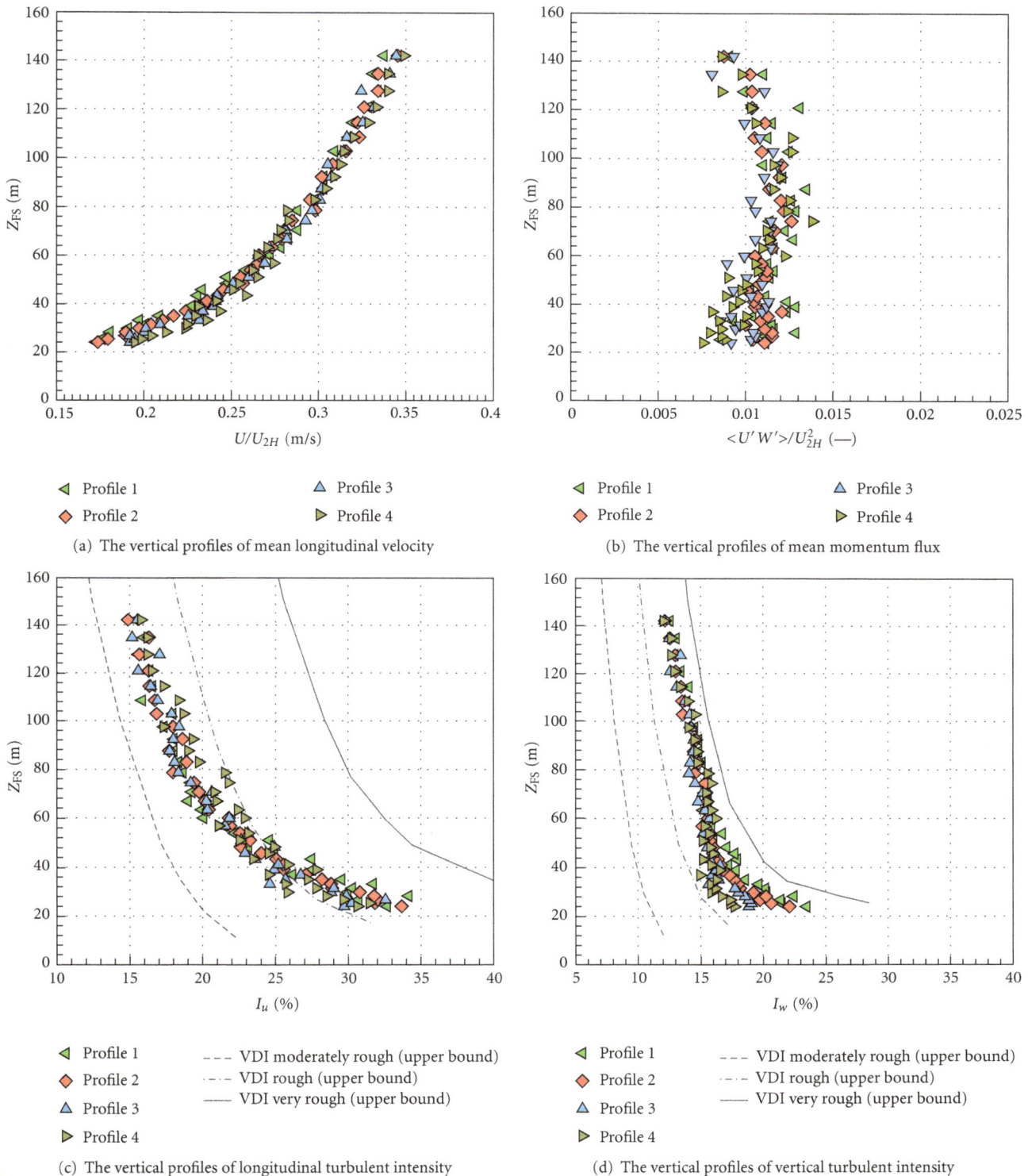

FIGURE 5: Boundary layer characteristics above the urban area model.

3.2. Mean Concentration Fields. The dimensionless concentration for a point source was obtained from the formula published in VDI [20]:

$$C^* = \frac{CU_{2H}H^2}{Q}, \quad (3)$$

where C means the measured concentration in and Q is a source emission volume flow.

Values of computed dimensionless concentration for five angles of the approach flow directions are plotted in Figure 8. A roughly symmetrical concentration field was

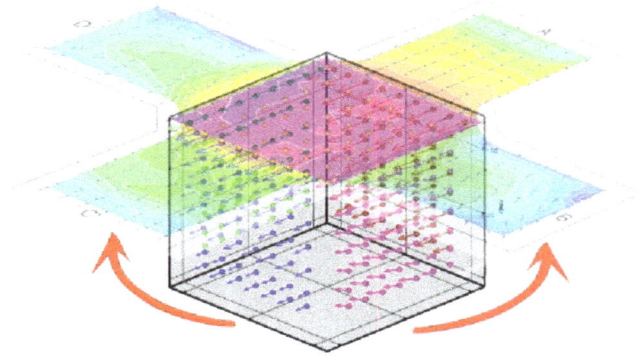

FIGURE 6: The transformation of the measured three-dimensional grid to an horizontal plane.

formed by $\varphi = 0°$ (Figure 8(a)), but notice slightly higher concentration in alongwind street (Cut D) in comparison with the left crosswind street (Cut B). There was almost zero concentration on the top of the intersection, which indicated weaker advective vertical transport of pollution than in the following cases.

We observed that a quite small change of the angle of the approach flow caused a radical change in concentration field by $\varphi = 5°$ (Figure 8(b)). An obvious deformation of the concentration field is probably caused mainly by street canyon vortices with horizontal axis (Cut A). The decrease in concentration in the right crosswind street (Cut D) was measured.

Transport of the majority of the tracer gas from the alongwind street (Cut A) to the left crosswind street (Cut B) was obvious by $\varphi = 15°$ (Figure 8(c)). Consequently there was almost zero concentration in Cut D. We measured the lowest concentrations in the intersection area in this case. We got a very similar concentration field for $\varphi = 30°$ (not shown).

We observed an overall increase in concentration by $\varphi = 45°$ (Figure 8(d)). There was an enhanced transport of the tracer gas to the right crosswind street up the approach wind (Cuts C and D) compared with case $\varphi = 15°$. It was probably caused by a small vortex with vertical axis at the leeward wall of this street. There was an area of significant concentration at the top of the intersection.

As we expected, the highest concentrations were measured at the ground levels in all cases and at the leeward wall of the street with the source (Cut A).

3.3. Advective Scalar Flux Fields.
The dimensionless advective scalar fluxes were computed from the average measured data to quantify advective spreading of pollutants within the studied intersection, see similar approach in Belcher [8], Robins [10]. We computed horizontal dimensionless advective fluxes using forms

$$\frac{C^*U}{U_{2H}}, \quad \frac{C^*V}{U_{2H}}, \quad (4)$$

where C^* is the mean dimensionless concentration of the tracer gas, U and V are the mean horizontal velocity com-

ponents of the flow. Vertical dimensionless advective flux given was given by

$$\frac{C^*W}{U_{2H}}, \quad (5)$$

where W is the mean vertical velocity of the flow. Results were obtained for all five values of the angle of the approach flow $\varphi = 0°, 5°, 15°, 30°,$ and $45°$.

The dimensionless advective scalar fluxes expressed a rate of emissions spreading through an unit area. Computed fluxes characterized the advective transport of pollution with the following convention of signs: the positive sign means the flux outwards and the negative sign means the flux inwards the studied intersection.

Values of computed fluxes for five angles of the approach flow directions are plotted in Figure 9. We can observe quite an asymmetrical flux field by $\varphi = 0°$ (Figure 9(a)). There is a higher flux into the right crosswind street (Cut D) than into the left crosswind street (Cut B). As we mentioned this was probably caused by minor geometrical deviations of the model. However, it means very strong sensitivity of scalar fluxes to the geometry of the model and approach flow direction. Notice a negative, that is, downward, flux at the top.

A roughly reversely spread flux field was formed by $\varphi = 5°$ (Figure 9(b)) compared to the case of $\varphi = 0°$. We could see a significant transport into the left crosswind street (Cut B).

A noticeable overall decrease in the flux in case of $\varphi = 15°$ was observed (Figure 9(c)). The lowest fluxes were measured in this case within the studied area. Emissions were transported mainly to the left crosswind street (Cut B). There was an area of the positive flux on the right side at the top of the intersection. We got similar flux field for $\varphi = 30°$ but with higher flux values.

A spreading of emissions mostly to the left side still predominated by $\varphi = 45°$ (Figure 9(d)). There was an increase in the flux especially in the left crosswind street (Cut B). There was mostly a positive flux at the top of the intersection.

3.4. Turbulent Scalar Flux Fields.
The dimensionless vertical turbulent scalar fluxes were computed from synchronised vertical velocity and concentration signals using eddy-correlation method, Arya [23], Stull [24].

The used Matlab postprocessing script synchronised simultaneously acquired vertical velocity and concentration data using the maximum of correlation between both signals. The synchronised time series were shifted by an average of 15 ms. This shift expressed the delay between a suck of the sample into the intake of the FFID probe tube and the moment of the sample analysing in the probe. The value of the shift agrees with very similar experimental setup published by Contini et al. [19].

The dimensionless vertical turbulent scalar flux is given by

$$\frac{\langle c^{*\prime}w^\prime \rangle}{U_{2H}}, \quad (6)$$

where $\langle \cdots \rangle$ mean a time average, $c^{*\prime}$ and $w^\prime$ indicate fluctuations of dimensionless concentration and vertical

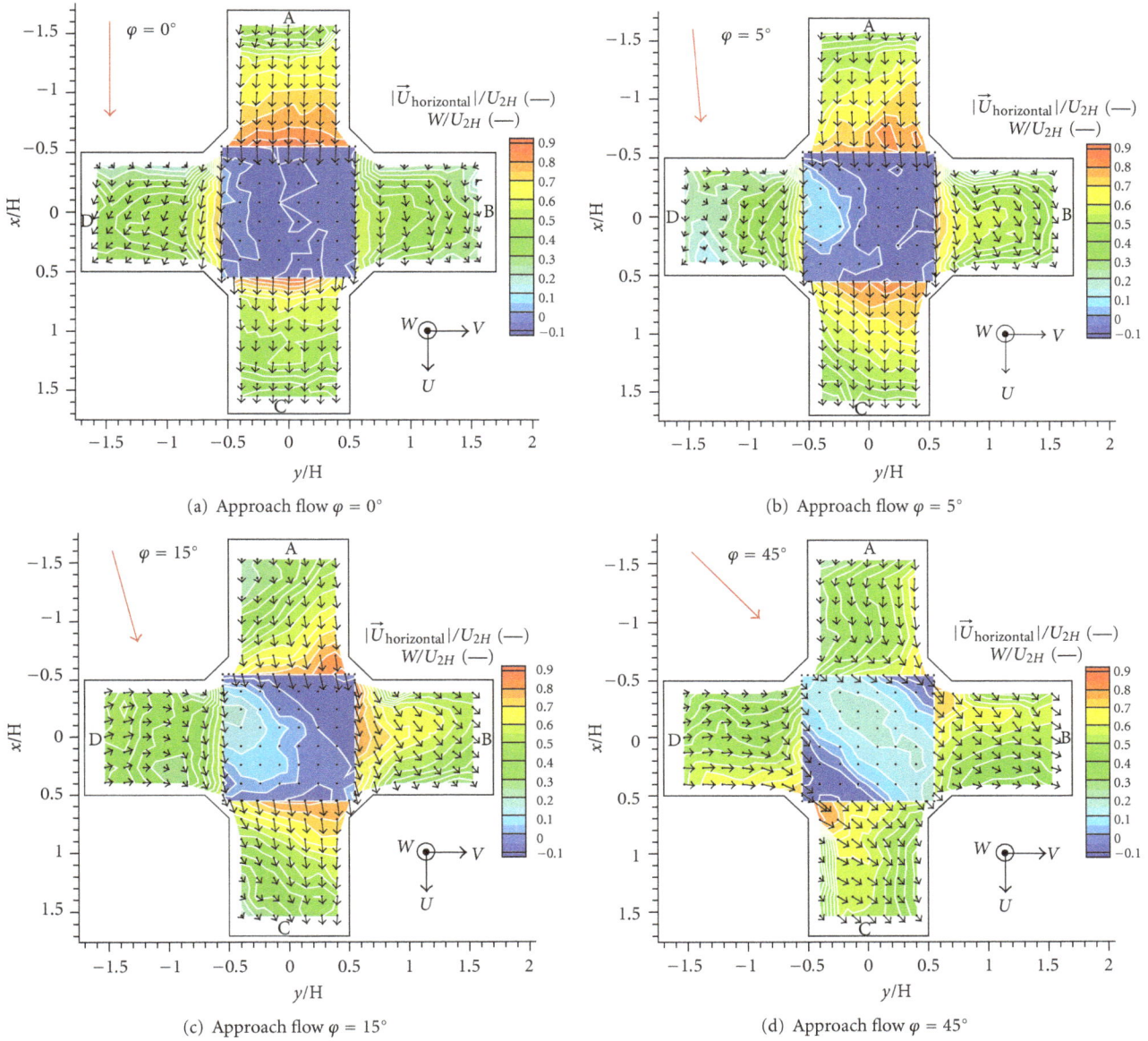

(a) Approach flow $\varphi = 0°$

(b) Approach flow $\varphi = 5°$

(c) Approach flow $\varphi = 15°$

(d) Approach flow $\varphi = 45°$

FIGURE 7: Dimensionless velocity fields for four angles of the approach flow direction.

velocity. Similar approach to turbulent transport computing was published in Jurčáková et al. [25].

Computed dimensionless vertical turbulent fluxes express a rate of emissions spreading through a unit area by turbulent transport with the same convention as mentioned above.

Values of determined vertical turbulent fluxes for the four approach flow directions are plotted in Figure 10. We measured relatively flat turbulent flux field by angle $\varphi = 0°$, but, compared with the advective flux, there is a positive turbulent transport of pollution, compare Figures 10(a) and 9(a).

In case $\varphi = 15°$, there are significantly positive values on the upwind side of the area (Figure 10(c)). The observed phenomenon became stronger by angle $\varphi = 45°$ (Figure 10(d)).

We estimated a significant turbulent transport of pollution near the leeward side of the buildings, see the upper part of Figures 10(a) and 10(b).

In comparison with the advective transport, the turbulent fluxes are positive in every case. The turbulent fluxes magnitude achieved almost two times the advective fluxes magnitude in the roof top level plane above the studied intersection.

3.5. *Quadrant Analysis.* We focused on the turbulent flow in vertical direction situated in the horizontal plane at the roof top level above the intersection in this part.

The first step to investigate the turbulent processes in strongly turbulent flow is the quadrant analysis, Kellnerová

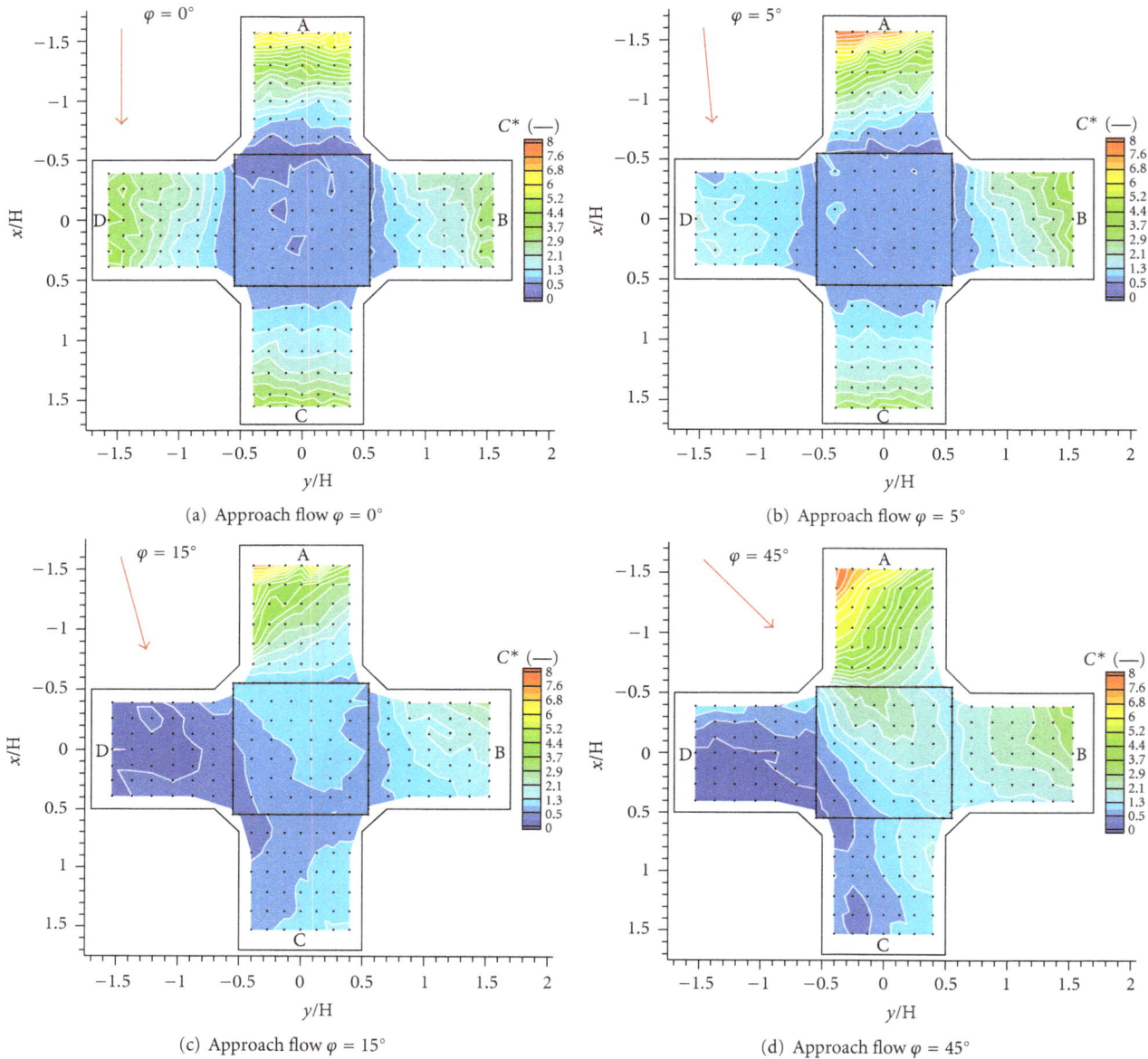

(a) Approach flow $\varphi = 0°$

(b) Approach flow $\varphi = 5°$

(c) Approach flow $\varphi = 15°$

(d) Approach flow $\varphi = 45°$

FIGURE 8: Dimensionless concentration fields for four angles of the approach flow direction.

et al. [26], Feddersen [27]. We applied this analysis to the velocity fluctuation time series to obtain contributions of the vertical flux of longitudinal momentum $\langle u'w' \rangle$ from particular quadrants defined as

1st quadrant "outward interaction" ($u' > 0$, $w' > 0$),

2nd quadrant "sweep" ($u' > 0$, $w' < 0$),

3rd quadrant "inward interaction" ($u' < 0$, $w' < 0$),

4th quadrant "ejection" ($u' < 0$, $w' > 0$).

These definitions are illustrated by Figure 11. The particular contribution from ith quadrant to the total momentum flux $\langle u'w' \rangle$ is given by

$$S_i = \frac{\langle u'w' \rangle_i N_i}{N_{\text{total}}}, \qquad (7)$$

where $\langle u'w' \rangle_i$ is the average stress and N_i is the number of events in the ith quadrant, number of all measured events is N_{total}.

The relative contribution R of the prevailing event to the total momentum flux is given by

$$R = \frac{S_{\max}}{\sum S_i} 100\%, \qquad (8)$$

where $S_{\max}$ is the particular contribution from the dominant event.

Relative contributions R of dominant events for four angles of the approach flow directions are plotted in Figure 12. As you see, ejections and sweeps are the prevailing events. Ejections characterize the upward transport of

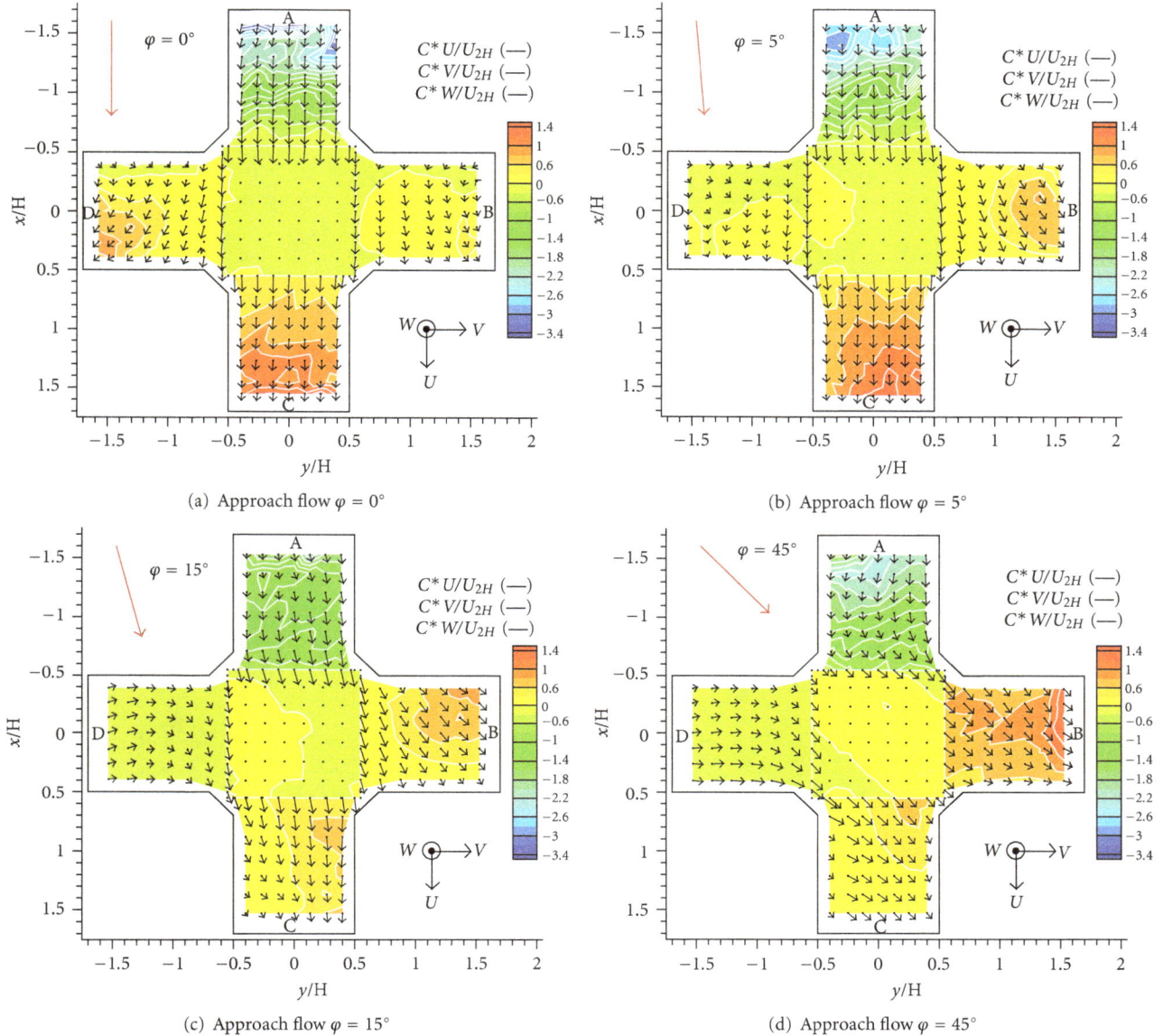

FIGURE 9: Horizontal and vertical dimensionless advective flux of passive contaminant with horizontal velocity vectors for four angles of the approach flow direction.

longitudinal momentum deficit, sweeps correspond to the downward transport of longitudinal momentum excess.

Ejections and sweeps were detected for the approach flow direction $\varphi \lesssim 5°$ with relatively small relative contribution to the mean momentum flux (Figures 12(a) and 12(b)). Large areas of sweeps with high contribution increased for higher angles $\varphi \gtrsim 15°$ caused probably by increasing magnitude of longitudinal velocity (Figures 12(c) and 12(d)).

We applied described quadrant analysis to the synchronized vertical velocity and concentration signals. In this case, particular quadrants are defined as

1st quadrant "outward interaction" ($c' > 0$, $w' > 0$),

2nd quadrant "sweep" ($c' > 0$, $w' < 0$),

3rd quadrant "inward interaction" ($c' < 0$, $w' < 0$),

4th quadrant "ejection" ($c' < 0$, $w' > 0$).

These definitions are illustrated by Figure 13.

Relative contributions R of dominant events for four angles of the approach flow directions are plotted in Figure 14. We observed outward interactions as the dominant event with high relative contribution for the approach flow angles $\varphi \sim 0°-15°$ (Figures 14(a)–14(c)). Inward interaction became dominant in part of the grid for the approach flow angles $\varphi \sim 45°$ but with low relative contribution (Figure 14(d)).

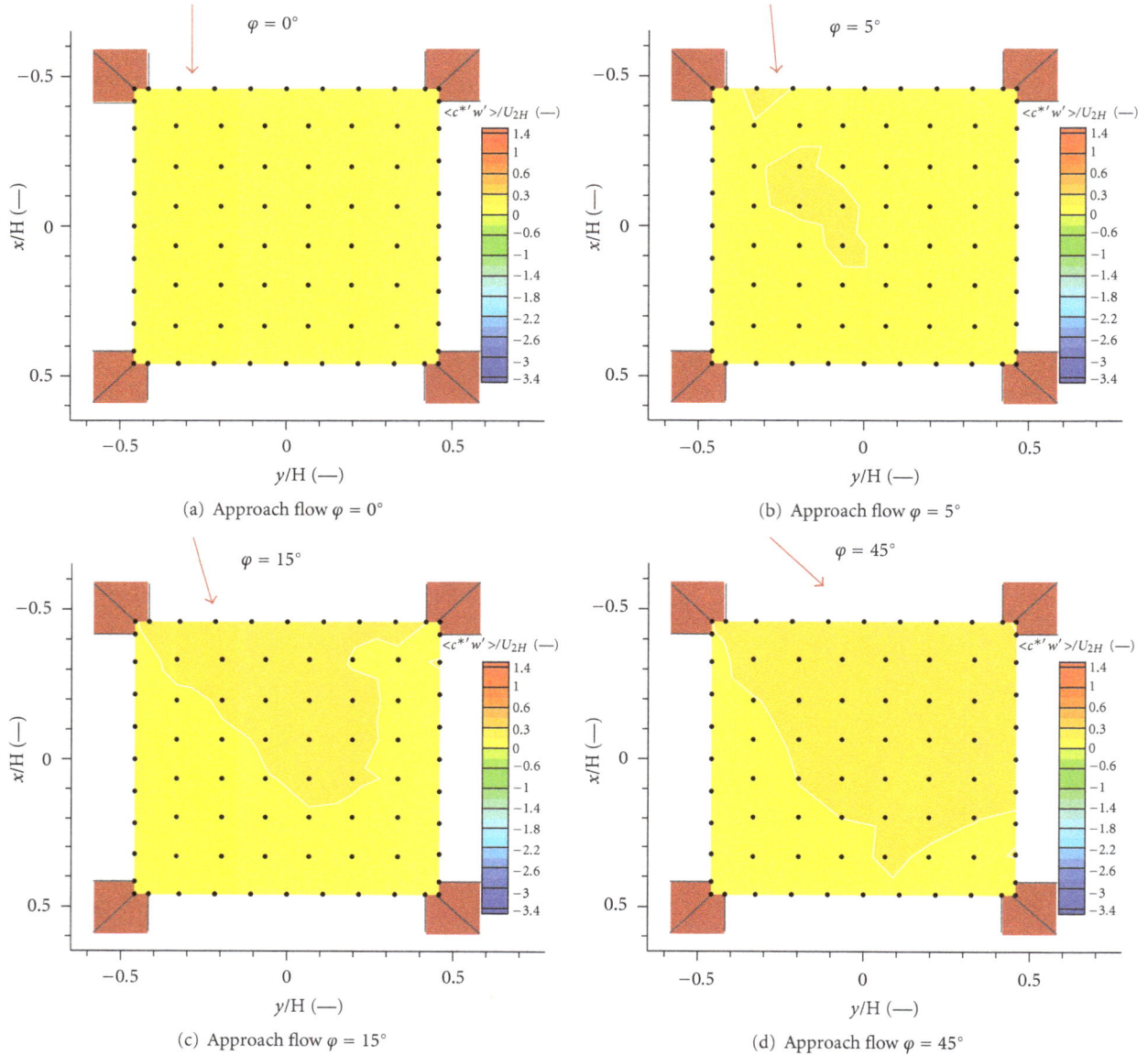

(a) Approach flow $\varphi = 0°$

(b) Approach flow $\varphi = 5°$

(c) Approach flow $\varphi = 15°$

(d) Approach flow $\varphi = 45°$

FIGURE 10: Vertical dimensionless turbulent scalar flux $\langle c^{*\prime} w' \rangle / U_{2H}$ for four angles of the approach flow direction.

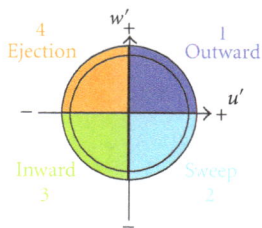

FIGURE 11: The scheme of event definitions used in velocity fluctuation quadrant analysis.

4. Conclusion

The described wind tunnel experiment quantified traffic pollutant dispersion within the X-shaped intersection in an idealized symmetrical urban area depending on the direction of the approach flow. The tracer gas is emitted into the urban area from the point source simulating "pollution hotspot," the place with higher emission of traffic pollution situated near a junction.

Velocity and concentration measurements were done by the building Reynolds number in the interval of Townsend hypothesis validity. We found out very complex flow and dispersion pattern within street canyons and high sensitivity to the approach flow direction. We determined a significant influence of the street canyon arrangements to the horizontal velocity in lower parts of the canyons at vertical levels of $z \lesssim 0.5H$. The highest concentration of pollution occurred at the bottom levels of streets.

Computed dimensionless advective scalar fluxes of contaminant showed spreading of pollution mostly within

(a) Approach flow $\varphi = 0°$

(b) Approach flow $\varphi = 5°$

(c) Approach flow $\varphi = 15°$

(d) Approach flow $\varphi = 45°$

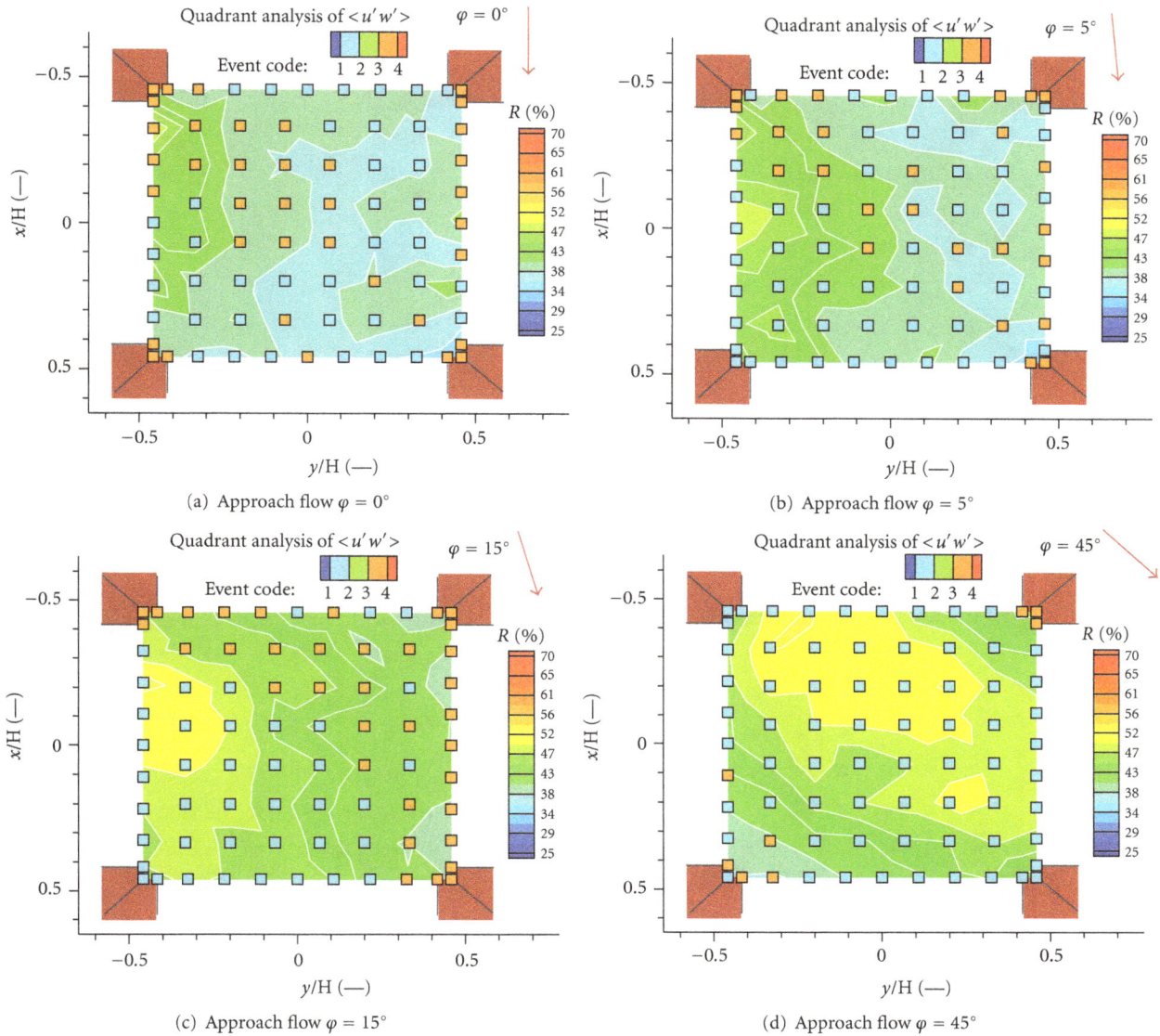

FIGURE 12: Relative contributions R of dominant event to the total momentum flux $\langle u'w' \rangle$ for four angles of the approach flow direction.

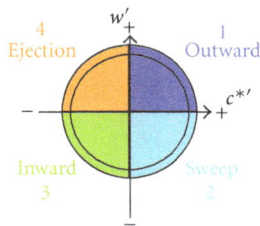

FIGURE 13: The scheme of event definitions used in turbulent flux quadrant analysis.

the alongwind street for flow almost parallel to the street canyon with pollution source. Spreading of pollution to the crosswind street down the wind was observed for approach flow diverging from orientation of the street canyon with pollution source. We determined the highest advective fluxes at the bottom parts of the street canyons.

A unique experimental setup for simultaneous measurement of the flow velocity and the tracer gas concentration was designed and assembled, based on Fast-Response Flame Ionisation Detector and Laser Doppler Anemometer. Vertical turbulent scalar fluxes of passive contaminant were computed from obtained synchronized signals for a horizontal plane placed above the intersection.

Vertical turbulent fluxes magnitude reached two times higher magnitude of vertical advective fluxes in individual grid points. Determined vertical turbulent fluxes comprised significant and positive contribution to the vertical ventilation of the area. On the other side, horizontal advective fluxes magnitude reached up to four times higher magnitude of

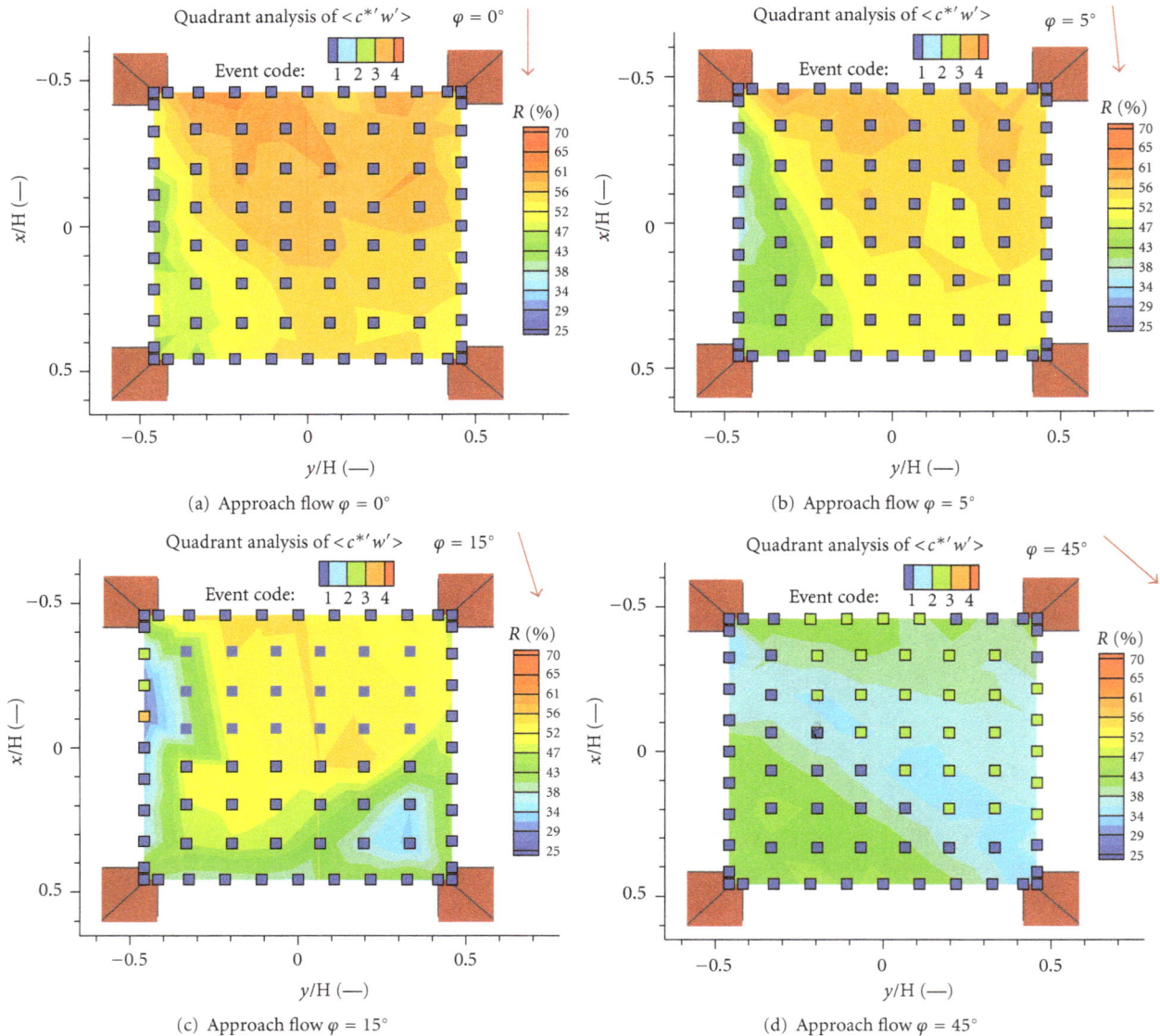

FIGURE 14: Relative contributions R of dominant event to the vertical turbulent scalar flux $\langle c^{*\prime} w' \rangle / U_{2H}$ for four angles of the approach flow direction.

vertical turbulent flux, so the contribution of the horizontal advective pollution transport to total ventilation is dominant in all the cases.

The best dispersive conditions in the studied intersection were measured for the approach flow angle $\varphi \approx 15°$. In this case we measured generally the lowest concentration in the studied area and the lowest scalar flux from the source to the intersection.

The quadrant analysis was applied to the velocity fluctuation signals determining the sweep as a dominant event in flow above the intersection. The relative contribution of the sweep invents to the momentum flux increased for approach flow diverging from orientation of the street canyon with pollution source.

The quadrant analysis was applied to the synchronized vertical velocity and concentration signals. We determined the outward interaction as a dominant event with high relative contribution to the vertical turbulent flux for flow almost parallel to the street canyon with pollution source. Inward interaction events became dominant for diverging flow but with small relative contribution. The flow in this case is strongly turbulent so that we investigated almost the same contribution to the vertical turbulent flux from all events.

The data set acquired from the experiment in the complex urban structure can be used for validations of numerical models of flow and dispersion in street scale or for comparisons of results obtained using these models. The data contains unique synchronized flow velocity and pollution concentration fluctuations signals in a high temporal resolution that can be used to verify pollution transport properties.

Acknowledgments

The authors would like to thank the Ministry of Education, Youth and Sports of the Czech Republic (Project AVOZ-20760514) and the Academy of Sciences of the Czech Republic (Project M100760901) for their financial support.

References

[1] J. Fenger, "Urban air quality," *Atmospheric Environment*, vol. 33, no. 29, pp. 4877–4900, 1999.

[2] R. N. Colvile, E. J. Hutchinson, J. S. Mindell, and R. F. Warren, "The transport sector as a source of air pollution," *Atmospheric Environment*, vol. 35, no. 9, pp. 1537–1565, 2001.

[3] G. Hoek, B. Brunekreef, A. Verhoeff, J. Van Wijnen, and P. Fischer, "Daily mortality and air pollution in the Netherlands," *Journal of the Air and Waste Management Association*, vol. 50, no. 8, pp. 1380–1389, 2000.

[4] F. Nyberg, P. Gustavsson, L. Järup et al., "Urban air pollution and lung cancer in Stockholm," *Epidemiology*, vol. 11, no. 5, pp. 487–495, 2000.

[5] P. Kastner-Klein and E. J. Plate, "Wind-tunnel study of concentration fields in street canyons," *Atmospheric Environment*, vol. 33, no. 24-25, pp. 3973–3979, 1999.

[6] M. Pavageau and M. Schatzmann, "Wind tunnel measurements of concentration fluctuations in an urban street canyon," *Atmospheric Environment*, vol. 33, no. 24-25, pp. 3961–3971, 1999.

[7] M. J. Brown, H. Khalsa, M. Nelson, and D. Boswell, "Street canyon flow patterns in a horizontal plane: measurements from the Joint URBAN 2003 field experiment," in *Proceedings of the 5th Symposium on the Urban Environment*, pp. 55–66, Canada, August 2004.

[8] S. E. Belcher, "Mixing and transport in urban areas," *Philosophical Transactions of the Royal Society A*, vol. 363, no. 1837, pp. 2947–2968, 2005.

[9] J. F. Barlow and S. E. Belcher, "A wind tunnel model for quantifying fluxes in the urban boundary layer," *Boundary-Layer Meteorology*, vol. 104, no. 1, pp. 131–150, 2002.

[10] A. Robins, "DAPPLE (dispersion of air pollution and its penetration into the local environment) experiments and modelling," *HPA Chemical Hazards and Poisons Report*, vol. 13, pp. 24–28, 2008.

[11] A. Scaperdas and R. N. Colvile, "Assessing the representativeness of monitoring data from an urban intersection site in central London, UK," *Atmospheric Environment*, vol. 33, no. 4, pp. 661–674, 1999.

[12] X. Wang and K. F. McNamara, "Effects of street orientation on dispersion at or near urban street intersections," *Journal of Wind Engineering and Industrial Aerodynamics*, vol. 95, no. 9–11, pp. 1526–1540, 2007.

[13] W. Dabberdt, W. Hoydysh, M. Schorling, F. Yang, and O. Holynskyj, "Dispersion modeling at urban intersections," *Science of the Total Environment*, vol. 169, pp. 93–102, 1995.

[14] M. Claggett, J. Shrock, and K. E. Noll, "Carbon monoxide near an urban intersection," *Atmospheric Environment Part A*, vol. 15, no. 9, pp. 1633–1642, 1981.

[15] P. Sharma and M. Khare, "Modelling of vehicular exhausts—a review," *Transportation Research Part D*, vol. 6, no. 3, pp. 179–198, 2001.

[16] L. Soulhac, V. Garbero, P. Salizzoni, P. Mejean, and R. J. Perkins, "Flow and dispersion in street intersections," *Atmospheric Environment*, vol. 43, no. 18, pp. 2981–2996, 2009.

[17] A. S. Tomlin, R. J. Smalley, J. E. Tate et al., "A field study of factors influencing the concentrations of a traffic-related pollutant in the vicinity of a complex urban junction," *Atmospheric Environment*, vol. 43, no. 32, pp. 5027–5037, 2009.

[18] D. J. Hall and M. A. Emmott, "Avoiding aerosol sampling problems in fast response flame ionisation detectors," *Experiments in Fluids*, vol. 10, no. 4, pp. 237–240, 1991.

[19] D. Contini, P. Hayden, and A. Robins, "Concentration field and turbulent fluxes during the mixing of two buoyant plumes," *Atmospheric Environment*, vol. 40, no. 40, pp. 7842–7857, 2006.

[20] VDI Verein Deutcher Ingenieure, *Physical Modelling of Flow and Dispersion Processes in the Atmospheric Boundary Layer—Application of Wind Tunnels*, Beuth, Berlin, Germany, 2000.

[21] A. A. Townsend, *A Structure of Turbulent Shear Flow*, Cambridge University Press, New York, NY, USA, 1976.

[22] R. N. Meroney, M. Pavageau, S. Rafailidis, and M. Schatzmann, "Study of line source characteristics for 2-D physical modelling of pollutant dispersion in street canyons," *Journal of Wind Engineering and Industrial Aerodynamics*, vol. 62, no. 1, pp. 37–56, 1996.

[23] S. P. Arya, *Air Pollution Meteorology and Dispersion*, Oxford University Press, New York, NY, USA, 1999.

[24] R. B. Stull, *An Introduction to Boundary Layer Meteorology*, Kluwer Aademic Publishers, Dordrecht, The Netherlands, 1988.

[25] K. Jurčáková, O. Massaki, and J. Zbyněk, "Contribution of advective and turbulent mass transfers to the ventilation of urban canopy," in *Proceeding of the 7th Asia-Pacific Conference on Wind Engineering*, Taipei, Taiwan, 2009.

[26] R. Kellnerová, L. Kukačka, and Z. Jaňour, "Quadrant analysis of boundary layer above pitched and flat roofs," *Acta Technica*, vol. 54, no. 4, pp. 401–413, 2009.

[27] B. Feddersen, *Wind tunnel modelling of turbulence and dispersion above tall and highly dens urban roughness*, Ph.D. thesis, ETH Zrich, 2004.

7

Particulate Matter and Health Risk under a Changing Climate: Assessment for Portugal

Daniela Dias, Oxana Tchepel, Anabela Carvalho, Ana Isabel Miranda, and Carlos Borrego

Centre for Environmental and Marine Studies and Department of Environment and Planning, University of Aveiro, 3810-193 Aveiro, Portugal

Correspondence should be addressed to Daniela Dias, danieladias@ua.pt

Academic Editor: Costas Varotsos

The potential impacts of climate-induced changes in air pollution levels and its impacts on population health were investigated. The IPCC scenario (SRES A2) was used to analyse the effects of climate on future PM10 concentrations over Portugal and their impact on short-term population exposure and mortality. The air quality modelling system has been applied with high spatial resolution looking on climate changes at regional scale. To quantify health impacts related to air pollution changes, the WHO methodology for health impact assessment was implemented. The results point to 8% increase of premature mortality attributed to future PM10 levels in Portugal. The pollution episodes with daily average PM10 concentration above the current legislated value $(50\,\mu g \cdot m^{-3})$ would be responsible for 81% of attributable cases. The absolute number of deaths attributable to PM10 under future climate emphasizes the importance of indirect effects of climate change on human health.

1. Introduction

Climate change affects human health by a combination of direct and indirect processes. Thus, the abrupt change of temperatures leading to heat waves or cold spells has become widespread, causing fatal illnesses, such as heat stress or hypothermia, as well as increasing death rates from heart and respiratory diseases. According to the World Health Organization (WHO), the statistics on mortality and hospital admissions show that death rates increase during extremely hot days, particularly among very old and very young people living in cities. In Portugal, during the European heat wave of 2003, a total of 2399 excessive deaths were estimated which implied an increase of 58% over the expected deaths [1].

The indirect effects of climate change on human health are related, among others, to the changes in air pollution levels under future climate. Thus, changes in the temperature, humidity, wind, and precipitation that may accompany future climate can deeply impact air quality because of induced changes in the transport, dispersion, and transformation of air pollutants at multiple scales [2, 3]. According to Sheffield et al. [4], climate change could cause an increase in regional summer ozone-related asthma emergency department visits for children aged 0–17 years of 7.3% across the New York metropolitan region by the 2020s. When population growth is included, the projections of morbidity related to ozone were even larger. The authors also highlighted that the use of regional climate and atmospheric chemistry models makes possible the projection of local climate change health effects for specific age groups and specific disease outcomes.

The potential impact of climate change on particulate matter (PM) is of major concern because their concentrations are most likely to increase under a changing climate [5–7] and because future changes in particulate matter concentrations are likely the most important component of changes in mortalities attributable to air pollution in future scenarios [8]. Over the last few decades, human exposure to particulate air pollution has been associated with human mortality and morbidity, as well as a broad range of negative health outcomes at levels usually experienced by populations due to short- and long-term exposure to particulate matter [9–14]. The European directive (2008/50/CE) revised the limit values for PM10 (particulate matter with an aerodynamic diameter less than or equivalent to $10\,\mu m$) previously defined by the

Framework Directive (1999/30/EC) and set up new quantitative standards for PM2.5 (particulate matter with an aerodynamic diameter less than or equivalent to 2.5 μm). Nevertheless, PM threshold levels to which exposure does not lead to adverse effects on human health have not yet been identified and given that there is a substantial inter-individual variability in exposure and in the response, it is unlikely that any standard or guideline value will lead to a complete protection for every individual against all possible adverse health effects of particulate matter [15].

For Portugal, studies show frequent exceedances of EU directive targets for air quality [16]. WHO has recently identified that Portugal is one of the 80 countries that exceed the reference values for particulate matter [17]. In addition, particulate emissions decreased in most European countries between 1990 and 2008 except for Portugal, Bulgaria, Romania, Malta, Finland, Denmark, Latvia, and Spain, where increases were recorded [18]. However, studies focusing on the health impacts of air quality in Portugal are very few. Several studies concerning the impact of meteorological factors on human health and the first attempt to relate air pollution levels and morbidity for Portugal have been published [1, 19–22]. The authors [20] highlight that under future climate the meteorological conditions will be more favourable for high ozone levels (low wind speed and high temperature) that could lead to impacts on human health. Recently, a number of studies on quantitative impact assessment of air pollution on mortality in Portuguese cities have emerged [23, 24] providing information on the association of current pollution levels with adverse health effects.

The main aim of the current study is to quantify the potential impact of short-term exposure to PM10 on population health under future climate. For this purpose, climate change scenario simulated with high temporal and spatial resolution is combined with health impact assessment (HIA). Air pollution modelling for the future scenario is performed assuming no changes in the PM10 precursor emissions in comparison with the reference situation thus allowing quantification of the climate change effect independently from the other factors that affect the pollution levels. The present study provides quantitative information on forecast of the health impact attributable to air pollution under a changing climate relevant for climate change mitigation and health policies.

2. Material and Methods

The potential impact on climate-induced human health effects caused by changes in PM10 concentrations over the continental Portugal is investigated using combined atmospheric and impact assessment modelling. The study is implemented in two main steps: (i) numerical simulation of PM10 concentrations over Portugal under the IPCC SRES A2 scenario and (ii) estimation of the number of deaths attributable to the changes in PM10 levels in the atmosphere under climate change.

To quantify the health impact related with air pollution changes, the WHO methodology [25] was adapted and ap-

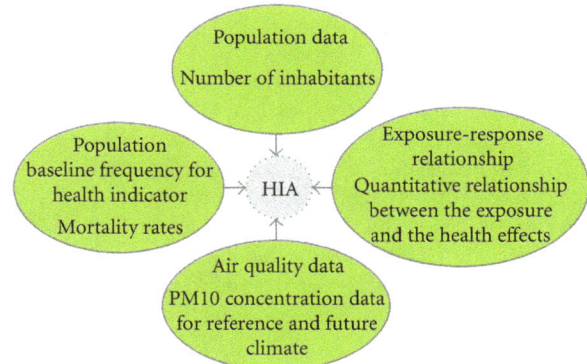

FIGURE 1: Schematic representation of the input information required by the health impact assessment performed in this study.

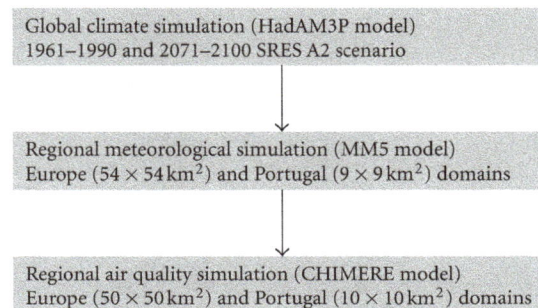

FIGURE 2: Schematic representation of the air quality numerical simulation.

plied to the study area using the input information schematically presented in Figure 1.

2.1. Air Quality Modelling under Climate Change. The air quality modelling was performed for a reference and a future climate scenarios first at the European scale and then over Portugal [26]. For this purpose, global climate simulations provided by the HadAM3P model were used to drive the air quality modelling system as represented in Figure 2. The climate conditions for 1961–1990 are considered to characterize the reference situation, and predictions for 2071–2100 are used for the future climate in accordance with the IPCC SRES A2 scenario [27]. This scenario is considered to be the highest emission scenario and the carbon dioxide (CO_2) concentrations reaching 850 ppm by 2100. In this sense, we are assessing the worst scenario with regard to air quality changes.

The air quality modelling system is based on the chemistry transport model CHIMERE [28, 29] forced by the mesoscale meteorological model MM5 [30]. The MM5/CHIMERE modelling system has been widely applied and validated in several air quality studies over Portugal [31–33] showing performance skills within the range found in several model evaluation studies using different air quality models [34, 35]. The MM5/CHIMERE modelling system has already been used in several studies that investigated the impacts of climate change on air pollutants levels over Europe [36] and specifically over Portugal [26]. The MM5 mesoscale model is a nonhydrostatic, vertical sigma coordinate model designed to simulate mesoscale atmospheric circulations. The selected MM5 physical options were based on the already performed validation and sensitivity studies over Portugal [37] and over the Iberian Peninsula [38]. A detailed description of the selected simulation characteristics is presented in [26]. The MM5 model generates the several meteorological fields required by the CHIMERE model, such as wind, temperature, water vapour mixing ratio, cloud liquid water content, 2 m temperature, surface heat and moisture fluxes, and precipitation.

CHIMERE is a tri-dimensional chemistry-transport model, based on the integration of the continuity equation for the concentrations of several chemical species in each cell of a given grid. It was developed for simulating gas-phase chemistry [28], aerosol formation, transport, and deposition [29, 39] at regional and urban scales. CHIMERE simulates the concentration of 44 gaseous species and 6 aerosol chemical compounds. In addition to the meteorological input, the CHIMERE model needs boundary and initial conditions, anthropogenic emission data, and the land use and topography characterization. The modelling system was firstly applied at the European scale (with 50×50 km^2 resolution) and then over Portugal using the same physics and a simple one-way nesting technique, with 10×10 km^2 horizontal resolution. The European domain covers an area from 14 W to 25 E and 35 N to 58 N. Over Portugal, the simulation domain goes from 9.5 W to 6 W and 37 N to 42.5 N [26]. The vertical resolution of CHIMERE model consists of eight vertical layers of various thicknesses extending from ground to 500 hPa. Lateral and top boundaries for the large-scale run were obtained from the LMDz-INCA (gas species) [40] and GOCART (aerosols) [41] global chemistry-transport models, both monthly mean values. The same boundaries conditions were used for both scenarios, since the objective is to only change the meteorological driver forcing. For the Portugal domain, boundary conditions are provided by the large-scale European simulation.

The CHIMERE model requires hourly spatially resolved emissions for the main anthropogenic gas and aerosol species. For the simulation over Europe, the anthropogenic emissions for nitrogen oxides (NO$_x$), carbon monoxide (CO), sulphur dioxide (SO$_2$), nonmethane volatile organic components (NMVOC) and ammonia (NH$_3$) gas-phase species, and for PM2.5 and PM10 are provided by EMEP (Co-operative Programme for Monitoring and Evaluation of the Long-range Transmission of Air Pollutants in Europe) [42] with a spatial resolution of 50 km. The national inventory INERPA was used over the Portugal domain [32].

Reference and the IPCC SRES-A2 climate scenario over Europe and over Portugal were simulated by dynamical downscaling using the outputs of HadAM3P [43], as initial and boundary conditions to the MM5 model. The MM5 model requires initial and time-evolving boundary conditions for wind components, temperature, geopotential height, relative humidity, surface pressure, and also the specification of SSTs. Carvalho et al. [26] discuss the global model HadAM3P and the MM5 ability to simulate the present

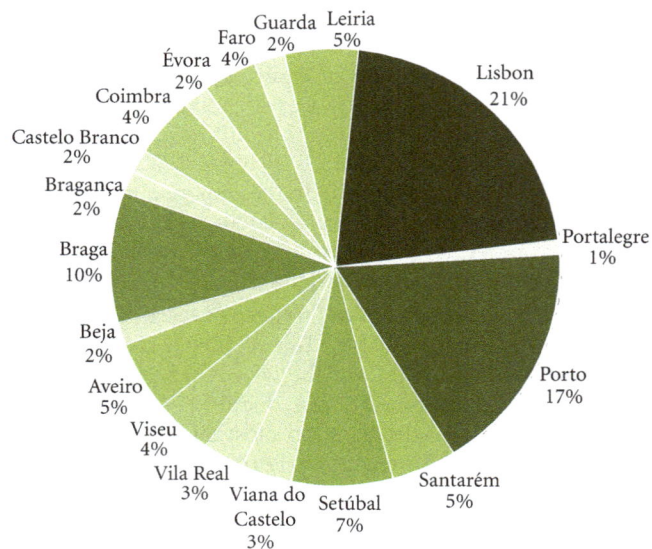

FIGURE 3: Distribution of demographic data by district in 2001.

climate. The HadAM3P was selected to drive the MM5 model because a previous work [44] has already concluded that the HadAM3P accurately reproduces the large-scale patterns, namely, the 500 hPa fields. The 500 hPa height reflects a broad range of meteorological influences on air quality. The authors concluded that the HadAM3P is able to capture the mean patterns of the circulation weather types. The obtained results give confidence to use the HadAM3P outputs as initial and boundary conditions for regional simulations.

To evaluate the influence of climate change on air quality, the anthropogenic emissions were kept constant (to the year 2003) in the simulations for the future climate and were not scaled in accordance with the IPCC SRES A2 scenario. This idealized regional model simulation provides insight into the contribution of possible future climate changes on the 3D distribution of particulate matter concentrations. The MM5/ CHIMERE simulations were conducted from May 1st to October 30th for the reference year (1990) and for the future scenario year (2100). Both simulations had the same chemical boundary conditions. Following this methodology, it is possible to analyse the changes caused by climate change only. In Carvalho et al. [26], a detailed analysis of the MM5/ CHIMERE modelling system application under climate change has been presented and validated.

2.2. Population Analysis. Population size, composition, and health status were analysed for the study area as important elements required for the health impact assessment. According to National Institute of Statistics, the resident population in Portugal in 2001 was 9,869,343 inhabitants [45]. Lisbon and Porto are emphasized as the most densely populated agglomerations representing about 38% of total national population (Figure 3).

The distribution of population by age groups is presented in Figure 4 stressing different proportion between active and older population for each district.

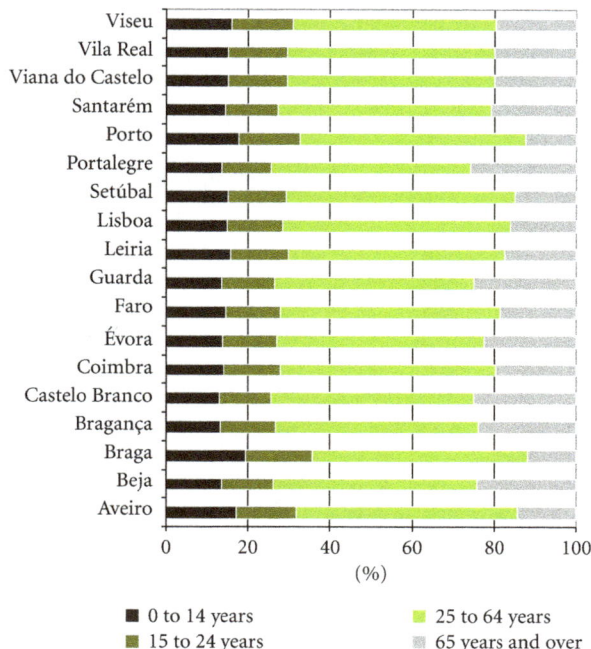

FIGURE 4: Distribution of population by age group for each Portuguese district in 2001.

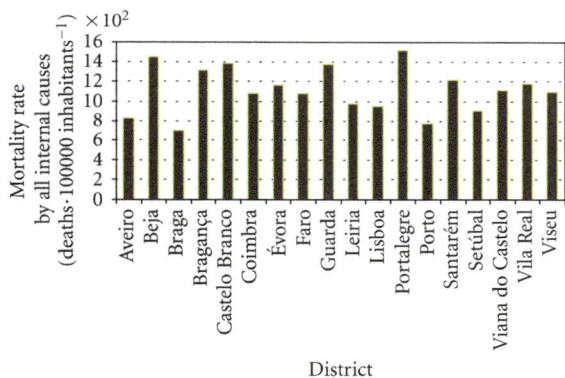

FIGURE 5: Annual mortality rate by all internal causes for each Portuguese district (deaths·100000 inhabitants^{-1}) [46].

The health indicator considered in this study includes all-cause mortality (except external causes) (ICD-10 codes A00-R99) expressed as daily mortality rates in the number of deaths per 100000 inhabitants. Figure 5 presents the distribution of annual mortality rate by district based on DGS [46].

As could be seen, there is not a homogeneous distribution of mortality rate by the districts in Portugal. In general, the highest mortality rate by all internal causes is observed for the regions with higher proportion of older population as presented previously in Figure 4. Although, the Lisbon district indicates greater mortality rate than Porto with main difference in the mortality rate for age group 25–64 years (Figure 6).

2.3. Health Impact Assessment. A methodology to quantify health effects is conducted in terms of number of cases at-

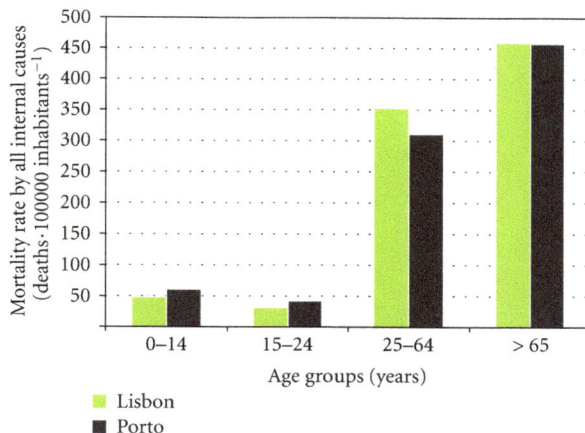

FIGURE 6: Annual mortality rate by all internal causes in Lisbon and Porto districts by age groups.

tributable to air pollution that may be prevented by reducing current levels of PM10 [25, 47]. An estimate of attributable deaths (AD) is obtained from the average number of deaths ($\overline{y}$), the regression coefficient β provided by epidemiology-based exposure-response functions, and the difference between the daily average concentration ($\overline{x}$) and a reference value under a given scenario (x^*):

$$AD = \overline{y} \times \beta(\overline{x} - x^*). \qquad (1)$$

The EIS-PA model, developed by French Surveillance System on Air Pollution and Health as a support tool for automated and standardized health risk assessment [48], is used in this study to calculate the number of premature deaths prevented annually due to the reduction of PM to the selected "target" concentration. The results of EIS-PA model application provide estimates of the health outcomes related to short-term (1 or 2 days) exposure.

The exposure-response function, expressed as Relative Risk (RR) per 10 μg·m^{-3}, from epidemiological studies recommended by the European study [47] was adopted, considering the Relative Risk (RR) of 1.006 (95% CI (1.004–1.008)) for all-cause mortality (except external causes) to assess the effects on human health associated with the very short-term PM10 exposure (1 or 2 days) [49].

The time series of PM10 concentrations for future climate scenario together with demographic data and specific health indicators were considered in accordance with the Apheis guidelines [47] and used as input in the EIS-PA model [48]. The health impact assessment is implemented for two air pollution scenarios: (i) a simulation for current climate (year 1990) and projected 2100 PM10 levels under the IPCC SRES A2 scenario; (ii) for the air pollution reduction scenario considering the legislation limit values of daily average 50 μg·m^{-3} recently revised by the Directive 2008/50/CE and proposed in the latest review of "Air Quality Guidelines" from WHO [15] as the reduction "target" level.

dif-TEMP-July

dif-RH-July

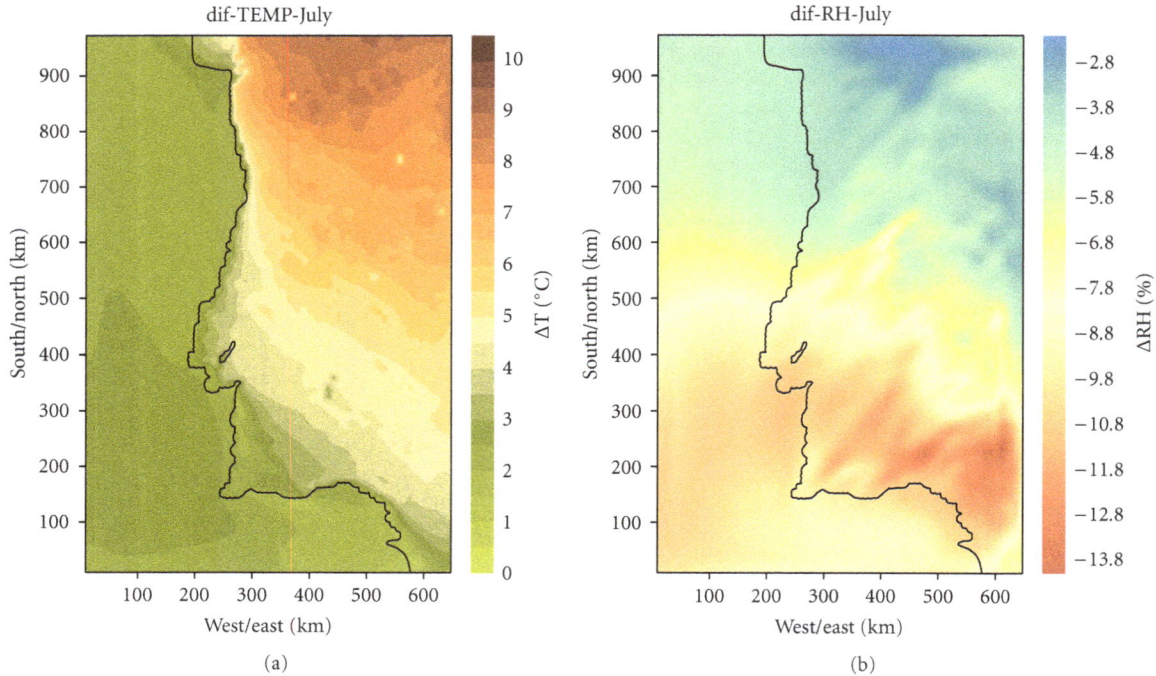

(a)

(b)

FIGURE 7: Temperature (°C) and relative humidity (%) differences between future and reference climates simulated with the MM5 model across Portugal for July.

3. Results and Discussion

In this section, the estimated PM10 levels and health impact for both climate scenarios are analysed. The results obtained for short-term exposure (1 or 2 days), expressed as a number of attributable cases by all internal causes mortality, are presented and discussed. The increased number of attributable cases between the future and current pollution levels and the potential number of attributable cases prevented annually by reducing future PM10 concentrations to the legislation limit value ($50 \, \mu g \cdot m^{-3}$) are also investigated.

3.1. Particulate Matter Levels under the IPCC SRES A2 Scenario. The simulated temperature increases under future climate almost reach 8.5°C over mid and southern Europe during the warm period of May–October [26]. These projections are in accordance to Rowell [50] who predicted that in winter the largest warming occurs over eastern Europe, up to 7°C, and in summer temperatures rise by 6–9°C south of about 50°N.

In Figure 7, an example of the projected climatic changes over Portugal is presented for July showing the largest temperature increases over the northwestern part of Portugal reaching almost 10°C. Relative humidity (RH) will decrease significantly all over Portugal. The changes in the meteorological fields (temperature, RH, wind, boundary layer) will influence the pollutants dispersion and transformation in the atmosphere.

Wind speed, mixing height, and relative humidity are the meteorological variables believed to mostly influence PM concentrations. Stagnant conditions are thought to correlate with high PM concentrations, as they allow particulates to

accumulate near the earth's surface. Although high wind speeds can increase ventilation, they are normally correlated with high PM concentrations because they allow the resuspension of particles from the ground, as well as long-range transport of particulates between regions. High PM concentrations are normally associated with dry conditions due to increased potential to resuspension of dust, soil, and other particles. Figure 8 presents the average PM10 levels over Portugal over the simulation period for both climates based on hourly data provided by the air quality model.

For the overall simulation period, the maximum averaged PM10 levels increase from $60 \, \mu g \cdot m^{-3}$ to $72 \, \mu g \cdot m^{-3}$. In addition, over Porto and Lisbon regions, the area affected by higher concentrations also increases in future climate (Figure 8).

Additionally to the changes in the average pollution levels, the frequency distribution of the PM10 concentrations is also very important for the human health studies. In Figure 9, an example for the most affected regions of Porto and Lisbon is presented providing information on the frequency of pollution episodes under the two climate scenarios.

The frequency distribution of the PM10 concentrations for both climatic scenarios emphasizes that Lisbon and Porto districts present an elevated number of days with high PM10 levels in comparison with the legislation limit value for the daily average PM10 concentration of $50 \, \mu g \cdot m^{-3}$ that cannot be exceeded more than 35 times per year. Moreover, the climate-driven effect on PM10 levels will be more noticeable in Porto district leading to significant increase in the number of days with high daily average concentration.

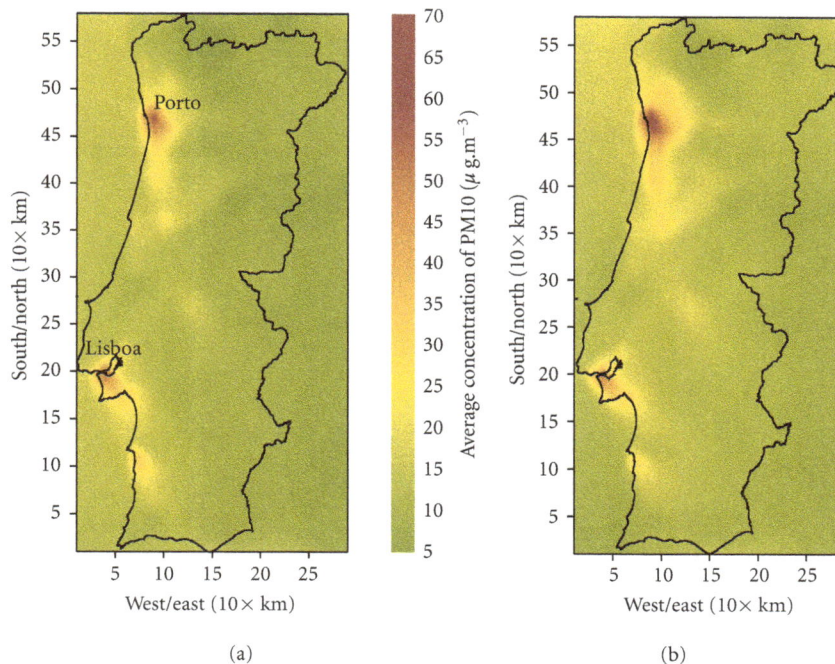

FIGURE 8: Average concentration of PM10 (μg$\cdot$m^{-3}) for the simulated period (from May to October) for (a) current and (b) future climate scenario.

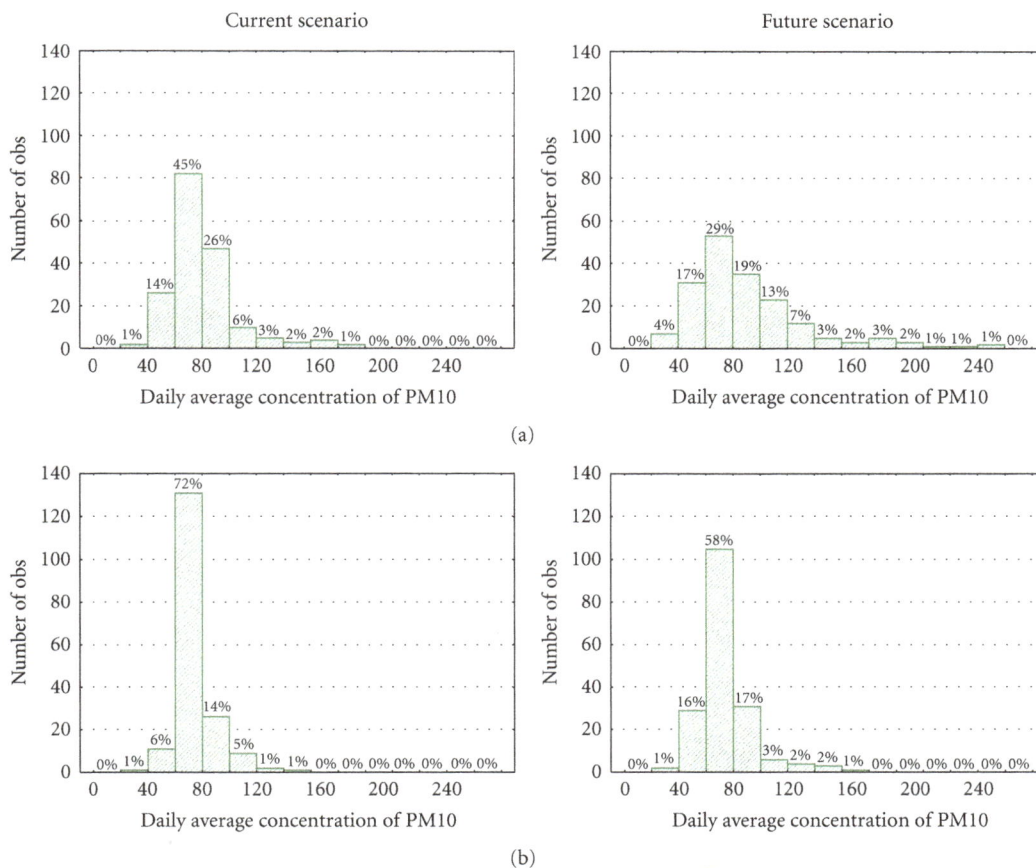

FIGURE 9: Frequency distribution of the PM10 concentrations for both climatic scenarios over the regions of (a) Porto and (b) Lisbon.

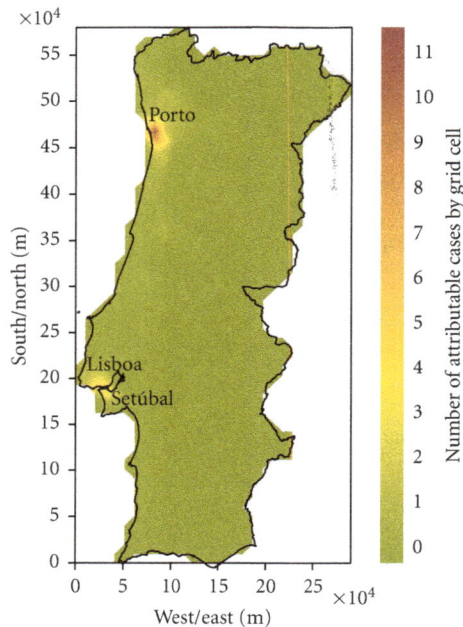

FIGURE 10: Spatial distribution of the increased number of attributable cases estimated by grid cell (10×10 km^2) related to short-term PM10 exposure for future climate.

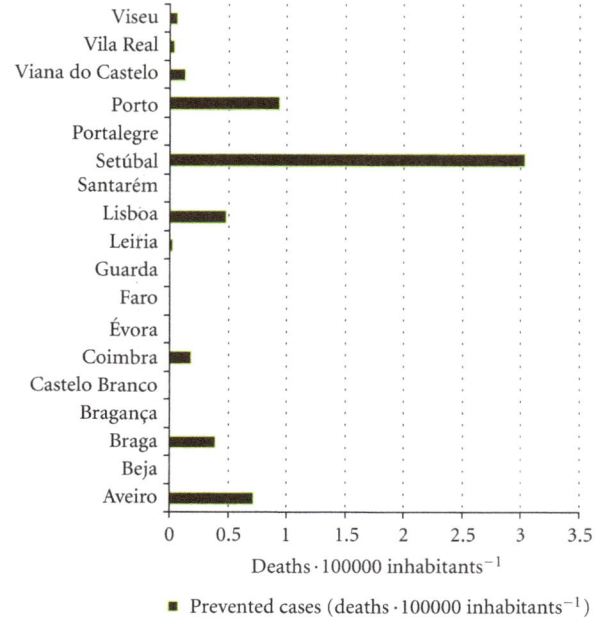

FIGURE 11: Prevented cases considering the fulfilment of the legislated value (deaths $\cdot$ 100000 inhabitants^{-1}).

3.2. Prognosis of Health Impact: Future versus Current Pollution Levels. The health impact assessment based on the estimated changes in PM10 between the future and reference climate shows some locations with no significant increment in the number of attributable cases to short-term PM10 exposure while other locations show important increase in PM10-induced premature mortality (Figure 10). Since the number of estimated attributable cases depends on both air quality and the number of the inhabitants exposed, air quality changes in the densely populated areas of the country have a greater effect than air quality changes in less densely populated areas, in general. Modelling results suggest that worsened PM10 levels will coincide spatially with many of the most densely populated areas of the country (Figure 8).

As could be seen from Figure 10, the highest increase of the number of attributable cases under a future climate scenario would be expected in the Northern coastal region and Lisbon metropolitan area achieving a maximum augment of 11 cases by grid cell. The results presented in Table 1 highlight that the changes on the PM10 concentrations lead to a significant increase in the number of deaths in the future for most districts, especially those with the larger urban areas. additionally the Lisbon district is characterised by larger population size and the current mortality rate is higher, and the Porto district is the most affected (about 31% of total national deaths), reaching two times higher values than expected for the Lisbon district due to different prognosis of future pollution levels for these areas.

On the other hand, South of Portugal presents the lowest changes in the average mortality rate (Faro district: 0.9 (95% CI 0.6–1.2)) since the PM10 concentrations projected for 2100 will not increase significantly in comparison with the current pollution levels. At national level, about 203 (95% CI 137–271) more premature deaths per year are projected for 2100 in comparison to the current scenario due to indirect effect of climate change.

3.3. Prognosis of Health Impact: Future Pollution versus Legislation. Additionally to the impact assessment based on prognosis of future pollution, the benefit for human health related with potential reduction of PM10 to the legislation limit value (daily average concentration of $50 \, \mu g \cdot m^{-3}$) was analysed. The number of prevented cases for all internal causes mortality attributed to the short-term (1 or 2 days) exposure is quantified considering that no exceedances to the limit value will occur. The results for each district are presented in Figure 11.

Porto district will be the greatest benefited in case of the legislated value fulfilment that is possible to achieve with implementation of additional policy measures such as emission reductions. Therefore, if no air quality exceedances will occur, about 50 premature deaths related to PM10 exposure may be avoided annually, which corresponds to four times higher values than prevented cases estimated for the Lisbon district. As expected, this fact is related with highest increase in air pollution levels predicted for Porto in future climate.

A more detailed analysis of the results obtained for the Porto area in terms of the number of attributable cases associated with different levels of exposure to PM10 is presented in Figure 12.

Although in Porto district average PM10 concentrations above $120 \, \mu g \cdot m^{-3}$ will occur in 13% of days, they are responsible for 50% of deaths attributable to air pollution. Thus emphasizing the greatest impact associated with "high pollution" days, despite their low frequency.

TABLE 1: Increase of mortality attributable to PM10 pollution levels under the climate scenario in comparison with the reference situation. Values presented in parenthesis correspond to the 95% confidence interval (CI).

District	Mortality rate average and 95% CI (deaths per 100000 inhabitants)	Annual mortality average and 95% CI (deaths)
Aveiro	2.6 (1.7–3.5)	13 (9–18)
Beja	1.7 (1.1–2.2)	3 (2-3)
Braga	1.9 (1.3–2.6)	19 (12–25)
Bragança	2.0 (1.3–2.6)	3 (2–4)
Castelo Branco	1.7 (1.1–2.2)	3 (2–4)
Coimbra	2.5 (1.7–3.4)	11 (7–15)
Évora	1.4 (0.9–1.9)	3 (2-3)
Faro	0.9 (0.6–1.2)	3 (2–4)
Guarda	1.8 (1.2–2.5)	4 (3–5)
Leiria	1.6 (1.1–2.2)	8 (6–11)
Lisbon	1.3 (0.8–1.7)	26 (17–35)
Portalegre	1.8 (1.2–2.4)	2 (2-3)
Porto	3.7 (2.5–5.0)	62 (41–83)
Santarém	1.8 (1.2–2.3)	8 (5–11)
Setúbal	1.9 (1.2–2.5)	13 (9–18)
Viana do Castelo	2.4 (1.6–3.2)	8 (5–11)
Vila Real	1.9 (1.3–2.6)	6 (4–7)
Viseu	2.0 (1.3–2.6)	8 (5–11)
National	2.1 (1.4–2.8)	203 (135–271)

4. Conclusions

In this study, a quantitative assessment of the impact of climate change on human health related with short-term exposure to PM10 has been performed using combined atmospheric and impact assessment modelling. The modelling results obtained for the continental region of Portugal revealed that climate change alone will deeply impact the PM10 levels in the atmosphere. All the Portuguese districts will be negatively affected but negative effects on human health are more pronounced in major urban areas. The short-term variations in the PM10 concentration under future climate will potentially lead to an increase of 203 premature deaths per year in Portugal. The Porto district is the most affected in terms of occurrence of number of days with higher concentrations, consequently leading to the most significant increase in premature deaths that correspond to approximately 8% increase of its current mortality rate by all internal causes.

The pollution episodes with daily average PM10 concentration above the current legislated value ($50\,\mu g \cdot m^{-3}$) would be responsible for 81% of attributable cases. Although "high pollution" days have low frequency, they show the greatest impact and highlight the significant contribution of pollution peaks to acute exposure. Thus, the reduction of "high pollution" days with daily average concentration above $120\,\mu g \cdot m^{-3}$ projected to the Porto district will avoid about 50% of premature deaths attributable to air pollution.

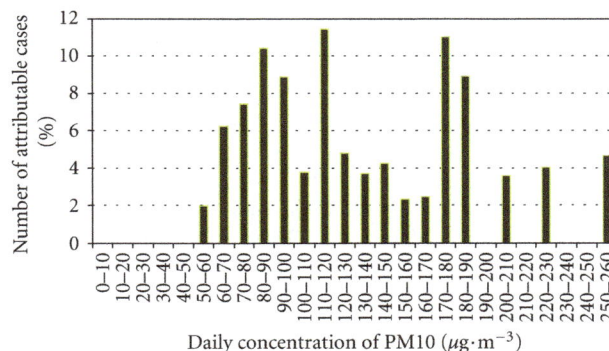

FIGURE 12: Distribution of the number of attributable cases (%) by PM10 concentration classes in Porto.

Although the hypothetical situation of what would happen if the predicted future climate conditions will occur in 2100 and assuming that PM10 precursor emissions and population maintain constant, the information provided in this study suggests that climate-driven changes on air pollutants and human health could be substantial. Therefore, additional efforts should be made to improve on this type of modelling approach in order to support local and wider-scale climate change mitigation and adaptation policies.

Acknowledgment

The authors thank the Portuguese Foundation for Science and Technology for the Ph.D. Grant of D. Dias (SFRH/BD/47578/2008).

References

[1] R. M. Trigo, A. M. Ramos, P. J. Nogueira et al., "Evaluating the impact of extreme temperature based indices in the 2003 heatwave excessive mortality in Portugal," *Environmental Science and Policy*, vol. 12, no. 7, pp. 844–854, 2009.

[2] S. M. Bernard, J. M. Samet, A. Grambsch, K. L. Ebi, and I. Romieu, "The potential impacts of climate variability and change on air pollution-related health effects in the United States," *Environmental Health Perspectives*, vol. 109, no. 2, pp. 199–209, 2001.

[3] NRC, *Global Air Quality*, National Academy Press, Washington, DC, USA, 2001.

[4] P. E. Sheffield, K. Knowlton, J. L. Carr, and P. L. Kinney, "Modeling of regional climate change effects on ground-level ozone and childhood asthma," *American Journal of Preventive Medicine*, vol. 41, no. 3, pp. 251–257, 2011.

[5] P. L. Kinney, "Climate change, air quality, and human health," *American Journal of Preventive Medicine*, vol. 35, no. 5, pp. 459–467, 2008.

[6] J. G. Ayres, B. Forsberg, I. Annesi-Maesano et al., "Climate change and respiratory disease: European Respiratory Society position statement," *European Respiratory Journal*, vol. 34, no. 2, pp. 295–302, 2008.

[7] D. J. Jacob and D. A. Winner, "Effect of climate change on air quality," *Atmospheric Environment*, vol. 43, no. 1, pp. 51–63, 2009.

[8] J. J. West, S. Szopa, and D. A. Hauglustaine, "Human mortality effects of future concentrations of tropospheric ozone," *Comptes Rendus*, vol. 339, no. 11-12, pp. 775–783, 2007.

[9] N. Kunzli, R. Kaiser, S. Medina et al., "Public-health impact of outdoor and traffic-related air pollution: a European assessment," *The Lancet*, vol. 356, no. 9232, pp. 795–801, 2000.

[10] R. Anderson, A. Atkinson, J. L. Peacock, L. Marston, and K. Konstantinou, *Meta-Analysis of Time-Series and Panel Studies on Particulate Matter and Ozone (O3)*, Document EUR/04/5042688, WHO Task Group. WHO Regional Office for Europe, Copenhagen, Denmark, 2004.

[11] H. R. Anderson, R. W. Atkinson, J. L. Peacock, M. J. Sweeting, and L. Marston, "Ambient particulate matter and health effects: publication bias in studies of short-term associations," *Epidemiology*, vol. 16, no. 2, pp. 155–163, 2005.

[12] C. A. Pope and D. W. Dockery, "Health effects of fine particulate air pollution: lines that connect," *Journal of the Air and Waste Management Association*, vol. 56, no. 6, pp. 709–742, 2006.

[13] E. Samoli, R. Peng, T. Ramsay et al., "Acute effects of ambient particulate matter on mortality in Europe and North America: results from the APHENA study," *Environmental Health Perspectives*, vol. 116, no. 11, pp. 1480–1486, 2008.

[14] K. Katsouyanni, J. M. Samet, H. R. Anderson et al., "Air pollution and health: a European and North American approach (APHENA)," *HEI Research Report*, no. 142, pp. 5–90, 2009.

[15] World Health Organization (WHO), *Air Quality Guidelines for Particulate Matter, Ozone, Nitrogen Dioxide and Sulfur Dioxide—Global Update 2005 Summary of Risk Assessment*, Geneva, Switzerland, 2006.

[16] EEA, "Spatial assessment of PM10 and ozone concentrations in Europe (2005)," Tech. Rep. 1/2009, 2009.

[17] World Health Organization (WHO), Database: outdoor air pollution in cities, 2011, http://www.who.int/phe/health_topics/outdoorair/databases/en/index.html.

[18] EEA, National emissions reported to the Convention on Long-range Transboundary Air Pollution (LRTAP Convention), 2010, http://www.eea.europa.eu/data-and-maps/data/national-emissions-reported-to-the-convention-on-long-range-transboundary-air-pollution-lrtap-convention-4.

[19] P. J. Nogueira, J. M. Falcão, M. T. Contreiras, E. Paixão, J. Brandão, and I. Batista, "Mortality in Portugal associated with the heat wave of August 2003: early estimation of effect, using a rapid method," *Euro Surveillance*, vol. 10, no. 7, pp. 150–153, 2005.

[20] E. Casimiro, J. Calheiros, F. D. Santos, and S. Kovats, "National assessment of human health effects of climate change in Portugal: approach and key findings," *Environmental Health Perspectives*, vol. 114, no. 12, pp. 1950–1956, 2006.

[21] C. A. Alves and C. A. Ferraz, "Effects of air pollution on emergency admissions for chronic obstructive pulmonary diseases in Oporto, Portugal," *International Journal of Environment and Pollution*, vol. 23, no. 1, pp. 42–64, 2005.

[22] C. A. Alves, M. G. Scotto, and M. D. C. Freitas, "Air pollution and emergency admissions for cardiorespiratory diseases in lisbon (Portugal)," *Quimica Nova*, vol. 33, no. 2, pp. 337–344, 2010.

[23] O. Tchepel and D. Dias, "Quantification of health benefits related with reduction of atmospheric PM10 levels: implementation of population mobility approach," *International Journal of Environmental Health Research*, vol. 21, no. 3, pp. 189–200, 2011.

[24] P. Garrett and E. Casimiro, "Short-term effect of fine particulate matter (PM2.5) and ozone on daily mortality in Lisbon, Portugal," *Environmental Science and Pollution Research*, vol. 18, no. 9, pp. 1585–1592, 2011.

[25] World Health Organization (WHO), *Quantification of Health Effects of Exposure to Air Pollution*, Document EUR/01/5026-342, WHO Regional Office for Europe, Copenhagen, Denmark, 2001.

[26] A. Carvalho, A. Monteiro, S. Solman, A. I. Miranda, and C. Borrego, "Climate-driven changes in air quality over Europe by the end of the 21st century, with special reference to Portugal," *Environmental Science and Policy*, vol. 13, no. 6, pp. 445–458, 2010.

[27] N. Nakicenovic, J. Alcamo, G. Davis et al., *IPCC Special Report on Emissions Scenarios*, Cambridge University Press, Cambridge, UK, 2000.

[28] H. Schmidt, C. Derognat, R. Vautard, and M. Beekmann, "A comparison of simulated and observed ozone mixing ratios for the summer of 1998 in Western Europe," *Atmospheric Environment*, vol. 35, no. 36, pp. 6277–6297, 2001.

[29] B. Bessagnet, A. Hodzic, R. Vautard et al., "Aerosol modeling with CHIMERE—preliminary evaluation at the continental scale," *Atmospheric Environment*, vol. 38, no. 18, pp. 2803–2817, 2004.

[30] G. A. Grell, J. Dudhia, and D. R. Stauffer, "A description of the fifth-generation Penn State/NCAR Mesoscale Model (MM5)," Tech. Rep. NCAR/TN-398+STR, The National Center for Atmospheric Research, Boulder, Colo, USA, 1994.

[31] A. Monteiro, R. Vautard, C. Borrego, and A. I. Miranda, "Long-term simulations of photo oxidant pollution over Portugal using the CHIMERE model," *Atmospheric Environment*, vol. 39, no. 17, pp. 3089–3101, 2005.

[32] A. Monteiro, A. I. Miranda, C. Borrego, R. Vautard, J. Ferreira, and A. T. Perez, "Long-term assessment of particulate matter using CHIMERE model," *Atmospheric Environment*, vol. 41, no. 36, pp. 7726–7738, 2007.

[33] C. Borrego, A. Monteiro, J. Ferreira et al., "Procedures for estimation of modelling uncertainty in air quality assessment," *Environment International*, vol. 34, no. 5, pp. 613–620, 2008.

[34] R. Vautard, P. H. J. Builtjes, P. Thunis et al., "Evaluation and intercomparison of Ozone and PM10 simulations by several chemistry transport models over four European cities within the CityDelta project," *Atmospheric Environment*, vol. 41, no. 1, pp. 173–188, 2007.

[35] R. Stern, P. Builtjes, M. Schaap et al., "A model inter-comparison study focussing on episodes with elevated PM10 concentrations," *Atmospheric Environment*, vol. 42, no. 19, pp. 4567–4588, 2008.

[36] S. Szopa, D. A. Hauglustaine, R. Vautard, and L. Menut, "Future global tropospheric ozone changes and impact on European air quality," *Geophysical Research Letters*, vol. 33, no. 14, Article ID L14805, 2006.

[37] A. C. Carvalho, A. Carvalho, I. Gelpi et al., "Influence of topography and land use on pollutants dispersion in the Atlantic coast of Iberian Peninsula," *Atmospheric Environment*, vol. 40, no. 21, pp. 3969–3982, 2006.

[38] J. Fernández, J. P. Montávez, J. Sáenz, J. F. González-Rouco, and E. Zorita, "Sensitivity of the MM5 mesoscale model to physical parameterizations for regional climate studies: annual cycle," *Journal of Geophysical Research D*, vol. 112, no. 4, Article ID D04101, 2007.

[39] R. Vautard, B. Bessagnet, M. Chin, and L. Menut, "On the contribution of natural Aeolian sources to particulate matter concentrations in Europe: testing hypotheses with a modelling approach," *Atmospheric Environment*, vol. 39, no. 18, pp. 3291–3303, 2005.

[40] D. A. Hauglustaine, J. Lathière, S. Szopa, and G. A. Folberth, "Future tropospheric ozone simulated with a climate-chemistry-biosphere model," *Geophysical Research Letters*, vol. 32, no. 24, pp. 1–5, 2005.

[41] M. Chin, P. Ginoux, S. Kinne et al., "Tropospheric aerosol optical thickness from the GOCART model and comparisons with satellite and sun photometer measurements," *Journal of the Atmospheric Sciences*, vol. 59, no. 3, pp. 461–483, 2002.

[42] V. Vestreng, "Review and revision of emission data reported to CLRTAP," EMEP Status Report, 2003.

[43] R. G. Jones, J. M. Murphy, D. C. Hassel, and M. J. Woodage, "A high resolution atmospheric GCM for the generation of regional climate scenarios," Hadley Center Technical Note 63, Met Office, Exeter, UK, 2005.

[44] C. H. R. Anagnostopoulou, K. Tolika, P. Maheras, H. Kutiel, and H. A. Flocas, "Performance of the general circulation HadAM3P model in simulating circulation types over the Mediterranean region," *International Journal of Climatology*, vol. 28, no. 2, pp. 185–203, 2008.

[45] Instituto Nacional de Estatística (INE), *Recenseamento da População e da Habitação—Censos 2001*, Instituto Nacional de Estatística, Lisboan, Portugal, 2002.

[46] Direcção-Geral da Saúde (DGS), *Risco de Morrer em Portugal, 2001*, DSIA. Divisão de Epidemiologia, Lisboan, Portugal, 2003.

[47] Air Pollution and Health: A European Information System (APHEIS), "Health impact assessment of air pollution and communication strategy," 3rd year report, 2005.

[48] Institut de veille sanitaire (INVS), Évaluation de l'impact sanitaire de la pollution atmosphérique urbaine concepts et méthodes, 2000, http://www.invs.sante.fr/surveillance/psas9/.

[49] World Health Organization (WHO), *Health Aspects of Air Pollution–Answers to Follow-up Questions from CAFE. Report on a WHO Working Group*, Document EUR/04/5046026, WHO Regional Office for Europe, Copenhagen, Denmark, 2004.

[50] D. P. Rowell, "A scenario of European climate change for the late twenty-first century: seasonal means and interannual variability," *Climate Dynamics*, vol. 25, no. 7-8, pp. 837–849, 2005.

8

Possible Source of Intermediate Ions over Marine Environment

Sunil D. Pawar and V. Gopalakrishnan

Thunderstorm Dynamics Program, Indian Institute of Tropical Meteorology, Pune 411 008, India

Correspondence should be addressed to V. Gopalakrishnan, gopal@tropmet.res.in

Academic Editors: J. Dodson, N. R. Jensen, and G. O. Thomas

Measurements of small, intermediate and large ions made onboard ORV Sagarkanya over the Arabian Sea in May-June 2003 during Arabian Sea Monsoon Experiment (ARMEX) are reported here. The daily averaged values of small-, intermediate-, and large-ion concentrations measured for 36 days during this cruise have been used for analysis. The analysis shows a weak positive correlation of 0.14 between intermediate- and large-ion concentrations, which indicates that the sources of these two types of ions are different over ocean surface. The negative correlation is observed between small- and intermediate-ion concentration for entire period of cruise. In addition, it is seen that the intermediate-ion concentration shows a very good ($r = 0.58$) and significant positive correlation with sea surface pressure. Based on good negative correlation between small- and intermediate-ion concentrations and good positive correlation between intermediate-ion concentration and sea surface pressure, it has been proposed that attachment of small ions to the ultrafine particles transported from upper troposphere to marine boundary layer is the main source of intermediate ions over ocean surface. This study supports the idea that the main source of ultrafine particles over marine boundary layer (MBL) is entrainment of aerosol particles from the free troposphere.

1. Introduction

Atmospheric ions play an important role in determining the electrical state of the atmosphere. They determine not only the conductivity of air but also the influence of other atmospheric electricity parameters like air-earth current, atmospheric electric field, and space charge [1–3]. Cosmic radiation and radioactivity from ground surface are the major source of ionization over the ocean surface and land surface respectively. Air molecules attach themselves to primary ion pairs (single-charged positive ion and free electron) by ionization processes to form "cluster ions" known as small ions. Consequently, large and intermediate ions are formed by the attachment of small ions to uncharged aerosol particles. The main characteristic of air ion is electrical mobility, which is a function of its mass and size. In recent years, it has been recognized that, under some suitable environmental conditions, small ions grow to form ultrafine aerosol particles by the process known as ion-induced nucleation process [4–7]. Laakso et al. [8] have shown that the ion-induced nucleation process takes place when

temperature, relative humidity (RH), concentrations of preexisting particles, and rate of ion production are favorable. Such conditions favorable for initiating ion-induced nucleation process exist in the upper troposphere and lower stratosphere [6]. Hõrrak et al. [4] and Vana et al. [9] have reported ultrafine particles bursts at three stations in Estonia and Finland. They attribute these bursts to the ion-nucleation process. Burst of nucleation mode particles has been reported over marine boundary layer also by Covert et al. [10] and Kamra et al. [11]. However, their contribution to ion nucleation has not been estimated.

In recent years, it has been recognized that atmospheric ions not only play important role in determining the electrical state of atmosphere but also can affect the aerosol and cloud properties and ultimately climate [5, 12–14]. Many studies have shown that electrical charge on aerosols can affect their dynamical properties. For example, coagulation rates and removal rate of aerosols by droplets can be greatly affected by electrical charge on aerosols, Clement et al., 1995 [15]. Moreover, many studies have shown that small ions can provide a source of atmospheric cloud condensation nuclei,

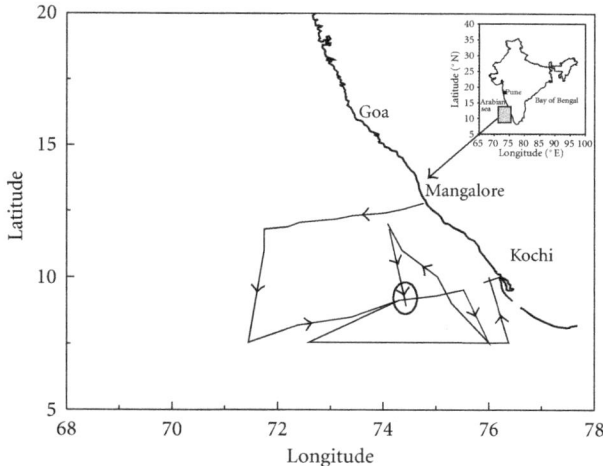

FIGURE 1: Cruise track of ORV Sagarkanya during the period from May 16, 2003 to June 19, 2003.

which indicates a potential effect on clouds and ultimately climate [5, 13]. This has been supported by observations which show that variations in cosmic rays on the scales of days and years influence global cloudiness [16, 17]. Harrison and Carslaw [14] have emphasized that a mechanism linking cosmic ray ionization and cloud properties cannot be excluded and that there are established electrical effects on aerosol and cloud microphysics. They also emphasize the need of further work includes measurements of cloud, droplet, and aerosol charging and ion-aerosol conversion, together with modeling of the electrical aspects of nonthunderstorm cloud microphysics.

We report here our measurements of atmospheric ions classified as small, intermediate, and large ions based on mobility, made over Arabian Sea in May-June 2003 during Arabian Sea Monsoon Experiment (ARMEX). These are analyzed with respect to the observed sea surface pressure to examine the source of intermediate ions in the marine boundary layer.

2. Instrumentation

Hirsikko et al. [18] list various techniques used by many researchers for measurements of atmospheric ions. For example, Mitisen et al. [19] used integral air ion counters for their measurements at Tartu, Estonia. Kulmala et al. [20], Iida et al. [21], and Vanhanen et al. [22] used Condensation Particle Counter-Based Differential Mobility Analyser for measuring the air ions of diameter down to 1 nm. We used an ion counter similar to one described by Dhanorkar and Kamra [23], is used to measure the concentration of positive atmospheric ions in three different size categories based on their mobility. The ion counter consists of three Gerdien's condensers. A common fan is used to suck the air through these three condensers. The three condensers are designed to measure the concentrations of small, intermediate, and large ions in the mobility ranges of $>0.77 \times 10^{-4}\,\mathrm{m^2\,V^{-1}\,s^{-1}}$, $1.21 \times 10^{-6}\text{--}0.77 \times 10^{-4}\,\mathrm{m^2\,V^{-1}\,s^{-1}}$, and $0.97 \times 10^{-8}\text{--}1.21 \times$

$10^{-6}\,\mathrm{m^2\,V^{-1}\,s^{-1}}$, respectively. The ion counter is installed on the balloon-launching platform of ORV Sagarkanya at ~9 m above sea level with its intake perpendicular to the direction of ship's motion. Care has been taken to see that ship's chimney exhaust does not contaminate the measurements by positioning the inlet of apparatus in upwind of ship's exhaust. However, the data obtained during the periods when the measurements are visibly observed to be contaminated are not included in the analysis. Some of the observations made using this ion counter have already been reported in Pawar et al. [24].

3. Cruise Track and Meteorological Conditions

Figure 1 shows the cruise track of ship ORV Sagarkanya from May 16 to June 19, 2003. The cruise started from Mangalore, India (12.9°N, 74.8°E), on May 16, 2003 and sailed initially westwards and then southwards. The ship was kept stationary at around 9.2°N and 74.5°E from May 23 to June 7, 2003 to conduct time series observation of conductivity-temperature-depth (CTD). On June 7, ship started moving towards east and reached Kochi on June 19. Figure 2 shows the daily averaged values of different meteorological parameters recorded onboard the ship. The atmospheric temperature was generally around 30°C except for some rainy days when it decreased to around 27-28°C. The relative humidity varied from 65% to 80% during cruise period. Winds, which were calm and westerly initially, became gradually strong and changed their direction to southwesterly with the onset of south-west June 7, 2003. Wind speed was between 4 to 9 m sec^{-1} during the cruise period. Sea surface pressure varied from 1002 mb to 1008 mb during this period. We report here only the observations about 50 to 400 km away from the Indian coast.

4. Observations

4.1. Relationship with Small and Large Ions. Figure 3 shows the daily average values of small, intermediate and large ions measured during whole cruise period. As shown in this figure, small-ion concentration lies in the range 300 to 3000 cm^{-3} with an average of 974 cm^{-3}. On the other hand, large-ion concentrations show large variations over the cruise period and lie between 1000 and 10000 cm^{-3} with an average of 5528 cm^{-3}. Intermediate-ion concentrations also show large day-to-day variation, and the concentrations vary from 200 to 3000 cm^{-3} with an average concentration of 877 cm^{-3}. As shown in Figure 2, large-ion concentration shows overall decreasing trend with the progress of cruise and small-ion concentration shows increasing trend. However, intermediate-ion concentration does not show any trend during the cruise period. Further, the intermediate ions generally follow variations in large ions and vary inversely with small-ion concentration. It should be worth noting here that the inverse correlation between intermediate- and small-ion concentrations is enhanced for periods when large-ion concentration is less.

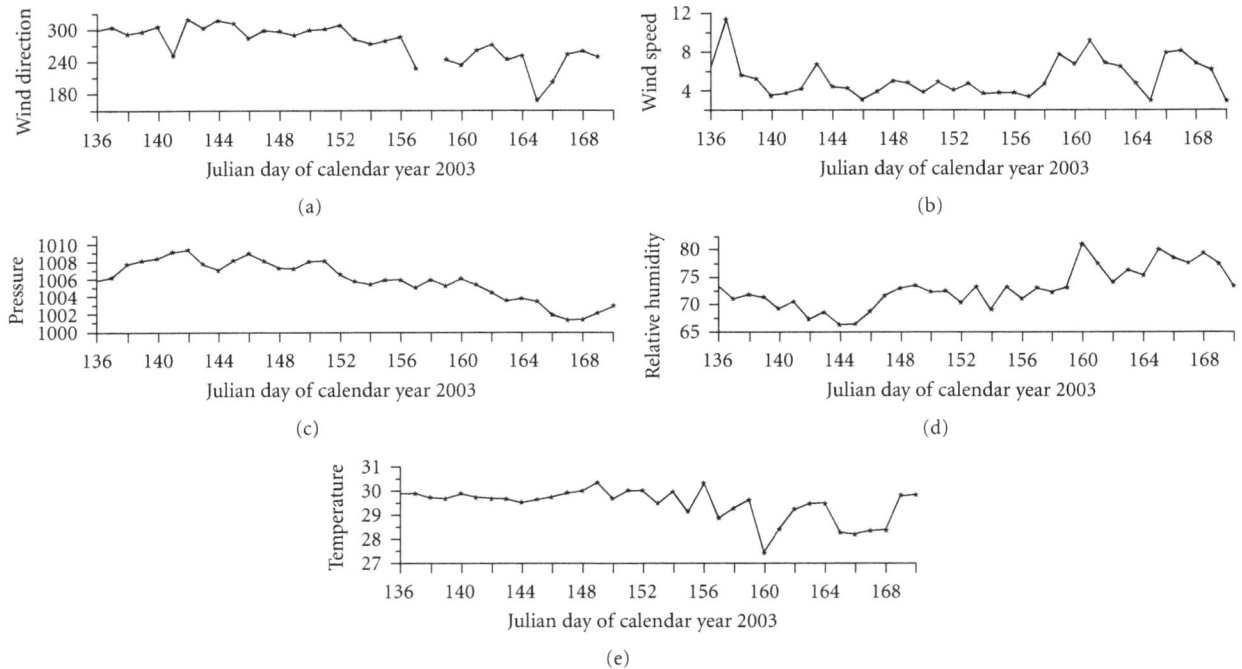

(a)

(b)

(c)

(d)

(e)

FIGURE 2: Daily averaged values of air temperature, relative humidity, atmospheric pressure, wind speed, and direction during the cruise period.

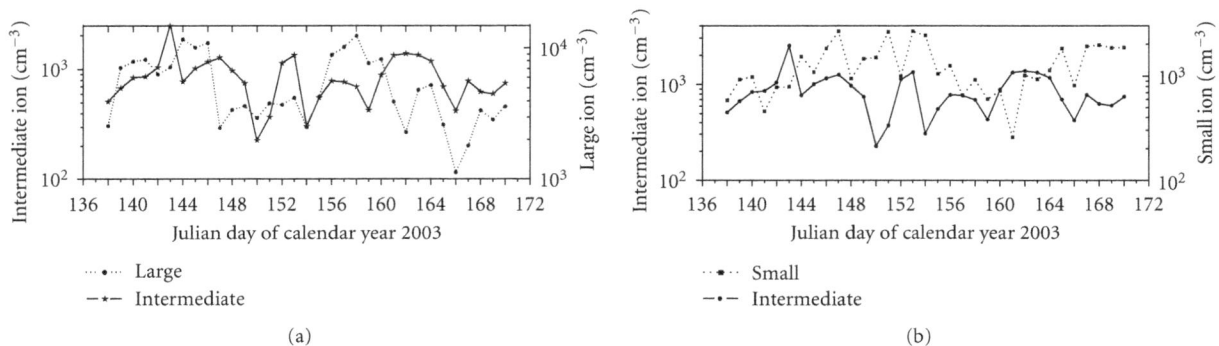

(a)

(b)

FIGURE 3: Daily averaged concentrations of small, intermediate, and large ions for the entire cruise period.

Figure 4 shows the scattered plot of intermediate-ion concentration versus small-ion (Figure 4(a)) and large-ion concentrations (Figure 4(b)). The correlation coefficient between intermediate ions and large ions is only 0.14, which indicates these two parameters are poorly correlated or sources and sinks of both types of ions can be different from each other over ocean surface. The inverse correlation between intermediate and small ions is also less ($r = -0.26$). The negative correlation between small ions and large ions is also not significant (Figure 5) which suggests that the attachment of small ions to uncharged aerosol particles may not be the main source of large ions over ocean surface. Further, Pawar et al. [24] have shown from National Centre for Environmental Prediction (NCEP) derived wind analysis that charged salt particles generated by wave breaking are the main source of this high concentration of large ions over this region.

4.2. Diurnal Variation. Dhanorkar and Kamra [23] have shown that all three types of ions, that is, small, intermediate, and large, show higher concentrations during night compared to day. They attributed this increased ion concentrations to the increased ionization rate during night time due to increased radioactive gasses. The observations of Hõrrak et al. [25] at Tahkuse observatory in Estonia show that the concentration of small ions shows maxima in the early morning hours and minima in the evening hours. They also observed that the size distribution of intermediate and light large ions in the range of 1.6–22 nm is strongly affected by nucleation bursts of nanometer particles. On the burst days, the maximum concentration of intermediate ions (1.6–7.4 nm) is about the noontime and that of light large ions (7.4–22 nm) about 2 hours later. Hõrrak et al. [4] observed that the concentration of intermediate ions is strongly correlated with temperature during the diurnal cycle. There are no

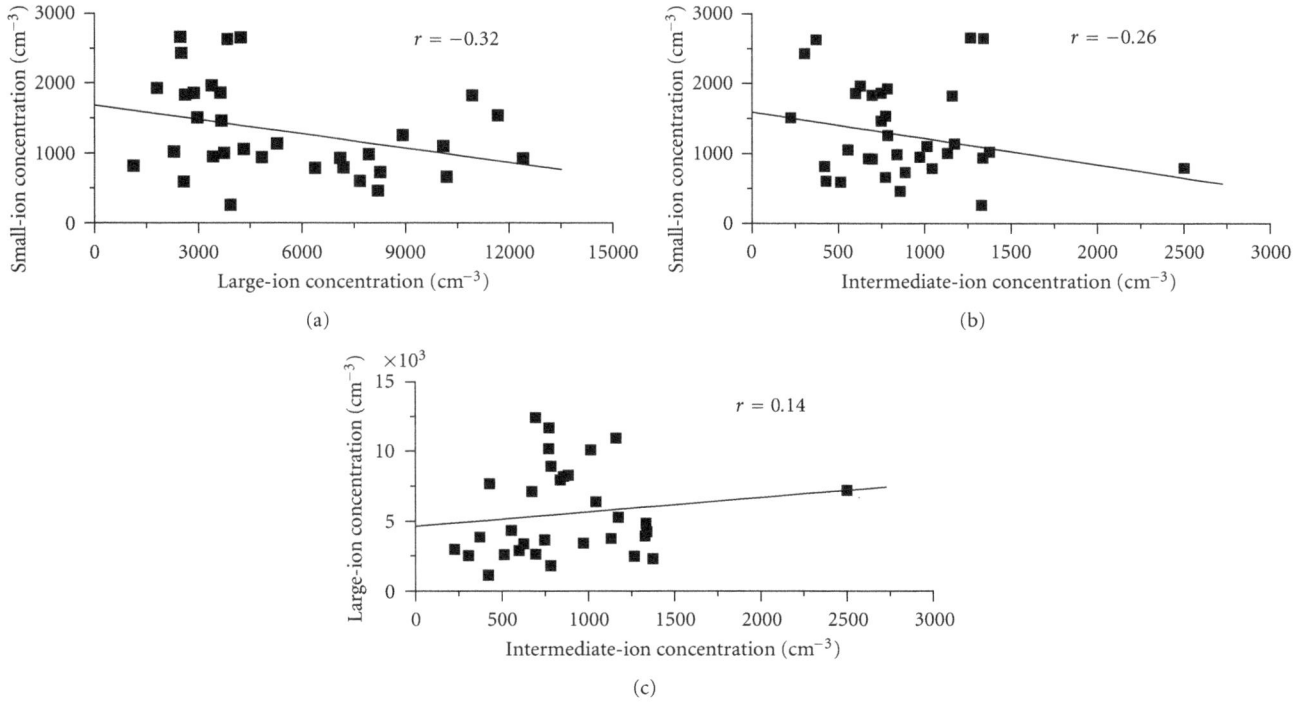

FIGURE 4: Scattered diagram of daily averaged (a) concentrations of large and small ions; (b) concentrations of intermediate and small ions; (c) concentrations of intermediate and large ions. Line of best fit is also shown.

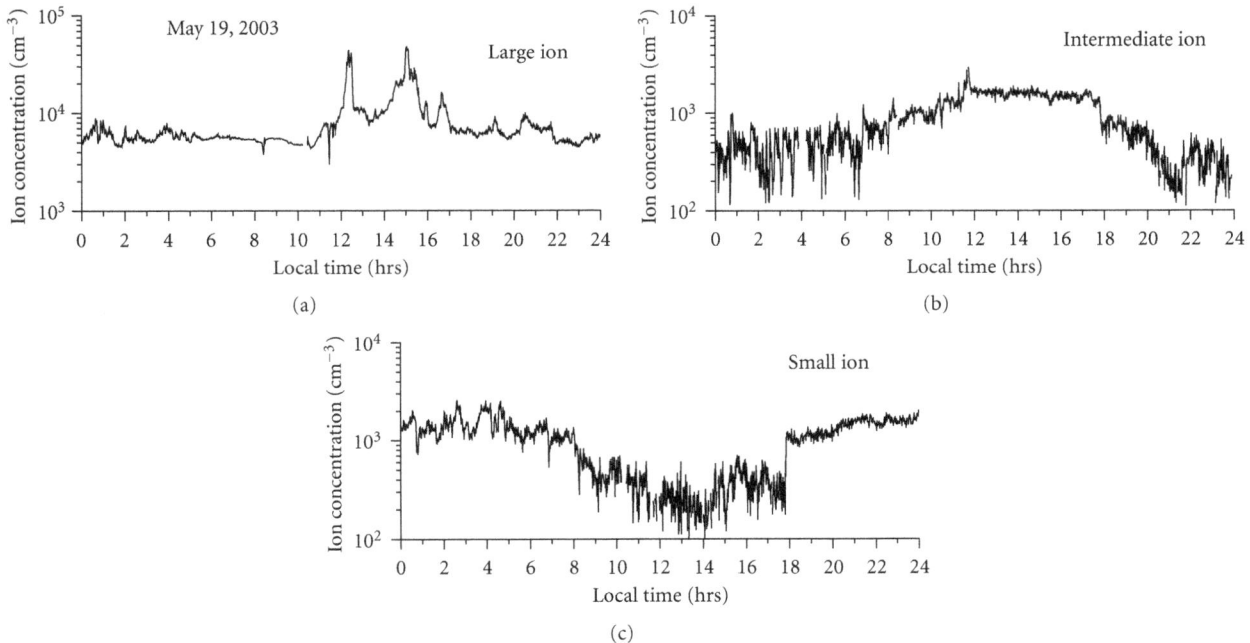

FIGURE 5: Diurnal variation of small, intermediate, and large ions on a fair weather day (May 19, 2003) observed during cruise period.

reports of diurnal variation of atmospheric ions over ocean. Figure 5 shows the diurnal variation of small, intermediate, and large in concentrations on a typical fair weather day, May 19, 2003. As shown in this figure, there is an increase in intermediate ions in the afternoon hours. The small-ion concentration shows decreases during same time. Large-ion concentrations also reach their maxima during the afternoon hours. Almost similar pattern of diurnal variation is seen on other fair-weather days also. The diurnal variation of associated meteorological parameters is shown in Figure 6. This day was a fair weather day with clear sky throughout the day. The winds were moderate and from west to

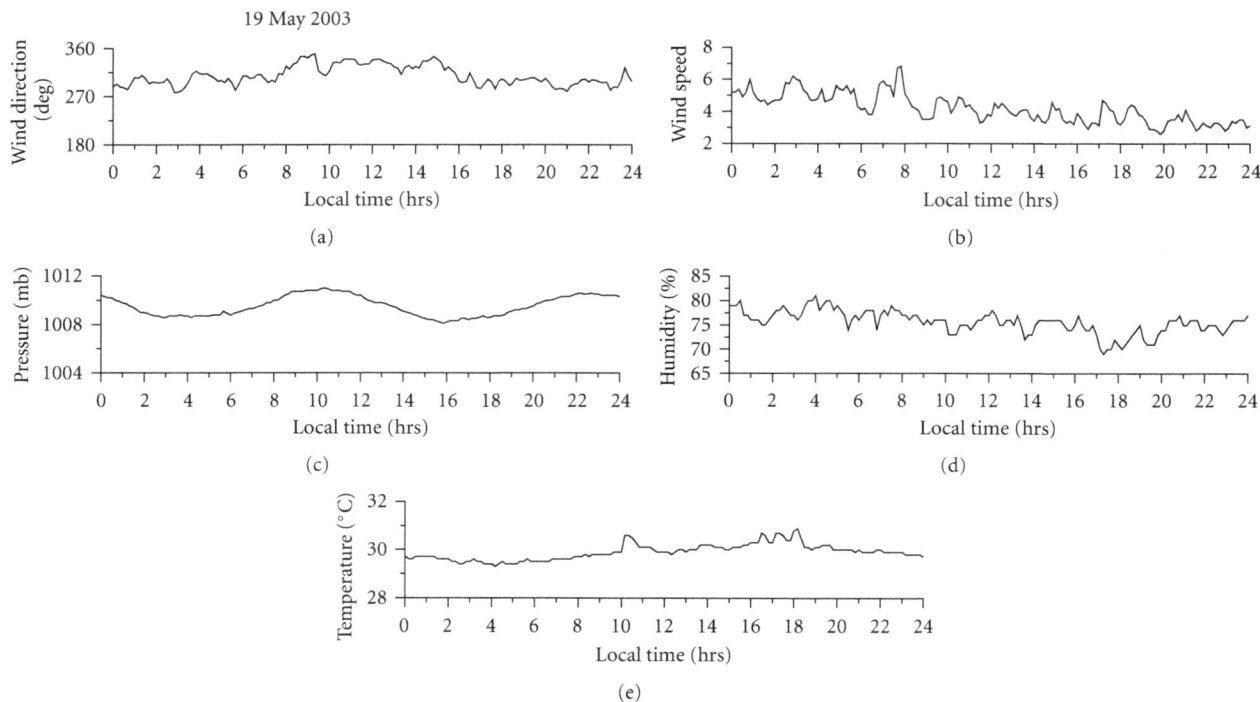

19 May 2003

(a)

(b)

(c)

(d)

(e)

FIGURE 6: Diurnal variation of meteorological parameters on May 19, 2003.

west-northwest direction. The diurnal variation of small-, intermediate-, and large-ion concentrations on nonfair weather day is shown in Figure 7. On that day, sky was cloudy and there was moderate rain in the morning hours. Figure 8 shows the associated meteorological parameters observed on that day. After the rainy spell, the atmospheric temperature was around 30°C and the relative humidity remained about 75%. There were winds exceeding 5 m sec^{-1} during the day and afterwards the winds were calm. As shown in Figure 7, intermediate-ion concentration does not show any diurnal variation as observed on fair weather day (Figure 6). There was a sharp increase in concentration of intermediate and large ions during rain. There was not much variation in small-ion concentration.

4.3. Relationship with Air Pressure. The main source of ultrafine particles over marine boundary layer is transported from free troposphere [10, 26]. The observations of Pawar et al. [27] show such a transport of ultrafine particles can be controlled by sea level pressure. Our observations also show a significant positive correlation between daily averaged values of intermediate-ion concentrations and sea surface pressure (Figure 9). Many studies [6, 28] have shown that, in the free troposphere because of favorable conditions like high ionization rate, low background aerosol concentration, and low temperature, large number of ultrafine particles form via ion-induced nucleation process. The good correlation observed between intermediate ions and sea surface pressure suggests that the transport from troposphere is the main source of observed intermediate ions. It should be noted

here that, in their life time of about few minutes, possibility of intermediate ions formed in the upper or middle troposphere reaching sea surface is much less. However, the possibility of transportation of intermediate ions formed in the lower troposphere or just above marine boundary layer to marine boundary layer cannot be ruled out.

5. Discussion

The attachment of small ions to ultrafine particles is considered as one of the dominant processes for formation of intermediate ions in the atmosphere [23]. Our observations, which show inverse relationship between small ions and intermediate ions, support this idea (Figure 4). However, correlation between these two parameters is low and statistically insignificant mainly because of presence of aerosols bigger than ultrafine particles, which can also act as a sink for small ions. The concentration of large ions can give some indication about the concentration of aerosols bigger than ultrafine particles. To illustrate the idea that the main source of intermediate ions is the attachment of small ions to ultrafine particles, we have chosen two periods, from May 26 to June 3 (Julian day 147 to 155) and from June 10 to June 18 (Julian day 162–170), when large-ion concentrations are less (Figure 3). As the large-ion concentration are low, we can assume that aerosols bigger than ultrafine particles will also be less during these periods. During these two periods, a good, statistically significant inverse correlation up to 96% confidence exists between small-ion and intermediate-ion concentration. From this good and statistically significant inverse correlation observed between intermediate- and

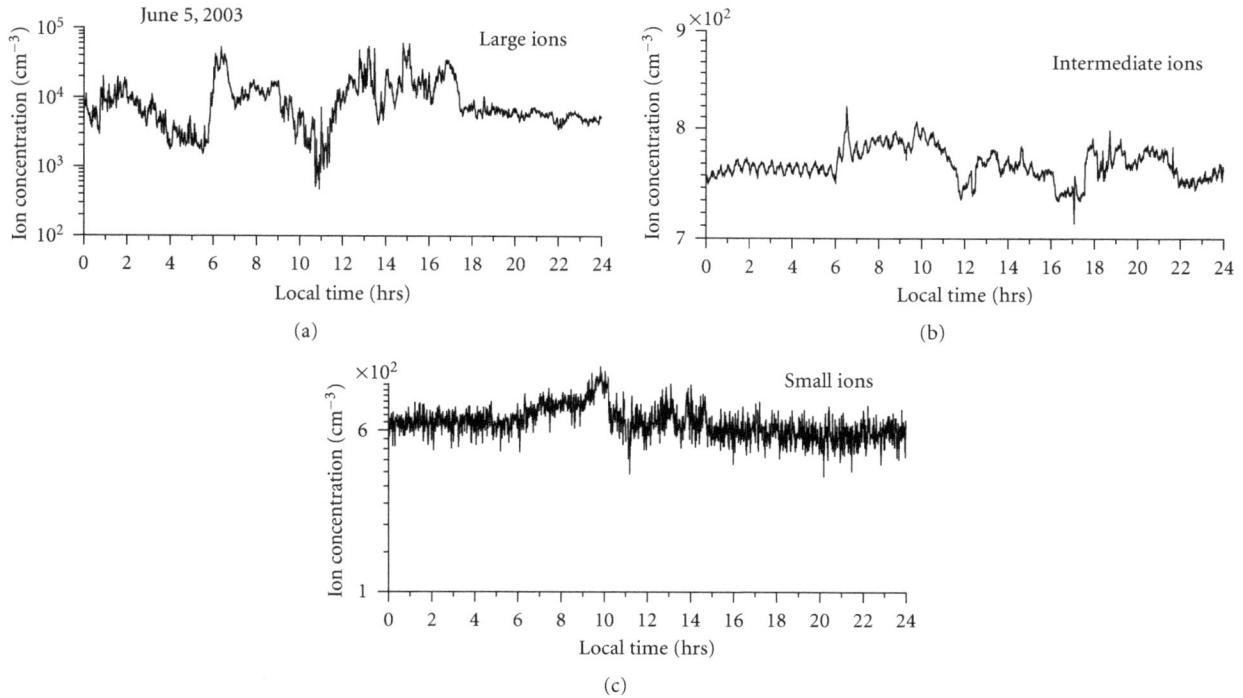

FIGURE 7: Diurnal variation of small, intermediate, and large ions on a cloudy day (June 5, 2003) observed during cruise period.

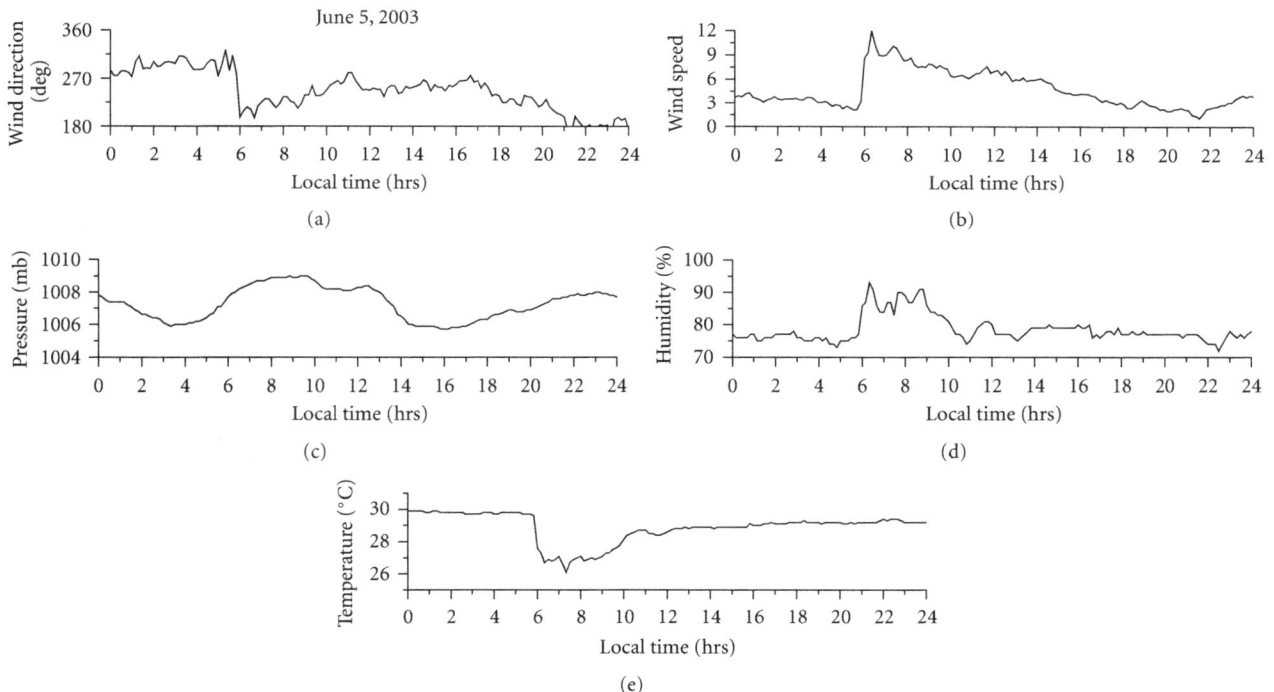

FIGURE 8: Diurnal variation of meteorological parameters on June 5, 2003.

small-ion concentrations, it is proposed that attachment of small ions to ultrafine particles can be the main source of intermediate ions over ocean surface. The typical diurnal variations observed on fair weather and cloudy days support this idea (Figures 5 and 7). Small and intermediate ions show significant inverse correlation on fair weather day, whereas, on cloudy days, the correlation is positive. The inverse correlation between small and intermediate ions on fair weather day suggests that small ions have depleted by newly formed ultrafine particles. It is interesting to note

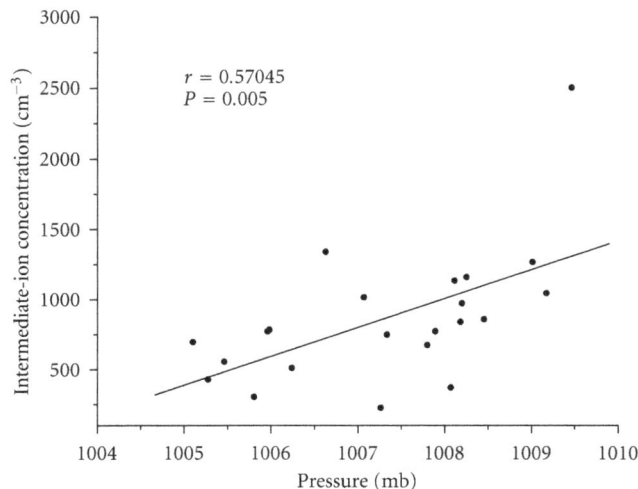

FIGURE 9: Scattered diagram of daily averaged intermediate-ion concentration and daily averaged sea surface pressure with line of best fit.

that, on cloudy day, the correlation between these ions is positive, even though correlation coefficient is small, this correlation is significant up to 99.9%. Such a positive correlation suggests that the source of these two types of ions can be same during rainy day. Our observations of typical diurnal variation of intermediate and small ions on fair weather days as shown in Figure 4(c) and significant inverse correlation observed between them also suggest that source of intermediate ions is the attachments of small ions to the newly formed ultrafine particles by photochemical reactions on sunny days.

The observations by Covert et al. [10] and Kamra et al. [11] and model calculations by Raes [26] and Yu et al. [29] show that the entrainment from free troposphere is a main source of the particles in nucleation mode in the marine boundary layer. Pawar et al. [27] have shown that the sea level pressure controls such transport of aerosol particles during subsidence motion from free troposphere to the marine boundary layer. The good positive correlation observed between surface pressure and intermediate ions (Figure 9) suggests that the one of the source of intermediate ions is attachment of small ions to the ultrafine particles transported from free troposphere.

The negligible correlation between large and intermediate ions suggests that source of these two types of ions can be completely different from each other. From the good positive correlation observed between the large-ion concentrations and area averaged NCEP-derived winds over Arabian sea, Pawar et al. [24] have suggested that the charged salt particles generated by wave breaking may be the main source of large ions over marine boundary layer.

6. Conclusions

Our observations strongly demonstrate that the attachment of small ions to the ultrafine particles is the main source of intermediate ions over marine boundary layer. The inverse

correlation observed between small ions and intermediate ions in the daily averaged as well as in one-minute averaged data supports this idea. Our observations also demonstrate that the ultrafine particles are being transported from upper troposphere to marine boundary layer. The diurnal variation of intermediate ion on fair weather day which shows increased concentration in the noon time suggests that new particle formation by photochemical reaction may be one of the source of ultrafine particles over marine boundary layer. The poor correlation between intermediate and large ions suggests that the sources of these two types of ions can be completely different over marine boundary layer. Poor correlation between large ions and small ions indicates that attachment of small ions to aerosol particles may not be the main source of formation of large ions over marine boundary layer.

Acknowledgments

The authors thank organizers of ARMEX for their participation in the experiment. They are also grateful to Professor G. S. Bhat for providing the meteorological data collected onboard Sagarkanya. They are grateful to Dr. A. K. Kamra for his support and constant encouragement.

References

[1] T. Ogawa, "Fair-weather electricity," *Journal of Geophysical Research*, vol. 90, no. 4, 1985.

[2] W. A. Hoppel, R. V. Anderson, and J. C. Willet, *Atmospheric Electricity in the Planetary Boundary Layer, Study in Geophysics-The Earth Electrical Environment*, National Academy Press, Washington, DC, USA, 1986.

[3] M. J. Rycroft, S. Israelsson, and C. Price, "The global atmospheric electric circuit, solar activity and climate change," *Journal of Atmospheric and Solar-Terrestrial Physics*, vol. 62, no. 17-18, pp. 1563–1576, 2000.

[4] U. Hõrrak, J. Salm, and H. Tammet, "Bursts of intermediate ions in atmospheric air," *Journal of Geophysical Research D*, vol. 103, no. D12, pp. 13909–13915, 1998.

[5] F. Yu and R. P. Turco, "From molecular clusters to nanoparticles: role of ambient ionization in tropospheric aerosol formation," *Journal of Geophysical Research D*, vol. 106, no. 5, pp. 4797–4814, 2001.

[6] S. H. Lee, J. M. Reeves, J. C. Wilson et al., "Particle formation by ion nucleation in the upper troposphere and lower stratosphere," *Science*, vol. 301, no. 5641, pp. 1886–1889, 2003.

[7] F. Yu and R. P. Turco, "Ultrafine aerosol formation via ion-mediated nucleation," *Geophysical Research Letters*, vol. 27, no. 6, pp. 883–886, 2000.

[8] L. Laakso, M. Kulmala, and K. E. Lehtinen, "Effect of condensation rate enhancement factor on 3-nm (diameter) particle formation in binary ion-induced and homogeneous nucleation," *Journal of Geophysical Research D*, vol. 108, no. 18, pp. 1–6, 2003.

[9] M. Vana, M. Kulmala, M. dal Maso, U. Horrak, and E. Tamm, "Comparative study of nucleation mode aerosol particles and intermediate air ions formation events at three sites," *Journal of Geophysical Research*, vol. 109, article D17201, 2004.

[10] D. S. Covert, V. N. Kapustin, P. K. Quinn, and T. S. Bates, "New particle formation in the marine boundary layer," *Journal of Geophysical Research*, vol. 97, no. 18, pp. 20581–20590, 1992.

[11] A. K. Kamra, P. Murugavel, and S. D. Pawar, "Measured size distributions of aerosols over the Indian Ocean during INDOEX," *Journal of Geophysical Research D*, vol. 108, no. 3, pp. 1–13, 2003.

[12] R. G. Harrison, "Cloud formation and the possible significance of charge for atmospheric condensation and ice nuclei," *Space Science Reviews*, vol. 94, no. 1-2, pp. 381–396, 2000.

[13] K. S. Carslaw, R. G. Harrison, and J. Kirkby, "Cosmic rays, clouds, and climate," *Science*, vol. 298, no. 5599, pp. 1732–1737, 2002.

[14] R. G. Harrison and K. S. Carslaw, "Ion-aerosol-cloud processes in the lower atmosphere," *Reviews of Geophysics*, vol. 41, no. 3, pp. 2–1, 2003.

[15] C. F. Clement, R. A. Clement, and R. G. Harrison, "Charge distribution and coagulation of radioactive aerosols," *Journal of Aerosol Science*, vol. 26, pp. 1207–1225, 1995.

[16] H. Svensmark and E. Friis-Christensen, "Variation of cosmic ray flux and global cloud coverage—a missing link in solar-climate relationships," *Journal of Atmospheric and Solar-Terrestrial Physics*, vol. 59, no. 11, pp. 1225–1232, 1997.

[17] M. C. Todd and D. R. Kniveton, "Changes in cloud cover associated with Forbush decreases of galactic cosmic rays," *Journal of Geophysical Research D*, vol. 106, no. 23, pp. 32031–32041, 2001.

[18] A. Hirsikko, T. Nieminen, S. Gagné et al., "Atmospheric ions and nucleation: a review of observations," *Atmospheric Chemistry and Physics*, vol. 11, no. 2, pp. 767–798, 2011.

[19] R. Matisen, F. Miller, H. Tammet, and J. Salm, "Air ion counters and spectrometers designed in Tartu University," *Acta et Commentationes Universitatis Tartuensis*, vol. 947, pp. 60–67, 1992.

[20] M. Kulmala, I. Riipinen, M. Sipilä et al., "Toward direct measurement of atmospheric nucleation," *Science*, vol. 318, no. 5847, pp. 89–92, 2007.

[21] K. Iida, M. R. Stolzenburg, and P. H. McMurry, "Effect of working fluid on sub-2 nm particle detection with a laminar flow ultrafine condensation particle counter," *Aerosol Science and Technology*, vol. 43, no. 1, pp. 81–96, 2009.

[22] J. Vanhanen, J. Mikkilä, K. Lehtipalo et al., "Particle size magnifier for nano-CN detection," *Aerosol Science and Technology*, vol. 45, no. 4, pp. 533–542, 2011.

[23] S. Dhanorkar and A. K. Kamra, "Diurnal and seasonal variations of the small-, intermediate-, and large-ion concentrations and their contributions to polar conductivity," *Journal of Geophysical Research*, vol. 98, no. 8, pp. 14895–14908, 1993.

[24] S. D. Pawar, D. Siingh, V. Gopalakrishnan, and A. K. Kamra, "Effect of the onset of southwest monsoon on the atmospheric electric conductivity over the Arabian Sea," *Journal of Geophysical Research D*, vol. 110, no. 10, Article ID D10204, pp. 1–8, 2005.

[25] U. Hõrrak, J. Salm, and H. Tammet, "Diurnal variation in the concentration of air ions of different mobility classes in a rural area," *Journal of Geophysical Research*, vol. 108, no. 20, p. 4653, 2003.

[26] F. Raes, "Entrainment of free tropospheric aerosols as a regulating mechanism for cloud condensation nuclei in the remote marine boundary layer," *Journal of Geophysical Research*, vol. 100, no. 2, pp. 2893–2903, 1995.

[27] S. D. Pawar, P. Murugavel, and D. M. Lal, "Effect of relative humidity and sea level pressure on electrical conductivity of air over Indian Ocean," *Journal of Geophysical Research D*, vol. 114, no. 2, Article ID D02205, 2009.

[28] J. Kazil, E. R. Lovejoy, M. C. Barth, and K. O'Brien, "Aerosol nucleation over oceans and the role of galactic cosmic rays," *Atmospheric Chemistry and Physics*, vol. 6, no. 12, pp. 4905–4924, 2006.

[29] F. Yu, Z. Wang, G. Luo, and R. Turco, "Ion-mediated nucleation as an important global source of tropospheric aerosols," *Atmospheric Chemistry and Physics*, vol. 8, no. 9, pp. 2537–2554, 2008.

9

Determination of Interesting Toxicological Elements in PM$_{2.5}$ by Neutron and Photon Activation Analysis

Pasquale Avino,[1] Geraldo Capannesi,[2] Francesco Lopez,[3] and Alberto Rosada[2]

[1] *Air Chemical Laboratory, DIPIA, INAIL Settore Ricerca, Via IV Novembre 144, 00187 Rome, Italy*
[2] *UTFIST-CATNUC, ENEA, Via Anguillarese 301, 00060 Rome, Italy*
[3] *Dipartimento di Agricoltura, Ambiente Alimenti (DIAAA), University of Molise, Via De Sanctis, 86100 Campobasso, Italy*

Correspondence should be addressed to Pasquale Avino; avino@unimol.it

Academic Editors: G.-C. Fang, J. Sun, and C. Varotsos

Human activities introduce compounds increasing levels of many dangerous species for environment and population. In this way, trace elements in airborne particulate have a preeminent position due to toxic element presence affecting the biological systems. The main problem is the analytical determination of such species at ultratrace levels: a very specific methodology is necessary with regard to the accuracy and precision and contamination problems. Instrumental Neutron Activation Analysis and Instrumental Photon Activation Analysis assure these requirements. A retrospective element analysis in airborne particulate collected in the last 4 decades has been carried out for studying their trend. The samples were collected in urban location in order to determine only effects due to global aerosol circulation; semiannual samples have been used to characterize the summer/winter behavior of natural and artificial origin. The levels of natural origin element are higher than those in other countries owing to geological and meteorological factors peculiar to Central Italy. The levels of artificial elements are sometimes less than those in other countries, suggesting a less polluted general situation for Central Italy. However, for a few elements (e.g., Pb) the levels measured are only slight lower than those proposed as air ambient standard.

1. Introduction

The distribution of elements in airborne particulate matter is fundamentally determined by resuspension of various substances of natural and/or artificial origin, by their type of circulation due to the meteorological events, and by the chemical behavior of the elements.

Anthropogenic activities introduce into the environment materials that give rise to increasing levels of many substances which may endanger the environmental quality and represent a hazard to human health.

Major attention has been given to those elements who are more liable to alter the environment and endanger human health.

Trace elements in airborne particulate are an important issue for the implications regarding the health effects of some elements (e.g., Cd, Hg, and Pb). Furthermore, the airborne particulate matter pollutant, especially the distribution and multielemental composition of fine particles with diameter <2.5 μm (PM$_{2.5}$), is considered one of the most difficult tasks in environmental chemistry for its complex composition and implications that complicate notably the behavior comprehension [1–5].

Since pollution from trace elements has been considered, the evaluation of background levels due to natural pathways of circulation seems the proper preliminary action to be undertaken. This study was therefore extended as far back as possible in time (from 1965 until 2000) in order to analyze the trend of airborne concentration of pollutant elements in connection with the industrial and lifestyle growth during the entire period.

Instrumental nuclear techniques are widely used in this field [4, 6–28] as they represent the most reliable method for analyzing trace and/or ultratrace elements in air particulate PM$_{2.5}$. Instrumental Neutron Activation Analysis (INAA) as well as Instrumental Photon Activation Analysis (IPAA) have been employed in this work to measure interesting toxicological elements. In particular, over metals such as As, Cd, Cr, Hg,

Pb, Sb, and Zn considered of greater health concern, other elements of relevant environmental interest or less considered previously were measured.

2. Experimental Part

2.1. Sampling. For particulate matter sampling a dichotomous sampler (mod. SA 241, Graseby-Andersen) operating at 16.7 L min^{-1} was used. This sampler has a PM10 size selective inlet and separates the aerosol into fine (aerodynamic diameter, D_a, <2.5 μm) and coarse (2.5 μm < D_a < 10 μm) fractions. Particulate matter was collected on polymethylpentene ringed, 2.0 μm pore size, 37 μm, Teflon membranes (Gelman, type R2PJ). This sampler has been designed as reference by USA-EPA. The PM$_{2.5}$ samples were stored in box at controlled conditions (atmosphere and temperature).

About 300 air samples have been collected in downtown Rome (41°53'54"N, 12°29'43"E), in an area characterized by high presence of anthropogenic activities. The sampling was 24-hour long for each filter. All the storage and handling sample treatment were carried out at the ENEA and INAIL laboratories.

2.2. INAA and IPAA Analyses. Among the different analytical methodologies available for element determination we used nuclear approach for its important analytical properties. In fact, the various analytical techniques (spectroscopy, electrochemistry, and chromatography) do not permit to have the maximum information because of their limitations. INAA is well known as reference analytical technique because all the experimental steps are totally traceable and there is absence of physical-chemical sample manipulation reducing the positive and/or negative artifacts formation. Furthermore, because of its high sensitivity, multielemental character allowing the determination of about 40 elements with a good Limit of Detection (LOD) (Table 1) and accuracy, the INAA has surpassed other instrumental methods for trace/ultratrace metal and rare earth element analysis: a comparative study [29], however, has pointed out that INAA is blank free and especially suitable for the analysis of reference materials [30]. Further, we used IPAA as a complementary technique for determining elements: in particular, Pb, Tl, and Zr which are difficult to determine by INAA, are important from toxicological and environmental point of view.

Table 1 shows the nuclear data (as product nuclide, half-life and energy peak emission) and LOD (expressed as ng m^{-3}) of each element investigated in this study by means of INAA and IPAA. It should be noted the very low LODs reached by nuclear techniques in relation with other analytical methodologies [29, 31].

INAA: samples, blank, and standards, put in nuclear-grade polyethyene cylinders (Kartell), were irradiated at a neutron flux of 2.6 × 10^{12} n × cm^{-2}× s^{-1} for 32.55 h in rotatory rack "Lazy Susan" of the nuclear reactor Triga Mark II of the ENEA-Casaccia Laboratories [32].

For the analysis primary and secondary standards were used. Primary standards (Carlo Erba, Milano, Italy) were As, Cd, Co, Cr, Cs, Fe, Hg, La, Ni, Sb, Se, Sm, and Zn whereas as secondary standards United States Geochemical Survey

TABLE 1: Nuclear data and LOD of elements by INAA and IPAA (m: minutes; h: hours; d: days; y: years). LODs calculated according to [30].

Element	Product nuclide	Half life	γ-Ray used (keV)	LOD (ng m^{-3})
INAA				
Ag	^{110m}Ag	253.0 d	657.8	0.06
Al	^{28}Al	2.3 m	1778.9	200
As	^{76}As	26.3 h	559.2	1
Au	^{198}Au	2.70 d	411.8	1
Ba	^{131}Ba	11.5 d	496.3	5
Br	^{82}Br	1.47 d	776.5	0.5
Ca	^{49}Ca	8.8 m	3083.0	200
Cd	^{115m}In	53.0 h	336.6	0.3
Ce	^{141}Ce	32.38 d	145.4	0.05
Cl	^{38}Cl	37.3 m	1642.0	100
Co	^{60}Co	5.272 y	1332.5	0.01
Cr	^{51}Cr	27.7 d	320.0	0.2
Cs	^{134}Cs	2.062 y	795.7	0.02
Eu	^{152}Eu	12.7 y	1408.0	0.01
Fe	^{59}Fe	45.1 d	1099.2	50
Hf	^{181}Hf	42.4 d	482.2	0.01
Hg	^{203}Hg	46.9 d	279.0	0.02
I	^{128}I	25.0 m	442.7	0.1
K	^{42}K	12.52 h	1524.7	50
La	^{140}La	40.27 h	1596.2	0.1
Mg	^{27}Mg	9.5 m	1014.1	50
Mn	^{56}Mn	2.58 h	1810.7	0.1
Mo	^{99}Mo	2.75 d	141.0	1
Na	^{24}Na	15.0 h	1368.4	100
Nd	^{147}Nd	11.1 d	531.0	1
Ni	^{58}Co	70.78 d	810.7	1
Rb	^{86}Rb	18.66 d	1076.7	1
Sb	^{122}Sb	2.70 d	564.0	0.2
Sc	^{46}Sc	83.85 d	889.2	0.003
Se	^{75}Se	120.4 d	264.6	0.1
Sm	^{153}Sm	1.948 d	103.1	0.04
Ta	^{182}Ta	115 d	1221.6	0.005
Tb	^{160}Tb	72.1 d	879.4	0.01
Th	^{233}Pa	27.4 d	311.8	0.1
Ti	^{51}Ti	5.8 m	320.0	200
U	^{239}Np	2.35 d	228.2	0.03
V	^{52}V	3.76 m	1434.4	1
W	^{187}W	24.0 h	685.7	0.01
Yb	^{175}Yb	4.21 d	396.1	0.01
Zn	^{65}Zn	243.8 d	1115.5	0.5
IPAA				
As	^{75}As	17.9 d	596.0	0.2
Pb	^{204}Pb	52.1 h	279.0	5
Tl	^{203}Tl	12.0 d	440.0	0.5
Zr	^{90}Zr	79.4 h	909.0	0.2

$$\begin{array}{c}
\hspace{3cm}\rightarrow t_r = 5/7\,\mathrm{d}\ \text{As, Au, Br, Cd, La, Mo, Sb, Sm, W}\\
(t_m = 9000\,\mathrm{s})
\end{array}$$

$$\boxed{\begin{array}{l}\text{Standard}\\ \text{and Sample}\end{array}}\ t_r = 24\,\mathrm{h}\ \xrightarrow{\hspace{1cm}}$$

$$\begin{array}{c}
\rightarrow t_r = 20/120\,\mathrm{d}\ \text{Ce, Co, Cr, Cs, Eu, Fe, Hf, Hg, Nd,}\\
\text{Ni, Rb, Sb, Sc, Se, Ta, Th, Yb, Zn}\\
(t_m = 86400\,\mathrm{s})
\end{array}$$

t_i = irradiation timetr
t_r = cooling timetm
t_m = measurement time

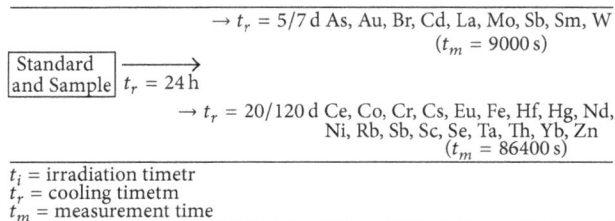

FIGURE 1: Scheme for INAA analysis of standards and samples.

(USGS) nn. 1, 4, 6 and the Coal Fly Ash (NIST) n. 1633a were used.

After irradiation, γ-ray spectrometry measurements of different durations (Figure 1) were carried out using a HPGe detector (FWHM 1.68 keV at 1332 keV) connected to a multichannel analyzer equipped with software packages for a γ-spectra analysis.

A first measurement series was performed 5/7 days after the end of irradiation with measurement times of 3000 and 9000 s for sample for determining As, Au, Br, Cd, La, Mo, Sb, Sm, and W [31]. The second series was performed 20/120 days after the end of irradiation with measurement times of 24–100 h for sample for determining Ce, Co, Cr, Cs, Eu, Fe, Hf, Hg, Nd, Ni, Rb, Sb, Sc, Se, Ta, Th, Yb, and Zn [31].

IPAA: samples, blank, and standards (NIST SRM 1571) were irradiated in the photon beam of the INFN Frascati National Laboratory Linear Accelerator (LINAC) at an average beam current of 40 μA, maximum electron energy of 300 MeV, and a W converter of 0.3 mm thickness.

Two series of measurements were carried out: after 36/70 hours Zr and Pb were measured for 2 hours whereas after 20 days from irradiation As and Tl were counted for 4 hours.

3. Results and Discussion

3.1. INAA Quality Assurance and Quality Control (QA/QC). An important task was devoted to the investigation of Quality Control (QC) and Quality Assurance (QA) of the methodology used in this study. For these goals we performed irradiations of primary and secondary reference standard materials (RSMs) for matching the entire analytical methodology and for minimizing and/or avoiding matrix effects, respectively. About the comparison between our data and certified values it should be noted that the very low levels of some elements ($<1\,\mu\mathrm{g\,g}^{-1}$) can affect the measurements in terms of precision and/or accuracy: for an element having mean value $<500\,\mu\mathrm{g\,g}^{-1}$, result can be considered as "good" if coefficient of variation (CV%) $<20\%$, "acceptable" if CV% between 20% and 30%, and "unsatisfactory" if CV% $>30\%$. Table 2 shows this analytical quality control of the INAA data. In particular, this control was performed through an intercomparison campaign for 14 elements promoted by IAEA on air filter samples among different worldwide laboratories (130) using both spectrochemical (colorimetry, fluorescence, X-ray fluorescence, infrared spectrometry, atomic absorption and emission spectrometry, ICP-AES, and ICP-MS), electrochemical (polarography, voltammetry), and nuclear (INAA,

TABLE 2: Results of INAA quality control on IAEA air filter samples ($\mu\mathrm{g\,g}^{-1}$). The "measured value" is the average of seven determinations on seven different replicates. s.d.: standard deviation; n.d.: not detected.

Element	Measured value Mean ± s.d.	Certified value Mean	Average value Mean	Average value 2σ (%)
As	4.9 ± 0.5	5.6	4.59	43
Au	1.26 ± 0.10	1.15	1.06	21
Ba	43.4 ± 0.5	53.8	39.05	40
Cd	10.6 ± 1.0	9.96	9.8	18
Co	1.3 ± 0.1	1.12	1.18	38
Cr	4.7 ± 0.8	5.6	4.8	13
Cu	51.6 ± 0.5	48.8	44.8	16
Fe	193 ± 17	207.9	200.1	8
Mn	31.2 ± 1.0	31.9	30.1	14
Mo	1.26 ± 0.2	1.14	1.56	70
Se	1.01 ± 0.10	1.06	1.01	11
U	0.78 ± 0.10	1.02	0.99	14
V	8.04 ± 0.35	8.00	7.2	16
Zn	132 ± 18	152	141	12

isotopic dilution, beta counting) analytical techniques. For each element in the table are reported our values determined by INAA ("measured value") and the "certified value": the last column ("average value") reports the value averaged among all the determinations performed by different laboratories interested in the measurements. As it can be noted, the agreement between INAA and the real value is quite good except for some elements such as Ba, U, and Zn. For Ba and U this discrepancies can be due to the difficulty to analyze such kind of elements even if for Ba our "measured value" and the "average value" are quite similar. For Zn the situation is little bit different: in fact, the "certified value" ($152\,\mu\mathrm{g\,g}^{-1}$) falls into the INAA value range ($132 \pm 18\,\mu\mathrm{g\,g}^{-1}$).

3.2. Particulate Matter Results. In Table 3 for each element both the summer and winter average levels are reported as well as the maximum and minimum values for each season measured in PM_{10} samples by INAA and IPAA. It may be noted that elements of artificial origin show winter concentration levels higher than the summer ones, probably owing to an enhanced production in the winter period; in contrast elements of natural origin show summer concentration levels higher than the winter ones, possibly as a consequence of an increased resuspension of soil matter in summer.

Such results are obviously strictly related to the general meteorological characteristics of Italy and therefore contain some peculiarity. If they are compared to similar results obtained in other countries under different meteorological conditions, it can be seen that they agree fairly well for the pollutant elements, whereas for most elements of natural origin there are sensible differences that may be related to the geomorphological and meteorological characteristics. In fact, the higher values found in this study for Al, Cs, Na, Rb, Th, Ti, U, and rare earths are to be related to the element content

TABLE 3: Seasonal element concentrations (average, min, and max levels expressed as ng m^{-3}) in PM$_{10}$ determined by INAA and IPAA([a]) in atmospheric particulate sampled in downtown Rome.

Element	Summer			Winter		
	Average	Min	Max	Average	Min	Max
Ag	0.22	0.1	0.5	0.25	0.1	0.6
Al	2900	500	5300	800	200	1600
As[a]	4	1	15	3	1	8
Ba	60	30	120	30	5	70
Br	40	20	70	50	10	140
Ca	2200	800	4200	1200	300	2000
Cd	0.4	0.3	0.9	0.65	0.3	2
Ce	5.6	0.7	10	2	0.2	5
Cl	1300	300	3900	1400	300	4700
Co	0.7	0.4	1.2	0.5	0.1	0.9
Cr	7	4	13	19	2	41
Cs	1.2	0.6	2.2	0.6	0.2	1.3
Eu	0.08	0.04	0.14	0.03	0.01	0.07
Fe	2200	1000	3600	1100	400	2700
Hf	0.36	0.15	0.62	0.13	0.01	0.30
Hg	0.14	0.05	0.36	0.12	0.03	0.20
I	6	3	8	6	3	10
K	1120	50	1900	320	50	1000
La	6	3	11	3	0.3	5
Mg	1090	50	2700	560	200	1200
Mn	40	15	60	35	10	90
Mo	0.5	0.2	0.6	2	0.3	7
Na	1500	300	7000	730	200	3500
Ni	1.7	1	5	6	3	13
Pb[a]	90	40	180	120	40	340
Rb	17	10	30	7	2	15
Sb	2	0.7	4	2	0.7	7
Sc	0.3	0.1	0.5	0.1	0.03	0.2
Se	0.6	0.2	1.1	0.4	0.2	1.3
Ta	0.03	0.01	0.05	0.01	0.005	0.02
Tb	0.055	0.1	0.10	0.024	0.01	0.05
Th	3	1	5	0.7	0.2	2.4
Ti	540	200	1050	265	200	500
Tl[a]	1	0.5	2	1.4	0.5	4
U	0.4	0.1	0.7	0.2	0.06	0.4
V	8	1	15	12	8	23
Zn	70	30	180	110	40	220
Zr[a]	50	10	90	20	5	50

TABLE 4: Grouping of elements in PM$_{10}$ according to the summer/winter ratio seasonal average.

Ratio >2	Ratio ~1	Ratio <1
Al, Ba, Ca, Ce, Cs, Eu, Fe, Hf, K, La, Mg, Na, Rb, Sc, Ta, Tb, Th, Ti, U, Zr	Ag, As, Br, Cl, Co, Hg, I, Mn, Sb, Se, Tl	Cd, Cr, Mo, Ni, Pb, V, Zn

of the volcanic rocks which are very widespread in Latium [33, 34].

As a very simple approach to the element behavior, they were grouped into three categories according to the value of

the ratio (R) of summer to winter average levels (Table 4). The first group includes elements whose R is greater than 2; the second group elements whose R is less than 2 but greater than 0.5; the last group includes elements whose R is less than 0.5.

Elements of natural origin are found in the first group only, while elements of both natural and anthropogenic origin are present in the second group. The third group includes only pollutant elements (Cd, Cr, Mo, Ni, Pb, V, and Zn).

After, we collected PM$_{2.5}$ aerosol samples in downtown Rome. Table 5 shows average concentration values, minimum and maximum levels and standard deviation of the elements determined in the PM$_{2.5}$ fraction, whereas the correlation

TABLE 5: Synoptic table (mean value, min–max values, and standard deviation) of elements concentration ($ng\,m^{-3}$) determined in $PM_{2.5}$ in downtown Rome (LOD: limit of detection; *expressed as $pg\,m^{-3}$; **expressed as $\mu g\,m^{-3}$).

| Element | PM$_{2.5}$ | | |
	Mean	Min–Max	St. Dev.
As	1.06	0.121–2.76	0.044
Au	0.009	0.000–0.050	0.012
Ba	3.76	1.91–6.45	2.38
Br	17.1	3.20–50.4	13.9
Ce	0.130	0.033–0.335	0.089
Co	0.167	0.077–0.331	0.065
Cr	3.03	1.29–6.40	1.30
Cs	0.047	0.004–0.124	0.037
Eu*	1.14	1.12–1.16	0.029
Fe**	0.074	0.005–0.212	0.059
Hf	0.018	0.006–0.032	0.010
Hg	0.818	0.195–2.12	0.655
La*	22.6	8.73–53.3	10.5
Mo	0.748	0.017–3.04	0.699
Nd	<LOD		
Ni	3.54	1.91–5.82	1.45
Rb	1.82	0.416–3.74	1.07
Sb	3.60	0.690–12.6	3.24
Sc*	3.14	0.208–7.49	2.41
Se	0.567	0.116–1.55	0.415
Sm*	3.88	0.208–7.78	2.13
Th	0.027	0.007–0.041	0.010
W	0.636	0.062–2.86	0.682
Yb	0.011	0.003–0.027	0.007
Zn	58.0	4.78–252	61.3

coefficients of the analyzed elements are reported in Table 6. It should be noted that many elements cannot be detected in these samples: the main reasons depend on both the granulometric size fraction, 2.5 μm, and the very low levels reached by some elements (e.g., Nd is below LOD). Basically, the concentration levels of the elements are low: the situation is good regarding the exposition to potential toxic elements. Furthermore, a wide scattering between the correlation coefficients can be noticed: only As, Co, Fe, Sc, Sb, and Se are highly correlated ($0.7 <$ correlation coefficient < 1, marked in bold in Table 6) whereas Au, Ba, Br, Ce, Cr, Cs, La, and Sm result scarcely correlated. This behavior can be expected considering the chemical-physical properties of the elements and the granulometric size ($<2.5\ \mu$m).

3.3. The Enrichment Factor (EF) Application. In order to investigate a retrospective study of elements in PM_{10} and their evolution in relationship with the natural or anthropogenic origins, Table 7 reports the levels of selected elements collected in the last 4 decades: the data obtained show a decrease ranging between 24% (Co) and 91% (La), except for Hg, Sb, and Se. This may be attributed to the technological

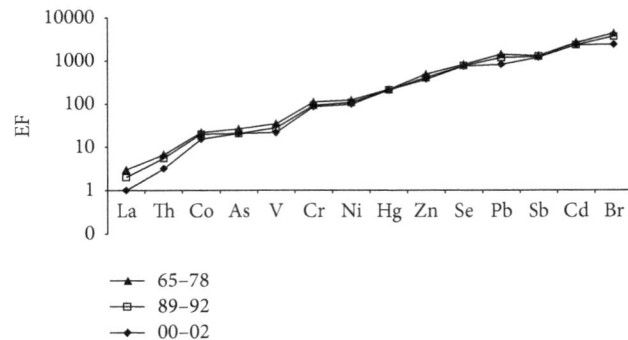

FIGURE 2: EF trend comparison of selected trace elements in PM10 fraction calculated using La as normalizing element.

growth during the entire period and to the adoption of antipollution system in domestic heating and in industrial plants. For a better knowledge of this evolution and, especially, of the element origin, we have calculated the enrichment factors (EFs) with respect to the element abundance in the upper continental crust. Elements with EF values much higher than 1 can be considered of noncrustal origin and may be attributed to long-transport phenomena from other natural and/or anthropogenic sources. The EFs have been calculated in agreement to the equation reported in Bergamaschi et al. (2002) [14] and La as normalizing element. Figure 2 shows the EF trend for selected elements in PM_{10} during four decades. As can be noted, all the elements may be attributed to long-range transport phenomena from other natural and/or anthropogenic sources: this behavior is common to all the period studied even if a very light decreasing trend can be evidenced from 1970 to 2002. Looking at Figure 2, three element groups can be identified according to their EFs: La and Th ranging between 1 and 5; Co, As, V, Cr, and Ni ranging between 1 and 100; Hg, Zn, Se, Pb, Sb, Cd, and Br ranging between 200 and 2500. Finally, some specific considerations can be extrapolated: the high EF values found for Br (and Pb as well) by both elaborations could be attributed to the use of leaded gasoline (cars with leaded gasoline are still present at the end of "nineties"); the sources of As, Pb, Sb, and Zn would be looked for among the various anthropogenic activities in the Rome area and particularly Sb and Zn could be attributed to traffic origin being essential components of antifriction alloys and car tires.

Finally, a same approach has been performed to elements investigated in the $PM_{2.5}$ fraction, even if no historical data series are available. Figure 3 reports the results obtained on the $PM_{2.5}$ fraction: Co, Cr, As, Zn, Hg, Sb, Se, and Br show EF values ranging between 5 and 8500, respectively. It should be noted that the EFs in $PM_{2.5}$ fraction are more elevated than in PM_{10} fraction, especially for Br, Se, Sb, and Hg: this could be due to the different granulometric size and the different penetrating abilities of such elements. This occurrence can be an index of the different bioavailability of an element series present in $PM_{2.5}$ fraction compared to PM_{10} fraction. As reported above, the higher EF value, found for Br and attributed to the use of leaded, is more evident in this fraction.

TABLE 6: Correlation coefficients for the trend of concentrations of analyzed elements present in PM$_{2.5}$.

As	Au	Ba	Br	Ce	Co	Cr	Cs	Fe	Hf	Hg	La	Mo	Rb	Sb	Sc	Se	Sm	Th	W	Zn	
	0.25	−0.80	**0.85**	0.37	**0.85**	0.65	0.50	**0.80**	0.51	−0.35	**0.70**	0.50	0.39	**0.83**	0.40	**0.80**	0.55	0.20	−0.19	0.37	As
		−0.97	0.58	0.51	0.35	0.06	0.54	0.33	**0.73**	0.35	0.01	0.10	**0.77**	0.48	0.01	0.56	0.08	−0.32	−0.19	0.16	Au
			−0.97	**−1.00**	**−0.98**	**−0.95**	**−0.89**	−0.65	**−1.00**	−0.43	**−0.81**	**−0.91**	**−1.00**	**−0.99**	0.62	**−0.94**	**−0.98**	**−0.98**	−0.11	−0.65	Ba
				0.61	**0.87**	0.54	0.57	**0.83**	**0.86**	−0.22	0.57	0.23	0.68	**0.95**	0.35	**0.89**	0.50	0.16	−0.17	0.58	Br
					0.51	0.39	0.29	0.46	0.59	0.15	0.41	0.15	**0.77**	0.64	0.05	0.66	0.10	−0.32	−0.14	0.50	Ce
						0.69	0.53	**0.81**	0.63	−0.22	**0.72**	0.24	**0.78**	**0.90**	0.35	**0.86**	0.53	0.25	0.10	0.60	Co
							0.09	0.67	−0.13	−0.34	0.66	0.43	0.45	0.61	0.29	**0.71**	0.45	0.04	−0.24	0.44	Cr
								0.44	**0.83**	0.15	**0.72**	0.60	0.13	0.43	0.19	0.18			−0.20	0.05	Cs
									0.39	−0.38	**0.73**	0.14	0.57	**0.85**	0.65	**0.79**	**0.72**	0.27	−0.29	0.50	Fe
										0.56	0.13	0.48	−0.05	**0.82**	0.26	**0.73**	0.46	0.26	0.01	0.55	Hf
											−0.34	−0.16	0.22	−0.17	−0.41	−0.21	−0.40	−0.37	−0.03	−0.04	Hg
												0.30	0.49	0.64	0.42	0.60	0.55	0.41	−0.15	0.26	La
													0.36	0.27	−0.07	0.40	0.12	−0.31	−0.21	0.03	Mo
														0.79	−0.34	**0.76**	−0.29	−0.57	−0.02	0.37	Rb
															0.41	**0.87**	0.53	0.15	−0.13	0.68	Sb
																0.24	**0.89**	0.47	−0.11	0.44	Sc
																	0.44	−0.06	−0.28	0.53	Se
																		0.39	−0.09	0.50	Sm
																			0.24	0.17	Th
																				0.07	W
																						Zn

TABLE 7: Levels (ng m^{-3}) of selected elements investigated along four decades in downtown Rome.

	PM$_{10}$		
	1965–78	1989–92	2000–02
As	4.00		1.35
Br	70	50	22
Cd	0.751		0.520
Co	0.498	0.523	0.379
Cr	16	2.3	7.28
Hg	0.131	0.092	1.07
La	2.04	0.803	0.188
Ni	8.01	1.72	4.54
Pb	270	172	92
Sb	1.99	2.13	9.22
Se	0.533	0.091	0.692
Th	0.911	0.723	0.229
V	14	4.82	4.02
Zn	190	28	80

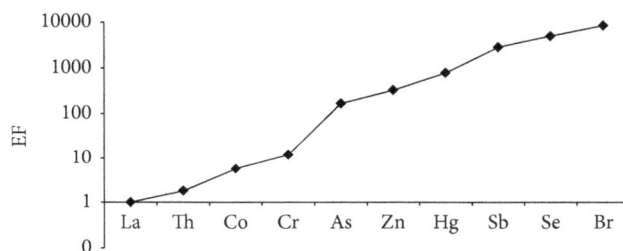

FIGURE 3: EFs of selected elements in PM$_{2.5}$ fraction calculated using La as normalizing element.

4. Conclusion

The experimentation has been addressed for getting the maximum analytical informative ability from the single sample determinations. The INAA and IPAA techniques have allowed to reach such elevated sensibility/accuracy levels to furnish discreet values for elements present at very low concentrations (trace and ultratrace levels). Particular attention has been devoted to reach elevated degrees of precision and accuracy for each determined element so that to increase its significance. In fact, for some elements such as Co, Se, and Th, this is the first determination in literature for Rome.

For some elements of geochemical origin (i.e., La, Th) the results reported can be considered representative of the urban area of Rome. The element concentrations determined in this study do not show a significant level of attention from a toxicological point of view taking into account that the simple measurements of the total airborne concentration of a metal may not be representative of its potential to participate in processes deleterious to health. Only few elements show a very good distribution between the two fractions whereas the greater part shows a distribution more elevated in the PM10 fraction than in the PM$_{2.5}$ fraction whereas EFs show an opposite trend. Finally, Co, Hg, and Zn (basically, three anthropogenic elements) show no predominant distribution between the two granulometric mass.

Acknowledgment

This work was supported under the Grant INAIL DIPIA P2/L2.

References

[1] D. W. Dockery, F. E. Speizer, D. O. Stram, J. H. Ware, J. D. Spengler, and B. G. Ferris Jr., "Effects of inhalable particles on respiratory health of children," *American Review of Respiratory Disease*, vol. 139, no. 3, pp. 587–594, 1989.

[2] D. W. Dockery and C. A. Pope III, "Acute respiratory effects of particulate air pollution," *Annual Review of Public Health*, vol. 15, pp. 107–132, 1994.

[3] R. M. Harrison and J. Yin, "Particulate matter in the atmosphere: which particle properties are important for its effects on health?" *Science of the Total Environment*, vol. 249, no. 1–3, pp. 85–101, 2000.

[4] L. Cao, W. Tian, B. Ni, Y. Zhang, and P. Wang, "Preliminary study of airborne particulate matter in a Beijing sampling station by instrumental neutron activation analysis," *Atmospheric Environment*, vol. 36, no. 12, pp. 1951–1956, 2002.

[5] P. Avino, G. Capannesi, and A. Rosada, "Heavy metal determination in atmospheric particulate matter by Instrumental Neutron Activation Analysis," *Microchemical Journal*, vol. 88, no. 2, pp. 97–106, 2008.

[6] N. K. Aras, W. H. Zoller, G. E. Gordon, and G. J. Lutz, "Instrumental photon activation analysis of atmospheric participate material," *Analytical Chemistry*, vol. 45, no. 8, pp. 1481–1490, 1973.

[7] R. E. Jervis and T. G. Pringle, "Aerosol characterization and apportionment using cascade impactors and activation analysis," *Journal of Radioanalytical and Nuclear Chemistry Articles*, vol. 123, no. 1, pp. 321–332, 1988.

[8] N. L. Chutke, M. N. Ambulkar, A. L. Aggarwal, and A. N. Garg, "Instrumental neutron activation analysis of ambient air dust particulates from metropolitan cities in India," *Environmental Pollution*, vol. 85, no. 1, pp. 67–76, 1994.

[9] L. Campanella, G. Crescentini, P. Avino, and A. Moauro, "Determination of macrominerals and trace elements in the alga Spirulina platensis," *Analusis*, vol. 26, no. 5, pp. 210–214, 1998.

[10] P. Avino, P. L. Carconi, L. Lepore, and A. Moauro, "Nutritional and environmental properties of algal products used in healthy diet by INAA and ICP-AES," *Journal of Radioanalytical and Nuclear Chemistry*, vol. 244, no. 1, pp. 247–252, 2000.

[11] E. Orvini, M. Speziali, A. Salvini, and C. Herborg, "Rare earth elements determination in environmental matrices by INAA," *Microchemical Journal*, vol. 67, no. 1–3, pp. 97–104, 2000.

[12] E. Rizzio, L. Bergamaschi, M. G. Valcuvia, A. Profumo, and M. Gallorini, "Trace elements determination in lichens and in the airborne particulate matter for the evaluation of the atmospheric pollution in a region of northern Italy," *Environment International*, vol. 26, no. 7-8, pp. 543–549, 2001.

[13] M. M. Farinha, M. C. Freitas, S. M. Almeida, and M. A. Reis, "Some improvements in air particulate matter analysis by INAA," *Radiation Physics and Chemistry*, vol. 61, no. 3–6, pp. 659–661, 2001.

[14] L. Bergamaschi, E. Rizzio, M. G. Valcuvia, G. Verza, A. Profumo, and M. Gallorini, "Determination of trace elements and evaluation of their enrichment factors in Himalayan lichens," *Environmental Pollution*, vol. 120, no. 1, pp. 137–144, 2002.

[15] M. C. Freitas, S. M. Almeida, M. A. Reis, and O. R. Oliveira, "Monitoring trace elements by nuclear techniques in PM10 and PM2.5," *Nuclear Instruments and Methods in Physics Research A*, vol. 505, no. 1-2, pp. 430–434, 2003.

[16] S. M. Almeida, M. A. Reis, M. C. Freitas, and C. A. Pio, "Quality assurance in elemental analysis of airborne particles," *Nuclear Instruments and Methods in Physics Research B*, vol. 207, no. 4, pp. 434–446, 2003.

[17] P. Avino, G. Capannesi, and A. Rosada, "Characterization and distribution of mineral content in fine and coarse airborne particle fractions by neutron activation analysis," *Toxicological and Environmental Chemistry*, vol. 88, no. 4, pp. 633–647, 2006.

[18] G. Capannesi, L. Diaco, A. Rosada, and P. Avino, "Investigation of trace and ultra-trace elements of nutritional and toxicological significance in Italian potable waters by INAA," *Journal of Radioanalytical and Nuclear Chemistry*, vol. 278, no. 2, pp. 353–357, 2008.

[19] G. Capannesi, A. Rosada, and P. Avino, "Elemental characterization of impurities at trace and ultra-trace levels in metallurgical lead samples by INAA," *Microchemical Journal*, vol. 93, no. 2, pp. 188–194, 2009.

[20] C. Seccaroni, N. Volante, A. Rosada, L. Ambrosone, G. Bufalo, and P. Avino, "Identification of provenance of obsidian samples analyzing elemental composition by INAA," *Journal of Radioanalytical and Nuclear Chemistry*, vol. 278, no. 2, pp. 277–282, 2008.

[21] P. Avino, G. Capannesi, L. Diaco, and A. Rosada, "Multivariate analysis applied to trace and ultra-trace elements in Italian potable waters determined by INAA," *Current Analytical Chemistry*, vol. 6, no. 1, pp. 26–36, 2010.

[22] G. Capannesi, A. Rosada, and P. Avino, "Radiochemical separation and anti-compton analysis of Ni, Sn, Te and Zn in lead standard reference materials at ultra-trace levels," *Current Analytical Chemistry*, vol. 6, no. 3, pp. 217–222, 2010.

[23] G. Buonanno, L. Stabile, P. Avino, and R. Vanoli, "Dimensional and chemical characterization of particles at a downwind receptor site of a waste-to-energy plant," *Waste Management*, vol. 30, no. 7, pp. 1325–1333, 2010.

[24] P. Avino, G. Capannesi, and A. Rosada, "Ultra-trace nutritional and toxicological elements in Rome and Florence drinking waters determined by Instrumental Neutron Activation Analysis," *Microchemical Journal*, vol. 97, no. 2, pp. 144–153, 2011.

[25] P. Avino, E. Santoro, F. Sarto, V. Violante, and A. Rosada, "Neutron activation analysis for investigating purity grade of copper, nickel and palladium thin films used in cold fusion experiments," *Journal of Radioanalytical and Nuclear Chemistry*, vol. 290, no. 2, pp. 427–436, 2011.

[26] P. Avino, G. Capannesi, M. Manigrasso, E. Sabbioni, and A. Rosada, "Element assessment in whole blood, serum and urine of three Italian healthy sub-populations by INAA," *Microchemical Journal*, vol. 99, no. 2, pp. 548–555, 2011.

[27] G. Capannesi, A. Rosada, M. Manigrasso, and P. Avino, "Rare earth elements, thorium and uranium in ores of the North-Latium (Italy)," *Journal of Radioanalytical and Nuclear Chemistry*, vol. 291, no. 1, pp. 163–168, 2012.

[28] P. Avino, G. Capannesi, L. Renzi, and A. Rosada, "Instrumental Neutron Activation Analysis and statistical approach for determining baseline values of essential and toxic elements," *Ecotoxicology and Environmental Safety*, vol. 92, pp. 206–214, 2013.

[29] C. Vandecasteele, "Activation analysis: present status in relation to other analytical techniques," *Mikrochimica Acta*, vol. 104, no. 1–6, pp. 379–389, 1991.

[30] L. A. Currie, "Limits for qualitative detection and quantitative determination: application to radiochemistry," *Analytical Chemistry*, vol. 40, no. 3, pp. 586–593, 1968.

[31] R. Djingova and I. Kuleff, "Chapter 5 Instrumental techniques for trace analysis," *Trace Metals in the Environment*, vol. 4, pp. 137–185, 2000.

[32] L. Di Palo, G. Focaccia, E. Lo Prato et al., "Il reattore RC-1 ad 1 MW del Centro Studi Nucleari della Casaccia. Caratteristiche generali e programmi di ricerca," *Energia Nucleare*, vol. 14, pp. 659–664, 1967.

[33] R. Djingova, S. Arpadjan, and I. Kuleff, "INAA and flame AAS of various vegetable reference materials," *Fresenius' Journal of Analytical Chemistry*, vol. 339, no. 3, pp. 181–186, 1991.

[34] E. Locardi and S. Sircana, "Distribuzione dell'uranio e del torio nelle vulcaniti quaternarie alcaline del Lazio settentrionale," *Rendiconti Della Società Mineralogica Italiana*, vol. 23, pp. 163–171, 1967.

10

A Study on Methane and Nitrous Oxide Emissions Characteristics from Anthracite Circulating Fluidized Bed Power Plant in Korea

Seehyung Lee,[1] Jinsu Kim,[2] Jeongwoo Lee,[1] and Eui-Chan Jeon[1]

[1] Department of Earth and Environmental Sciences, Sejong University, Seoul 143-747, Republic of Korea
[2] Cooperate Course for Climate Change, Sejong University, Seoul 143-747, Republic of Korea

Correspondence should be addressed to Eui-Chan Jeon, ecjeon@sejong.ac.kr

Academic Editor: Jürgen M. Lobert

In order to tackle climate change effectively, the greenhouse gas emissions produced in Korea should be assessed precisely. To do so, the nation needs to accumulate country-specific data reflecting the specific circumstances surrounding Korea's emissions. This paper analyzed element contents of domestic anthracite, calorific value, and concentration of methane (CH_4) and nitrous oxide (N_2O) in the exhaust gases from circulating fluidized bed plant. The findings showed the concentration of CH_4 and N_2O in the flue gas to be 1.85 and 3.25 ppm, respectively, and emission factors were 0.486 and 2.198 kg/TJ, respectively. The CH_4 emission factor in this paper was 52% lower than default emission factor presented by the IPCC. The N_2O emission factor was estimated to be 46% higher than default emission factor presented by the IPCC. This discrepancy can be attributable to the different methods and conditions of combustion because the default emission factors suggested by IPCC take only fuel characteristics into consideration without combustion technologies. Therefore, Korea needs to facilitate research on a legion of fuel and energy consumption facilities to develop country-specific emission factors so that the nation can have a competitive edge in the international climate change convention in the years to come.

1. Introduction

The Kyoto Protocol, which was adopted in 1997 at the third Conference of Parties, specified the GHG reduction targets and action plans for Annex I countries. The Bali Road Map was agreed upon at the 13th session of the Conference of Parties, held in December 2007, which encourage developing nations with making voluntary efforts to reduce GHG emissions. Accordingly, Korea established "Comprehensive Plans on Combating Climate Change" (2008~2012), to capitalize on the crisis of climate change to create new opportunities to further advance our nation. The nation will be proactive in joining international efforts to reduce GHG emissions, and at the same time will take an early response to climate change at home to minimize the burden of GHG emission reduction [1].

Burden sharing, with regard to GHG emission reduction, includes technology transfer and GHG emission trade, as well as the reduction of the absolute quantity of emissions. In this context, it is vital to have precise information on the amount of emissions produced and the emission sources to seek methods of reducing GHG emissions. In other words, only if we can quantitatively estimate the spatial distribution and time variance of emissions in different emission sources can we establish a concrete strategy to reduce emissions. First and foremost, we must establish an inventory of emissions based on credible data in order to achieve the goal of mandatory emission reduction [2].

In this regard, it is essential for Korea to secure statistics on basic greenhouse gases that reflect the nation's circumstances and reality. GHG emissions will vary depending on intrinsic factors including fuel type, fuel property, different types of boiler and control facilities, and the amount of fuel consumption. In particular, CH_4 and N_2O emissions are influenced by numerous additional factors, some of which are hidden, such as conditions of consumption and operation as well as technical elements [3]. Accordingly, IPCC recommends that nations apply country-specific or technology-specific emission factors rather than default emission factors of IPCC when assessing national GHG emissions [3]. However, Korea has used the default emission factors due to the lack of relevant research within the nation.

A Study on Methane and Nitrous Oxide Emissions Characteristics from Anthracite Circulating Fluidized Bed Power Plant in Korea

93

TABLE 1: Status of power generation of anthracite-fired fluidized bed plants and air-pollution control facilities (2005. 1.1.~12.31.).

Stack	Electric capacity (MW)	Electric generation				Control device	
		Gross generation (MWh)	Load factor (%)	Net generation (MWh)	Plant factor (%)	1st	2nd
1	200	1,149,264	63.7	1,027,830	65.42	ESP	Filter house
2	200	1,074,672	58.4	955,105	61.17	ESP	Filter house
Total	400	2,223,936		1,982,935			

Source: KEPCO(2006. 5.).

Therefore, it is fundamental to assess the emission factors befitting national circumstances in order to predict and estimate GHG emissions and establish emission reduction plans.

Although the anthracite coal-fired power plants in Korea consume a significant amount of domestic anthracite coal, there is virtually no research to assess emission factors on these power plants [4]. In particular, designated power plants in this study are one of the world's largest scaled anthracite coal-fired fluidized bed power plants [5].

Therefore, it should be categorized as a major GHG emission source in mapping out Korean GHG inventory as the plant uses anthracite produced in Korea and is exclusively a domestic technology. To this end, we need to develop Korean country-specific emission factors on fluidized power plants.

2. Selection of Facilities and Methods of Sample Extraction

2.1. Selection of Facilities. The research center designated first and second anthracite-fired fluidized bed power plants that use domestic low-grade anthracite. The details of the facilities are presented in Table 1. Each capacity of the facilities is 200 MW, respectively, and the total amount of power generation stood at approximately 2.2 TWh (as of 2005). An electric precipitator is installed as the first air pollution control facility, and a filter dust collector is installed as the second control facility.

The designated power plants are the only anthracite-fired power plants in Korea, which are run in the manner of circulating fluidized combustion, but not in the mixture combustion of pulverized coal and heavy oil. Annually, anthracite is consumed approximately 1.1 million tons, it means 25% of total domestic anthracite production, and 40% of total consumption from domestic anthracite-fired power plants. The combustion temperature of a fluidized bed is 900°C, significantly lower than pulverized coal combustion (1,200~1,500°C). The concentration of NO_x is 60 ppm, a figure much lower than 350 ppm, the effluent quality standard of NO_x.

2.2. Method of Sample Extraction. In this paper, EPA Method 18 (Figure 1) was applied when extracting GHG samples to assess the emission factor from GHG emitted from the plants [6]. A variety of tests were conducted to assess the temperature of flue gas, the amount of moisture, flow velocity, pressure, temperature, and others during sample extraction [7]. In order to reduce the margin of error, a one-liter Tedlar bag (SKC, US) was used for GHG sample extraction.

TABLE 2: Repeatability of elemental analysis for carbon (C) and hydrogen (H) in coal.

Times	Sample type	Injection weight (mg)	Element content (%)	
			C	H
1	BBOT	1.686	72.53	6.09
	Unknown[1]	1.832	72.90	6.04
	Absolute difference (%)		0.37	0.15
2	BBOT	1.686	72.53	6.09
	Unknown[1]	1.832	72.63	5.84
	Absolute difference (%)		0.10	0.25

[1] Unknown is that using the same BBOT sample but not inputting the information of the sulfanilamide element content.

Fuel samples were extracted from the sample extractor installed at a coal conveyor, which provides coal from the coal center to a boiler when replacing fuel. The samples for moisture measurement were extracted through two stages on each level in a swift manner. The rest of the samples were reduced through quartering, and then final samples were extracted. The fuel extraction was conducted at each power plant, and the extracted anthracite was the fuel used as the exhaust gases were extracted.

3. Analysis Methods

3.1. Fuel Analysis Methods

3.1.1. Element Contents Analysis Method. The contents of carbon and hydrogen are highly important in order to assess GHG emissions in the consumption of fossil fuel. It is essential to analyze elements of coal, because the fuel properties vary greatly depending on the origin. The same anthracite that is used for sample extraction in the research was analyzed by Thermo Finnigan-Flash EA 1112, USA, at a laboratory for analysis of Carbon, Nitrogen, Sulfur, Hydrogen, and other elements. A two-meter ParaQX column was used in the process. The reproducibility result of elemental analysis showed that the absolute difference of Carbon and Hydrogen was between 0.10%~0.37% and between 0.15%~0.25%, respectively, as shown in Table 2 when using BBOT (2,5-bis (5-tert-butyl-benzoxazolyl) thiophene) as samples.

3.1.2. Calorific Value Analysis Method. The calorific values of anthracite samples were analyzed with an automatic analyzer (IKA-C2000, Germany) at a laboratory. Injected

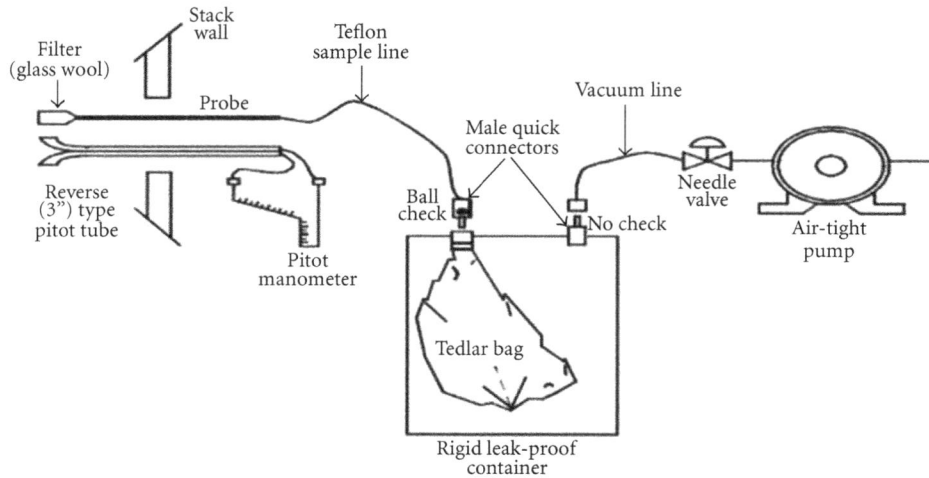

FIGURE 1: Diagram of greenhouse gas sampling system.

TABLE 3: Repeatability test of calorific analysis using benzoic acid.

Times	Mass of benzoic acid (g)	Calorific value (kcal/kg)
1	0.9998	6,316
2	0.9994	6,315
3	1.0002	6,317
4	0.9993	6,315
5	0.9989	6,314
	Mean	6,315
	S.D.	1.14
	RSD (%)	0.018
	S.E.	0.5

samples were quantified from an electronic scale (Mettler Toledo-AB204S, Switzerland) with 0.1 mg sensitivity. The water temperature of the analyzer was set at 25°C using a temperature control device (KV-500, Germany) to analyze calorific values in Isoperibolic Mode. Cooling water used for the analyzer was created with a pure water apparatus (Duplex-150H, Korea). The reproducibility of anthracite was analyzed five times. The reproducibility was exceptional, with relative standard error (RSE) standing at 0.008%, as shown in Table 3.

3.2. Exhausted Gas Analysis Method

3.2.1. Concentration of Methane and Nitrous Oxide Analysis Method. The concentration of CH_4 and N_2O was analyzed at a laboratory after extracting exhaust gases using a Tedlar bag. The analysis was conducted via gas chromatography (Model CP-3800, Varian, USA) in order to quantify the concentration of CH_4 and N_2O. Flame Ionization Detector (FID) and Electron Capture Detector (ECD) were connected to analyze CH_4 and N_2O. In this stage, one meter and three meters of Porapack QX 80/100 mesh columns (Stainless steel, external diameter of 3.175 mm, Restek) were used. The temperatures for the injector, oven, and detector were main-

tained at 120°C, 70°C, and 320°C, respectively. Ultrapure Nitrogen (99.9999%) was used as the carrier gas. Ten-port, six-port, and four-port gas switching valves were used when injecting samples in order to remove oxygen and moisture.

A calibration line was drawn up for each element before analysis in order to conduct a quantitative analysis for CH_4 and N_2O, and it was used for the concentration assessment. The calibration line was made based on five different samples with varied concentrations, ranging from 0.25–5 μmol/mol for CH_4. The calibration line for N_2O was drawn from five different samples with varied concentrations ranging from 0.5~10 μmol/mol. As a result, R^2 values of CH_4 and N_2O were 0.99977 and 0.99979, respectively, showing high relevance. Calibration results for CH_4 and N_2O were presented in Figures 2 and 3, respectively.

In order to confirm the reproducibility of CH_4 analysis, standard gas (RIGAS, KOREA) with a concentration of 1.1 μmol/mol was analyzed ten times repeatedly. A standard gas (RIGAS, KOREA) with a concentration of 1.0 μmol/mol was analyzed ten times repetitively for reproducibility confirmation of N_2O analysis. The result of reproducibility analysis was presented in Tables 4 and 5. The Relative Standard Error (RSE) of CH_4 and N_2O stood at 0.19340% and 0.57101%, respectively, displaying excellent reproducibility.

3.2.2. Moisture Content of Exhaust Gas Analysis Method. Moisture in the exhaust gas emitted from the power plants was measured from a moisture sampling apparatus (M-5, Astek Korea) and an electronic scale (Ohaus adventurer, USA). Heat rays were installed inside the tubes of the moisture sampling apparatus so that moisture in the exhaust gas can condense inside the tubes, maintaining a temperature at 120°C to extract moisture. In order to measure the amount of moisture, a certain amount of granular anhydrous calcium chloride was filled into a round-shape absorption bottle as a moisture absorbent connected to sampling tubes for GHG. The gases were measured down to two decimal places (EPA method 4) on an integrated

A Study on Methane and Nitrous Oxide Emissions Characteristics from Anthracite Circulating Fluidized
Bed Power Plant in Korea

95

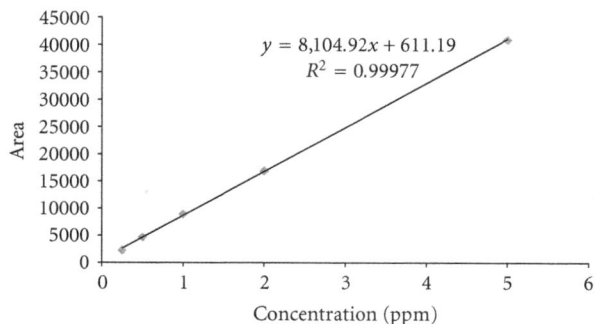

FIGURE 2: Result of calibration slope using CH_4 standard gas.

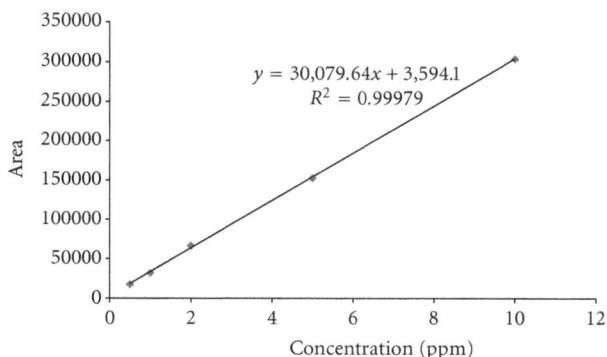

FIGURE 3: Result of calibration slope using N_2O standard gas.

TABLE 4: Repeatability test of concentration analysis using CH_4 standard gas.

Times	Concentration of CH_4 (μmol/mol)
1	1.11226
2	1.10572
3	1.10449
4	1.10930
5	1.09610
6	1.09919
7	1.09141
8	1.09956
9	1.09746
10	1.09437
Mean	1.10099
S.D.	0.00673
RSD (%)	0.61160
S.E.	0.00213
RSE (%)	0.19340

TABLE 5: Repeatability test of concentration analysis using N_2O standard gas.

Times	Concentration of N_2O (μmol/mol)
1	1.03801
2	1.01238
3	1.01570
4	0.98851
5	1.01171
6	0.98571
7	0.98551
8	1.00819
9	0.98588
10	0.98644
Mean	1.00180
S.D.	0.01809
RSD (%)	1.80570
S.E.	0.00572
RSE (%)	0.57101

3.2.3. Estimated Method of Emission Factor. The elemental analysis of fuel enables a reliable assessment of emission factors for CO_2. However, the emission factors for CH_4 and N_2O are susceptible to numerous variables of combustion conditions, such as combustion technology and its management. Therefore, it is hard to use the emission factors based on fuel analysis itself as a representative value [3]. Accordingly, emission factors for CH_4 and N_2O were assessed after precise measurement of the exhaust gas concentration from the power plants used in this research. There are four stages of work sheets for emission factor assessment through measurement. The measured concentration and flow rate of CH_4 and N_2O are inputted, and the unit is converted for the emission factor assessment in the first stage. The energy unit for fuel consumption is standardized in the second stage, and the amount of fuel consumption, power generation, and heat production are inputted in the third stage. In the fourth stage, the analysis for the calorific value of fuel is conducted, and the calorific value is inputted to assess CH_4 and N_2O emissions, and the same for CH_4 and N_2O emission factors in the fifth stage.

4. Results and Considerations

4.1. Analysis Results of Fuel Properties. The results from proximate analysis and elemental analysis for domestic anthracite used in the power plants are presented in Table 6. The carbon contents in the anthracite on dry basis stood at 65.35%, while hydrogen stood at 1.46%. The proximate analysis of volatile components, fixed carbon, and inherent moisture stood at 7.16%, 57.94%, and 4.42%, respectively. When comparing the results of proximate analysis and elemental analysis with the results of a previous anthracite analysis from different anthracite-fired plants, the carbon contents were similar at 62.26%. The hydrogen contents were nearly identical at 1.12% [8].

flow meter installed to a moisture sampling apparatus. After extracting the samples, the absorption bottle was weighed with a cap on it. The amount of moisture in the exhaust gases was measured, taking numerous factors into consideration including the differences of the bottle weight before and after extracting the samples, the flow rate of the samples, and the temperature of the gases.

TABLE 6: Results of proximate and elemental analysis of anthracite sampled at the power plants.

	Proximate analysis (air-dried basis), %				Elemental analysis (dry basis), %	
	IM	Ash	VM	FC	C	H
Stack 1	4.42	30.48	7.16	57.94	65.35	1.46
Stack 2	4.42	30.47	7.17	57.94	65.35	1.46

IM: Inherent moisture, VM: Volatile matter, FC: Fixed carbon, HHV: Higher heating value.

TABLE 7: Non-CO_2 concentration and exit condition of exhaust gas from stacks in the anthracite Fluidized Bed power plants.

Stack no.	CH_4 concentration (ppm)	N_2O concentration (ppm)	Temperature(°C) Stack	Ambient	Moisture (g/m^3)	Flow rate (m^3/hr)
1	2.22	3.72	151.40	32.00	63.8	566,898
2	1.41	2.78	151.00	32.00	63.8	626,667

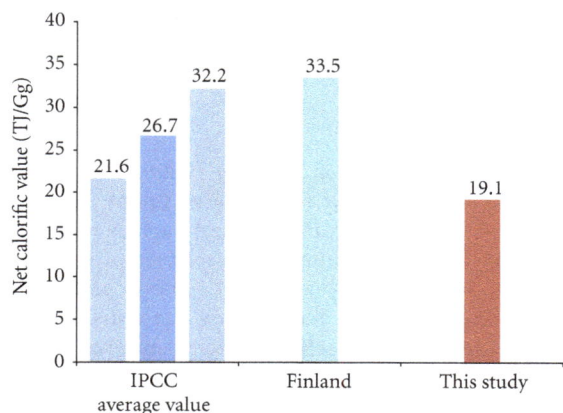

FIGURE 4: Comparison of net calorific value by this study and the IPCC.

Measuring calorific values, the calorific values as received basis with the application of total moisture displayed between 4,723–4,779 kcal/kg. The measured calorific values were compared with the standard calorific values of anthracite presented by the IPCC, as shown in Figure 4. Low calorific values of anthracite presented by IPCC G/L in 2006 are 26.7 TJ/Gg, and the margin of error within the 95% confidence interval is between 21.6~32.2 TJ/Gg. On the other hand, when the unit used in calculating calorific values of domestic anthracite from research-designated power plants is converted in order to compare with the values presented by the IPCC, the figure was found to be as low as at 19.1 TJ/Gg. The disparities of calorific values in different countries are attributable to the variances in the composition and origins of coal. These variables are decisive factors in determining the GHG emission factors; therefore, we need to develop a country-specific emission factors.

4.2. Analysis Results of Methane and Nitrous Oxide Emissions.
The numerous analysis results, which include the concentration of CH_4 and N_2O emissions, the temperature of exhaust gas, and the moisture and emissions measured from research-designated power plants, are shown in Table 7. The

average CH_4 concentration from the first exhaust pipe of the power plant was measured at 2.22 ppm, and the second exhaust pipe at 1.41 ppm. The average concentration of total emissions stood at 1.85 ppm. The average N_2O concentrations from each exhaust pipe were measured at a value between 2.78~3.72 ppm, and the average concentration of emissions from all facilities stood at 3.25 ppm. The research found that there were slight differences in the concentrations of CH_4 and N_2O emissions from some exhaust pipes, even though they used the same energy source. The disparities are likely attested to the concentrations of CH_4 and N_2O, which are more susceptible to circumstantial factors such as combustion conditions, fuel amount, and flow rate.

4.3. Estimated Results of Emission Factors of Methane and Nitrous Oxide Emissions.
In order to assess the CH_4 and N_2O emission factors of fluidized bed power plants that consume anthracite as their energy source, the low calorific values were calculated through fuel analysis. CH_4 and N_2O emission factors were assessed using the element's concentration out of exhaust gases extracted from exhaust pipes, the emission flow rate of Tele-monitoring system (TMS), and the amount of power generation. The assessed emission factors are presented in Table 8. The emission factors for CH_4 assessed from the anthracite-fired fluidized bed power plants in this research stood at 0.486 kg/TJ, 50% lower than the default emission factors for anthracite presented by the IPCC. Still, the assessed emission factors fell into the range of emission factors for CH_4 presented by the IPCC.

Emission factors for N_2O were assessed at 2.198 kg/TJ, 30% higher than the default emission factors for anthracite presented by the IPCC. The assessed emission factors for N_2O fell into the range of emission factors presented by the IPCC. The emission factors for CH_4 are 1 kg/TJ at circulating fluidized bed power plants with less than 5 MW facility capacity and 4 kg/TJ at plants with more than 5 MW. It shows that CH_4 and N_2O emissions can vary from power plants with the same method for combustion. The emission factors for N_2O presented by Finland are 30 kg/TJ, which is particularly distinctive from this research or the emission factors presented by the IPCC. Likewise, emission

A Study on Methane and Nitrous Oxide Emissions Characteristics from Anthracite Circulating Fluidized Bed Power Plant in Korea

97

TABLE 8: Non-CO_2 emission factors of anthracite fluidized Bed power plant investigated in this study.

	CH_4 emission factor (kg/TJ)	N_2O emission factor (kg/TJ)
This study	0.486	2.198
IPCC (2006)[1]	1 (0.3~3)	1.5 (0.5~5)
FINLAND (2007)[2]	4 (>5 MW)	30
	1 (<5 MW)	30

[1] Non-CO_2 default emission factor for anthracite.
[2] Non-CO_2 emission factor for coal-fired circulating fluidized bed power plants.

characteristics for CH_4 and N_2O can vary depending on different conditions, such as the country of origin or the method for combustion.

5. Conclusion

We selected circulating fluidized bed power plants as research facilities, which consume 25% of domestically produced anthracite. In order to analyze GHG emission factors of anthracite-fired circulating fluidized bed power plants, we assessed the calorific values and carbon contents for domestic low-grade anthracite through fuel analysis for domestic coal. We conducted an analysis on CH_4 and N_2O emission concentration from the stack of the plants. With the analysis results, we assessed GHG emission factors for CH_4 and N_2O.

The fuel analysis showed that the low calorific value for coal used at the research-designated facility was at 19.1 TJ/Gg. The analysis of CH_4 and N_2O concentration from exhausted gas showed that the average emission concentrations for CH_4 and N_2O from research-designated facilities were at 1.82 ppm and 3.25 ppm, respectively. The emission factors for CH_4 and N_2O from the same analysis stood at 0.486 kg/TJ and 2.198 kg/TJ, respectively. The assessed emission factors for CH_4 from anthracite-fired circulating fluidized bed power plants were 52% lower than 1 kg/TJ, the default emission factors for anthracite presented by the IPCC. The disparities were due to differences in the method for and the conditions of combustion. The emission factors for N_2O were assessed 46% higher than 1.5 kg/TJ, the emission factors presented by the IPCC. This disparity is attributable to the variance in method for combustion and other conditions because the default emission factors suggested by IPCC do not take combustion technologies into consideration. When calculate the national GHG emissions from these power plants using the emission factors in this study, CH_4 emissions could be 52% lower than emissions calculated by IPCC emission factor. But N_2O emissions should be 46% higher than emissions estimated used by emission factor of IPCC.

Therefore, it is vital to proactively promote research for developing country-specific emission factors on a wide variety of fuel and energy consumption facilities in order to secure a dominant position in future international negotiation on climate change convention.

Acknowledgments

This paper was supported by the Human Resources Development of the Korea Institute of Energy Technology Evaluation and Planning (KETEP) Grant funded by the Korea government Ministry of Knowledge Economy (no. 20100092).

References

[1] The Office for Government Policy Coordination, The 4th Comprehensive Plans on Combating Climate Change; pp. 1–15, 2007.

[2] Ministry of Environment Republic of Korea, Drew up GHG emissions inventory in the environmental sector and research and development on emission factors; pp. 3–15, 2006.

[3] IPCC, 2006 IPCC Guidelines for National Greenhouse Gas Inventories 2006.

[4] Korea Energy Economics Institute, *Yearbook of Energy Statistics*, Korea Energy Economics Institute, Uiwang, South Korea, 2006.

[5] Ministry of Commerce, Industry and Energy, Feasibility study on domestic fluidized bed expansion for utilizing domestic anthracite; pp. 3–6, 2000.

[6] US EPA, Promulgated method 18, 2001, http://www.epa.gov/ttn/emc/.

[7] G. D. Wight, *Fundamentals of Air sampling*, Lewis Publishers, Chelsea, Mich, USA, 1994.

[8] E.-C. Jeon, S.-J. Myung, J.-H. Jung et al., "Developed CO_2 emission factor for anthracite-fired thermal power plants," *Journal of Korean Society for Atmospheric Environment*, vol. 23, no. 4, pp. 440–448, 2007.

11

Investigating the Influence of Anthropogenic Forcing on Observed Mean and Extreme Sea Level Pressure Trends over the Mediterranean Region

Armineh Barkhordarian

Helmholtz-Zentrum Geesthacht Centre for Materials and Coastal Research, Max-Planck-Straße, 21502 Geesthacht, Germany

Correspondence should be addressed to Armineh Barkhordarian, armineh.barkhordarian@hzg.de

Academic Editors: L. Gimeno and C. Varotsos

We investigate whether the observed mean sea level pressure (SLP) trends over the Mediterranean region in the period from 1975 to 2004 are significantly consistent with what 17 models projected as response of SLP to anthropogenic forcing (greenhouse gases and sulphate aerosols, GS). Obtained results indicate that the observed trends in mean SLP cannot be explained by natural (internal) variability. Externally forced changes are detectable in all seasons, except spring. The large-scale component (spatial mean) of the GS signal is detectable in all the 17 models in winter and in 12 of the 17 models in summer. However, the small-scale component (spatial anomalies about the spatial mean) of GS signal is only detectable in winter within 11 of the 17 models. We also show that GS signal has a detectable influence on observed decreasing (increasing) tendency in the frequencies of extremely low (high) SLP days in winter and that these changes cannot be explained by internal climate variability. While the detection of GS forcing is robust in winter and summer, there are striking inconsistencies in autumn, where analysis points to the presence of an external forcing, which is not GS forcing.

1. Introduction

Applying formal detection methods, observed patterns of global December-to-February sea level pressure changes have been shown to be inconsistent with simulated internal variability but consistent, after rescaling its amplitude, with simulations driven by anthropogenic forcings. Therefore the SLP response to external forcing is considered to have been detected [1, 2]. However, the trend of SLP in the North Hemisphere has generally been found to be larger than that simulated trend in response to natural and anthropogenic influence [2]. Greenhouse gases, stratospheric ozone depletion, volcanic aerosol, and solar forcing have all been cited as having an impact on SLP [1, 3, 4].

Wang et al. [5] attempted to detect external influence on global SLP trends in each of the four seasons, but did not find detectable external influence on SLP in any season. Gillett and Stott [6] applied for the first time a formal attribution analysis and demonstrated that externally forced SLP trends are observed in all four seasons, with simulated and observed decreases in SLP at high latitudes and increases elsewhere. They found that the observed pattern of seasonal and zonal mean SLP changes is inconsistent with simulated internal variability and that the anthropogenic influence on SLP is therefore detectable, regardless of other natural influences. But when they separately considered the mid- and high-latitude regions of both hemispheres and the tropics and subtropics, they found that the external influence is only detectable in the low-latitude region, where models and observations show positive trends in SLP and where internal variability is low. In contrast, the anthropogenic influence could not be detected at mid- and high-latitude regions of either hemisphere.

Here, we apply a different approach to investigate whether greenhouse gases and tropospheric sulphate aerosols (GS)—the main human influences on climate—had a detectable influence on recently observed SLP changes over the Mediterranean region. This method has been earlier

Investigating the Influence of Anthropogenic Forcing on Observed Mean and Extreme Sea Level Pressure
Trends over the Mediterranean Region

99

applied to near-surface temperature [7] and precipitation [8]. The results obtained indicate that most likely (with less than 2.5% risk) GS forcing is a plausible explanation for the observed warming in this region and that present trends may be understood of what will come more so in the future, allowing for a better communication of the societal challenges to meet in the future [7]. However, in terms of precipitation, the expectation of future precipitation change over the Mediterranean is different from what we observe. Obtained results show that factors other than GS-forcing may have played a significant role in shaping the precipitation trends during the recent decades in this region. Therefore, recent observed changes cannot be used to illustrate the future expected changes of precipitation over the Mediterranean region [8].

In the present study we analyze mean sea level pressure (SLP) trends and investigate whether the observed changes are consistent with natural (internal) variability, and if not, whether they are consistent with what models simulate as response of SLP to GS forcing (greenhouse gases and sulphate aerosols, GS). Therefore we compare trends in observed SLP over the period from 1975 to 2004 with the response to GS forcing derived from the set of global climate model simulations provided through the World Climate Research Programmes (CMIP3, [9]).

Extremes of climate events are increasingly being recognized as key aspects of climate change. In view of this, we further investigate the role of GS forcing in the frequencies of extreme low and high mean SLP days. We used the P^{10} and the P^{90} percentiles of the distribution of daily SLP as a threshold to define the extreme daily values. Days with SLP less than the 10th percentile value are defined as low SLP days, and those greater than the 90th percentile are defined as high SLP days. We thus obtain the series of seasonal frequencies of the days with extreme SLP. This should allow it to be decided whether the temporal trend of the mean values of sea level pressure is due to the behaviour of its extreme values or not [10, 11].

For the first time this method is being applied to Mediterranean Sea level pressure. In contrast to other studies that have often focused on the winter season, we analyze the different seasons separately. The method used in this work allows us to deal with each model separately and present results for the 17 models individually.

The remainder of this paper is structured as follows. Details on the observational and model data are given in Section 2. The methodology, including estimating anthropogenic climate change signal, estimating natural (internal) variability, comparing the patterns of changes, and the method used to obtain the significance of pattern similarity statistics is presented in Section 3. The results including the detection of external forcing and the consistency with GS forcing at spatial mean level and at pattern level are shown in Section 4. The robustness of the results against observationally based internal variability is shown in Section 5. The fingerprint of NAO is discussed in Section 6. The results of analysis about the frequency of extreme SLP days are presented in Section 7. The summary and conclusions are in Section 8.

2. Observations and Model Data

In this study the Mediterranean area is defined as the region from 25°N to 50°N and 10°W to 40°E. We use the HadSLP2.0 noninterpolated gridded monthly sea level pressure observations for the period 1850–2004 [12]. We use global simulations with 17 coupled atmosphere-ocean general circulation models (AOGCMs) to estimate the response of of mean SLP to anthropogenic GS forcing. The simulations are included in the World Climate Research Programme's (WCRP's) Coupled Model Intercomparison Project (CMIP3, [9]). The simulations are driven by the IPCC SRES A1B scenario, ending with a CO_2 concentration of 720 ppm by the year 2100. The name of the models and the number of runs with each individual model are given in Table 1.

For changes in the frequency of extreme daily SLP, consistency analysis is based on the daily mean sea level pressure gridded reconstructions, EMSLP [13], available for the period 1850–2003. The daily SLP data in the CMIP3 archive is not completely available, and thus, to estimate the anthropogenic climate change signal of daily SLP extremes, we only use the INGV model [14], which is a global coupled ocean-atmosphere general circulation model (AOGCM), coupled with a high-resolution (7 Km) regional model of the Mediterranean Sea.

3. Methodology

Our analyses are based on the approach used in ([7] see also [15, 16]). In the first step, we investigate whether the externally forced, changes are detectable in the observed trends; therefore annual and seasonal observed mean SLP trends are compared with an estimation of the natural (internal) variability of the SLP trends (Section 3.3). If externally forced changes are detectable, in the second step we assess whether the observed trends are consistent with what models simulate as response of SLP to anthropogenic forcing (greenhouse gases and sulphate aerosols, GS). For this purpose, we compare trends in observed SLP over the period 1975–2004 with the anthropogenic (GS) climate change signal derived from a set of climate model simulations (Section 3.1). The comparison is carried out using several pattern-similarity statistics based on regression analysis and outlined in Section 3.2. Consistency of observed trend patterns with GS signal pattern is claimed in cases where the uncertainty range of regression indices, derived from control runs, does not include zero but includes unit scaling.

3.1. Anthropogenic Climate Change Signal Estimates. In this study two methods are used to estimate anthropogenic (GS) signal. On the one hand, we use time-slice experiment and define the anthropogenic climate change signal as the difference between the last decades of the 21st century (2071–2100) and the reference climatology (1961–1990). We assume a linear development from 1961–2140, and the resulting signal is scaled to mileibars of SLP change per year. Using well-separated time slices, 110 years in this study, has the advantage of increasing the signal-to-noise ratio and

TABLE 1: The seventeen coupled ocean-atmosphere models and the number of twentieth century runs with each individual model. The control integrations (runs) used in this study to estimate natural (internal) variability.

	Models (number of runs)	Number of years used form control integrations
1	BCCR-BCM2-0 (1)	250 (8 30-year segments)
2	CCCMA-CGCM3-1 (5)	1000 (33 30-year segments)
3	CCCMA-CGCM3-1-t63 (1)	350 (11 30-year segments)
4	CNRM-CM3 (1)	500 (16 30-year segments)
5	CSIRO-MK3-0 (1)	350 (11 30-year segments)
6	GFDL-CM2-0 (1)	500 (16 30-year segments)
7	GFDL-CM2-1 (1)	500 (16 30-year segments)
8	GISS-AOM (2)	250 (8 30-year segments)
9	INGV-ECHAM4 (1)	100 (3 30-year segments)
10	INMCM3-0 (1)	350 (11 30-year segments)
11	MIROC3-2-HIRES (1)	100 (3 30-year segments)
12	MIROC3-2-MEDRES (3)	500 (16 30-year segments)
13	MIUB-ECHO-G (3)	350 (11 30-year segments)
14	MPI-ECHAM5 (4)	500 (16 30-year segments)
15	MRI-CGCM2-3-2A (5)	350 (11 30-year segments)
16	NCAR-CCSM3-0 (7)	230 (7 30-year segments)
17	UKMO-HADGEM1 (1)	160 (5 30-year segments)
		Total: 206 non-overlapping 30-year segments

avoiding the need to average multiple models to get good signal estimates. Therefore, this method allows us to use different climate models and explicitly deal with individual models separately. A list of the climate models and the number of ensemble members of the individual models used in this study is given in Table 1.

On the other hand, we estimate the anthropogenic signal from model simulations, forced with estimates of historical anthropogenic forcing only (ANT), including greenhouse gases and sulphate aerosols (GS). The models used are BCCR-BCM2.0 (1 run), CNRM-CM3 (1 run), CSIRO-MK3.0 (1 run), GISS-AOM (2 runs), INGV-ECHAM4 (1 run), ECHAM5/MPI-OM (4 runs), CCCMA-CGCM3-1 (5 runs), CCCMA-CGCM3-T63 (1 run), and UKMO-HadGEM1 (1 run). Since the 20th century simulations generally finished in 2000, it was necessary to use outputs from SRES A1B scenario for the last 4 years of the simulation period. The multimodel ensemble mean is used here since it provides better signal estimate [17]. This averaging removes some traces of internal variability in the GS signal pattern. Since individual realizations can be regarded as being statistically independent, the uncertainty from internal variability in GS patterns decreases as $n^{-0.5}$ where n is the ensemble size [18]. Therefore, by using ensemble mean of 17 simulations in this study, the internal variability decreases and leads to increasing the signal-to-noise ratio by a factor of 4 in estimated anthropogenic patterns. This method is mainly used in this study to evaluate the robustness of the obtained results.

3.2. Comparing the Patterns of Change.

The comparison of observed and anthropogenic climate change signal patterns is carried out using four pattern similarity statistics. We examine the similarity between the two patterns by using uncentered and centred pattern correlation statistics (1) and (2)

$$UC(O,P) = \frac{\sum_{i=1}^{n} P_i \cdot O_i}{\sqrt{\sum_{i=1}^{n} P_i^2 \cdot \sum_{i=1}^{n} O_i^2}}, \tag{1}$$

$$CC(O,P) = \frac{\sum_{i=1}^{n} \left(P_i - \overline{P}\right) \cdot \left(O_i - \overline{O}\right)}{\sqrt{\sum_{i=1}^{n} \left(P_i - \overline{P}\right)^2 \cdot \sum_{i=1}^{n} \left(O_i - \overline{O}\right)^2}}. \tag{2}$$

The index subscript $i = 1,\ldots,n$ counts the spatial points. O_i and P_i refer to the observed and expected pattern of change, respectively. We also use uncentred and centred regression indices (3) and (4), which unlike the correlation statistics, also include information about the relative magnitude of the observed and model projected trend patterns. Uncentred statistics have the advantage of including information about the area mean change, while centred statistics focus on the pattern of spatial anomalies about the mean [19, 20]

$$UR(O,P) = \frac{\sum_{i=1}^{n} P_i \cdot O_i}{\sum_{i=1}^{n} P_i^2}, \tag{3}$$

$$R(O,P) = \frac{\sum_{i=1}^{n} \left(P_i - \overline{P}\right) \cdot \left(O_i - \overline{O}\right)}{\sum_{i=1}^{n} \left(P_i - \overline{P}\right)^2}. \tag{4}$$

In order to have a measure without the effect of spatial pattern information, we also compare the area-mean changes of observed and anthropogenic signal patterns. Trends in observations have been calculated using ordinary least squares linear regression.

Investigating the Influence of Anthropogenic Forcing on Observed Mean and Extreme Sea Level Pressure
Trends over the Mediterranean Region

101

3.3. Estimating Natural (Internal) Variability. We estimate natural (internal) variability from control integrations of CMIP3 climate models, which are preindustrial control experiments with all forcings held constant. A list of the climate models and the number of years used from control integrations to estimate the natural (internal) variability is given in Table 1. From 6,200-year control runs we draw 206 nonoverlapping 30-year segments to estimate chaotic variability from weather and other sources that would be present irrespective of any external influence on the climate system [18].

The quantiles of the pattern similarity indices of control run segments on the GS signal patterns are then used to test the null hypotheses that the pattern similarity statistics obtained between observation and modelled SLP response to GS arise by chance. We further test whether the similarity statistics are compatible with unit scaling, indicating consistency with GS forcing.

4. Results

Figure 1 displays the observed pattern of change of mean sea level pressure (SLP) during the period 1975–2004 in comparison with the projected GS signal pattern (ensemble mean of 17 models) based on the time-slice experiments, and Figure 2 compares the observed and projected (simulated) area-mean changes of SLP. In winter (DJF) the observed trends show an area of increased SLP centred over the Mediterranean, which extends to the north in spring (MAM).This area of high pressure is reproduced in response of SLP to GS forcing in all 17 models and, thus also in the multimodel mean. However, the climate models underestimate the magnitude of the SLP response in winter (see Figure 2). In summer (JJA) the observed trend pattern is characterized by negative trends, which is in agreement with response of SLP to GS forcing (Figures 1 and 2).

In autumn (SON), models suggest that anthropogenic forcing should have caused a small increase in mean SLP. However, the observed trend shows areas of pronounced decrease in mean SLP over the east and west of the Mediterranean (Figures 1 and 2), which contradicts 16 of the 17 climate change projections. As a consequence of these low-pressure areas, the Mediterranean region has unexpectedly experienced an upward trend in the amount of precipitation over the last few decades, which is strongly at odds with dry and stable conditions projected by climate change scenarios [8].

Figure 2 also displays the comparison between projected GS signal (derived from two well-separated time slices, 17 climate change projections) and simulated GS signal (derived from ensemble mean of 17 ANT simulations, forced with historical GS forcing only). This comparison reveals that the simulated and projected anthropogenic climate change signals are consistent, whereas in all seasons the simulated GS signal is within the range of projected GS signals derived from 17 models. In the following sections we first test the significance of the observed SLP trends and then apply the consistency analysis.

4.1. Are the Observed SLP Trends Derived from an Undisturbed Stationary Climate? Figure 2 shows observed seasonal and annual area mean changes of mean sea level pressure (SLP) over the period 1975–2004. The observed trend is likely not derived from an undisturbed stationary climate in cases where the 90% uncertainty ranges (red whiskers in Figure 2) based on control variability (Section 3.3) exclude zero. There is no significant trend in the annual area mean change of SLP, but at seasonal resolution the observations indicate highly enhanced seasonality over the Mediterranean region (Figure 2).

As shown by red whiskers in Figure 2, there is less than a 5% chance (one-sided test) that observed trends in winter, summer, and autumn are due to natural (internal) variability alone. Thus, observed trends cannot be explained by natural (internal) variability derived from 6,200-year control runs; externally forced changes are detectable in all seasons, except spring.

Therefore, in the second step we investigate whether the observed trends, which are found to be inconsistent with natural (internal) variability, are consistent with what models simulate as response of SLP to anthropogenic (GS) forcing. We separate the consistency process in two parts.The first is searching for projected-observed data correspondence at the spatial-mean level by using uncentered pattern similarity indices. The second is considering this also at the pattern level, by employing centered pattern similarity indices. The results of our analysis can be interpreted as follows: the GS forcing is detectable if the pattern similarity indices are significantly greater than zero,but not detected if the indices are negative or not significantly greater than zero. Consistency of observed trend patterns with GS signal pattern is claimed in cases where the uncertainty range of regression indices, derived from control runs, does not include zero but includes unit scaling.

4.2. Consistency of Spatially Averaged Changes. Here we investigate the consistency of observed trend pattern with GS signal patterns using uncentred pattern similarity statistics (1) and (3).

In winter the observed trend pattern shows a high pattern correlation with all the 17 GS signal patterns. The correlation coefficients are in the range of 0.55–0.96, and in most cases higher than 0.80 (see Table 2). The correlation coefficients between the observed and GS signal patterns are further compared with the distribution of the 90th percentile of correlation coefficients of 206 unforced trends (derived from 6,200-year control variability) with anthropogenic (GS) signal pattern. The comparison indicates that in winter all the 17 GS signal patterns yield a correlation with none of the 206 unforced trends as high as that with observed trends. Therefore the significantly (with probability of error of less than 5%) greater than zero high pattern correlation coefficients with all the 17 GS signal patterns lead us to conclude that the large-scale component of GS signal has a detectable influence in observed wintertime positive sea level pressure trend.

In spring the correlation coefficients are in the range 0.20–0.85 and with 12 of 17 models significantly (at 5%)

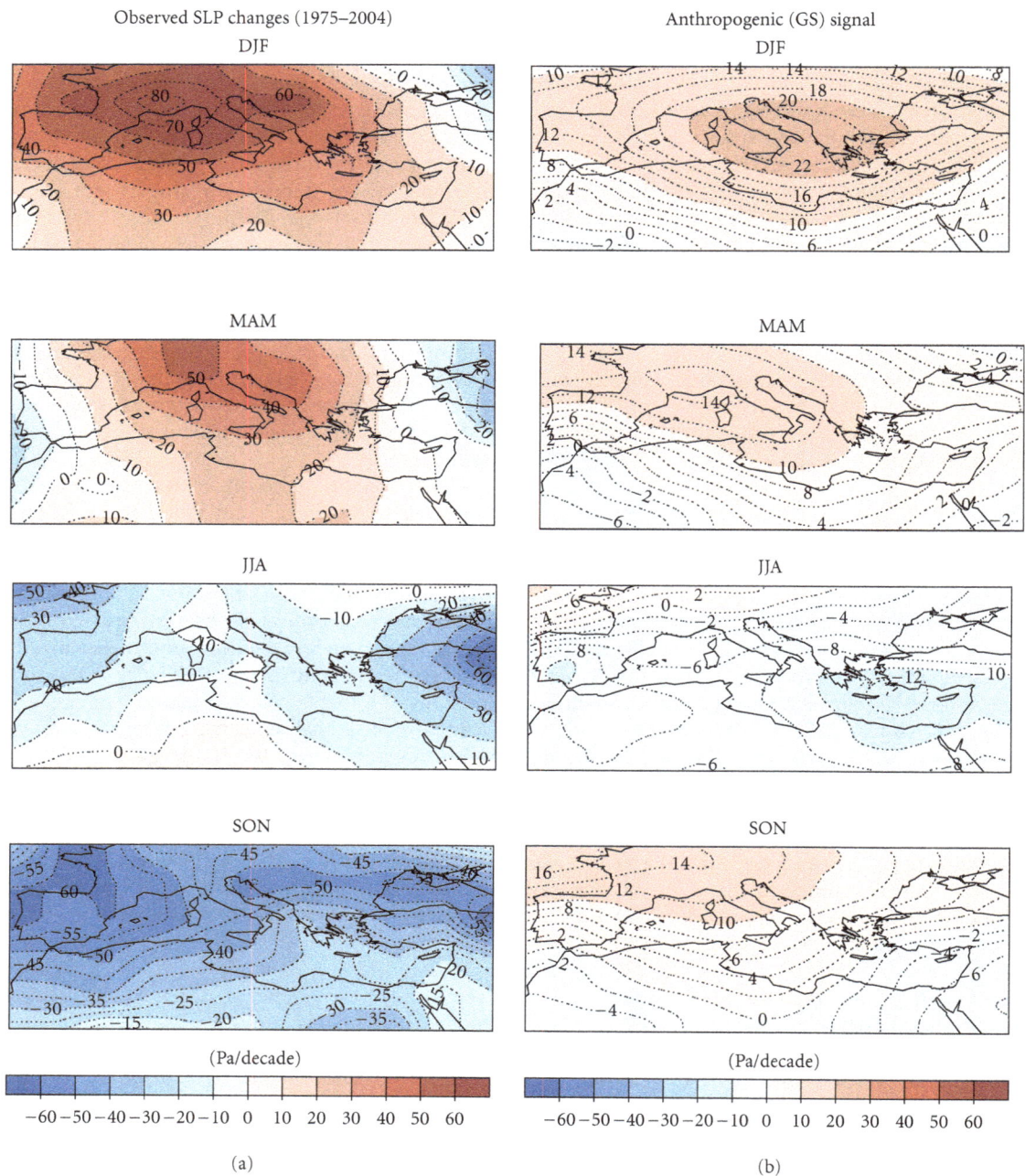

FIGURE 1: (a) Observed pattern on SLP derived from HadSLP2.0 dataset (1975–2004). (b) Projected GS signal pattern estimated from two well-separated time slices (2071–2100 minus 1961–90 mean scaled to change per decade) ensemble mean of 17 models from CMIP3 archive.

greater than zero. A smaller pattern correlation is found in summer; here indices are in the range of 0.47–0.78 and with 11 of 17 cases greater than the 90th percentile of pattern correlations of 206 control run segments. In autumn we found negative correlation with all 17 GS signal patterns (Table 2); however, indices are out of the range of the distribution of 90th percent %tile correlation coefficients of 206 unforced trends, pointing to the fact that externally forced changes are detectable in observed patterns, which are inconsistent with GS forcing.

We also use regression as a pattern, similarity measure, which unlike the correlation statistics, measures the relative magnitude of observed and simulated trend patterns. Figure 3 displays the uncentred regression coefficients of observed SLP changes into the GS signal patterns derived from the 17 model projections. In winter the uncentred regression indices are positive and significantly greater than zero (the uncertainty range of regression indices dose not include zero line) for all the 17 models, indicating that GS signal is detectable. In addition, the uncertainty range of the

Investigating the Influence of Anthropogenic Forcing on Observed Mean and Extreme Sea Level Pressure
Trends over the Mediterranean Region

103

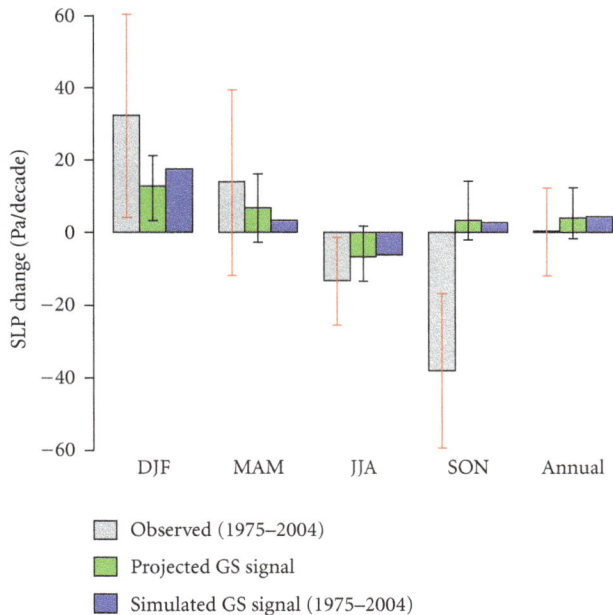

FIGURE 2: Observed seasonal and annual area mean changes of mean sea level pressure over the period 1975–2004 (grey bars) in comparison with the projected GS signals (GS) estimated from two well-separated time slices, according to SRES A1B scenario using 17 models (green bars) and the simulated GS signal based on ensemble mean of 17 ANT simulations conducted with 9 models forced with estimated historical GS forcing only (blue bars). The black whiskers indicate the spread of trends of 17 climate change projections used in this study. The red whiskers indicate the 90% uncertainty range of observed trends, derived from control variability (206 nonoverlapping 30-year segments derived from 6,200-year control runs).

regression indices includes unity in all cases, demonstrating that the observed trend pattern in winter is consistent with the GS signal patterns, derived from 17 climate change projections (with the probability of error of less than 5%).

In summer the 90% uncertainty range of the regression indices does not include the zero line either, but includes unit scaling for 12 of the 17 models (blue bars in Figure 3). Therefore, we conclude that there is less than 5% chance that natural (internal) variability rather than GS signal is responsible for the observed summertime SLP changes, and the observed trend pattern in summer is consistent in 12 of the 17 GS signal patterns.

In autumn, in contrast to other seasons, the negative but significantly greater than zero regression indices (blue bars in Figure 3) indicate that external forcings are detectable in the observed trend, but other forcings rather than GS signal are responsible for the observed unusual negative SLP trends in autumn.

Overall, significant uncentred correlation and uncentered regression indices clearly indicate that the combined large-scale and small-scale component of GS signal is detectable within all the 17 models in winter and within 12 of the 17 models in summer. In addition, the GS forcing is a plausible explanation of the observed positive trend in winter

and negative trend in summer (with probability of error of less than 5%).

We further investigate the robustness of the results against estimating the GS signal from ANT simulations outlined in Section 3.1. We use ensemble mean of 17 ANT simulations conducted with 9 models solely forced by historical anthropogenic forcing. Results indicates that in winter and summer the observed trends correlate significantly with the GS forcing pattern (at 5% level); correlation is 0.90 in winter and 0.78 in summer (see Table 2). When using uncentred regression indices (red bars in Figure 3), the influence of GS forcing is also detectable in winter at 5% level and in summer at 1% level. Therefore, obtained results in this section are robust against using simulated GS signal pattern.

4.3. Consistency of Anomaly Patterns. Our level of confidence in attributing observed changes to anthropogenic influence will be increased if we could demonstrate that even regional features of a model-projected anthropogenic signal, and not only the area average, are consistent with the observed changes. This is where centred statistics are useful since they focus on anomalies around the area mean [19, 20]. In this section we investigate the consistency of the observed and GS anomaly patterns, when the spatial mean is removed, using centred pattern similarity indices (2) and (4).

In winter the correlations between the observed and projected GS anomaly patterns are in the range 0.53–0.88. Significant test indicates that in winter 11 of 17 GS anomaly patterns yield a correlation with none of the 206 unforced trends as high as that with observed trends (with probability of error less than 5%). However, in summer we find no correlation between observed and projected GS anomaly patterns.

Figure 4 displays the centred regression indices and their respective uncertainty ranges. Analysis of the plots displayed in Figure 4 reveals that when subtracting the spatial mean and comparing the anomaly patterns, the small-scale component (spatial anomalies about the mean) of GS signal is detectable in winter with 11 of the 17 models. However, in summer the anomaly component of GS signal is hardly identifiable, suggesting that in summer the spatial-mean of GS signal is the dominant component.

Therefore, our consistency analysis at pattern level reveals that the small-scale component (spatial anomalies about the spatial mean) of GS signal is detectable only in winter within 11 of 17 models.

5. The Robustness of the Results against Observationally Based Internal Variability

Uncertainty in estimating climate variability plays a critical role to make inferences whether GS signal is detectable in the observed changes. This uncertainty is addressed here by further using an observationally based estimation of internal variability, instead of the estimation used in previous sections based on control simulations. We estimate interval variability by resampling the observed record (available from 1850 to 2004 based on HadSLP2.0 dataset). To preserve

TABLE 2: Seasonal uncentred pattern correlation coefficients of mean sea level pressure for 30-year trends from 1974 to 2005, compared to the trend of 17 anthropogenic (GS) signal patterns and ensemble mean of 17 ANT simulations conducted with 9 models, derived from the CMIP3 multimodel data set. The indices significantly greater than zero at 5% level are labelled with an asterisk.

	Models	DJF	MAM	JJA	SON
1	BCCR-BCM2-0	0.92*	0.70*	0.40	−0.95*
2	CCCMA-CGCM3-1	0.67*	0.63*	0.50*	−0.16
3	CCCMA-CGCM3-1-T63	0.55*	0.20	0.61*	−0.40*
4	CNRM-CM3	0.89*	0.85*	0.21	−0.94*
5	CSIRO-MK3-0	0.86*	0.55	0.67*	−0.69*
6	GFDL-CM2-0	0.87*	0.68*	0.30	−0.40*
7	GFDL-CM2-1	0.93*	0.78*	0.19	−0.20
8	GISS-AOM	0.73*	0.76*	0.39*	−0.74*
9	INGV-ECHAM4	0.91*	0.32	0.39	−0.77*
10	INMCM3-0	0.90*	0.62	0.67	−0.19
11	MIROC3-2-HIRES	0.81*	0.80*	0.64*	−0.71*
12	MIROC3-2-MEDRES	0.90*	0.78*	0.78*	0.10
13	MIUB-ECHO-G	0.96*	0.84*	0.47*	−0.59*
14	MPI-ECHAM5	0.90*	0.69*	0.65*	−0.18
15	MRI-CGCM2-3-2A	0.57*	0.81*	0.59*	−0.81*
16	NCAR-CCSM3-0	0.89*	0.84*	0.67*	−0.02
17	UKMO-HADGEM1	0.91*	0.73*	0.73*	−0.50
	GS signal (17 ANT Simulations)	0.90*	0.37*	0.78*	−0.51*

the serial correlation present in the observations, we use a method of re-sampling called z moving blocks bootstrap technique [21]. It consists in randomly drawing blocks of fixed length from the measured data, rather than single years, and concatenating them. The blocks can appear several times in the surrogate series. A good proportion of the original autocorrelation is thus preserved within each block. The disadvantage of this method is that the observed record is relativity short (124 years in this study) and may be affected by the response to external forcing; the advantage is that internal variability is produced from the real world itself not from model simulations.

The block length chosen for the moving blocks bootstrapping depends on the autocorrelation of the seasonal sea level pressure time series. This autocorrelation is different for each grid box and each season. Our analysis, based on a method suggested by Wiks [21], indicates that the average block length across the Mediterranean is 3.4, 2.2, 1.8, and 2.5 years for DJF, MAM, JJA, and SON, respectively. Thus, we draw 1000 30-year time series to estimate the variability of 30-year trends in a stationary climate.

Figure 5 displays the uncentred and centered regression indices with individual models and their 90% uncertainty ranges, derived from observationally based internal variability. Our results show that in spring the internal variability based on block bootstrapping is slightly smaller than variability based on control runs. Therefore, we are also able to detect the externally forced changes in spring (Figure 5). As shown in Figure 5 (left column), the GS signal is detectable and observed trends are consistent with GS signal patterns in 16 of the 17 models in winter, in 16 of the 17 models

in spring, and in 12 of the 17 models in summer (at 5% significant level). Analysis of anomaly patterns (Figure 5, right column) reveals that the smaller-scale component of GS signal is also detectable in 8 out of the 17 models in winter and in 13 of 17 models in spring, at 5% significant level (Figure 5).

Overall, our results are the same when using observationally based internal variability derived from block bootstrapping; however, the effect of GS signal, both at spatial mean level and at pattern level, is also detectable in observed springtime SLP changes, which was not detectable using estimation of internal variability based on control runs.

6. The Influence of the NAO

The Mediterranean climate is affected by the North Atlantic atmospheric circulation. The NAO is a measure of a pressure teleconnection and is defined as a large-scale alternation of atmospheric mass between regions of subtropical high pressure (the Azores high) and subpolar low pressure (the Icelandic low) in the North Atlantic. The NAO is the main feature driving the climate of western Mediterranean (see [22], and references therein), and the displacement of the Azores High, linked to NAO, leads to blocking episodes or to enhancement/weakening of the westerlies over Western Europe [23]. Ribera et al. [24] have shown that sea level pressure behavior over the Mediterranean region is dominated by QBO (quasi-periodic oscillation of the equatorial zonal wind between easterlies and westerlies in the tropical stratosphere) while NAO modulates the amplitude of sea level pressure. It is now recognized that models may underestimate the

Investigating the Influence of Anthropogenic Forcing on Observed Mean and Extreme Sea Level Pressure
Trends over the Mediterranean Region

105

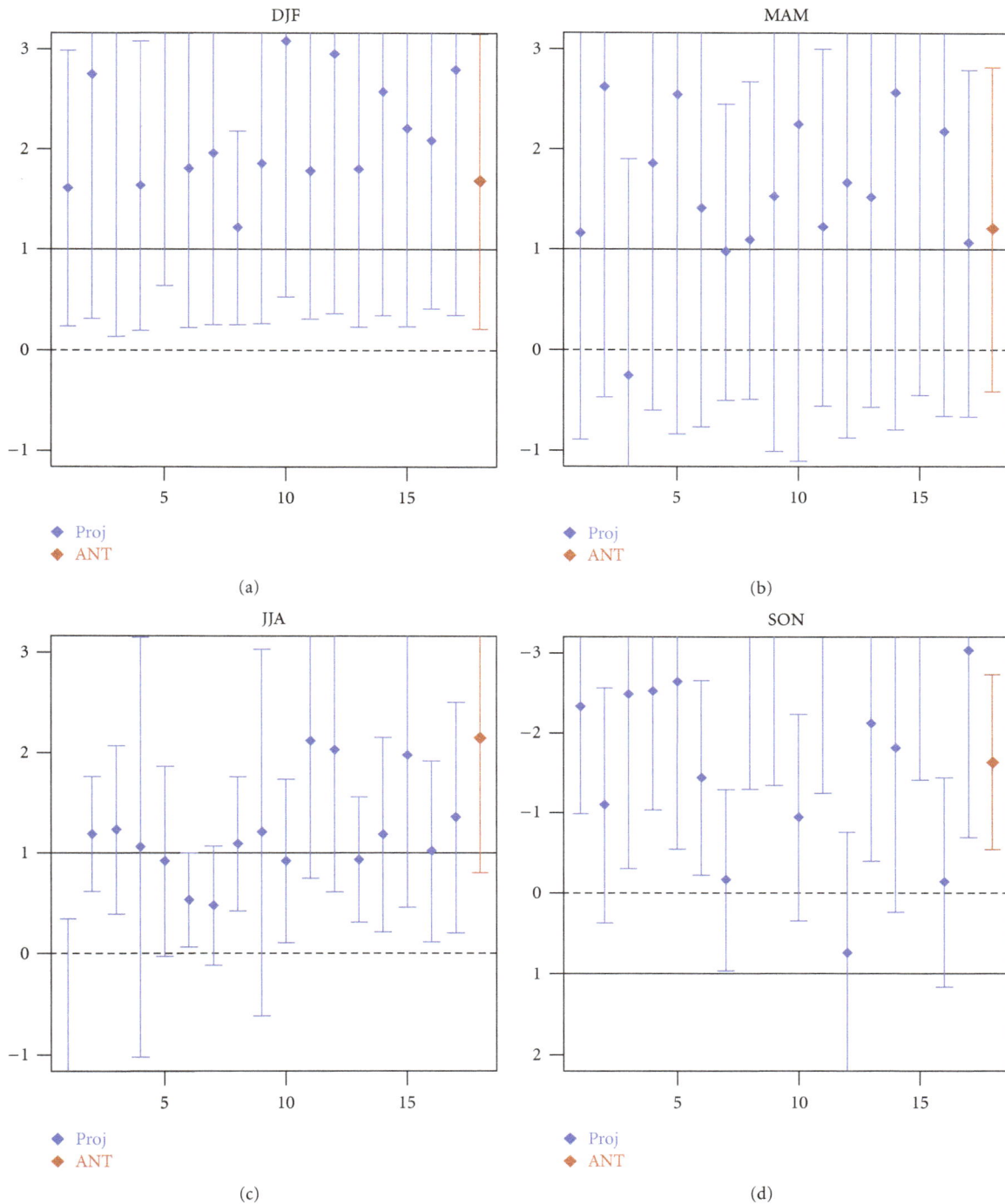

FIGURE 3: Results from consistency of spatially averaged changes. Blue bars: seasonal uncentred regression coefficients (y axes) of observed SLP changes onto the changes in response to GS forcing derived from 17 climate change projections (x axes) based on the SRES A1B scenario. Red bars: regression coefficient of observed SLP changes onto the simulated GS signal estimated from ensemble mean of 17 ANT simulations (conducted with 9 models). Whiskers show the 90% uncertainty ranges of regression coefficients derived from model-based estimates of natural (internal) variability (206 nonoverlapping 30-year segments derived from 6,200-year control runs). The solid lines mark regression indices equal to unity indicating consistency with GS forcing.

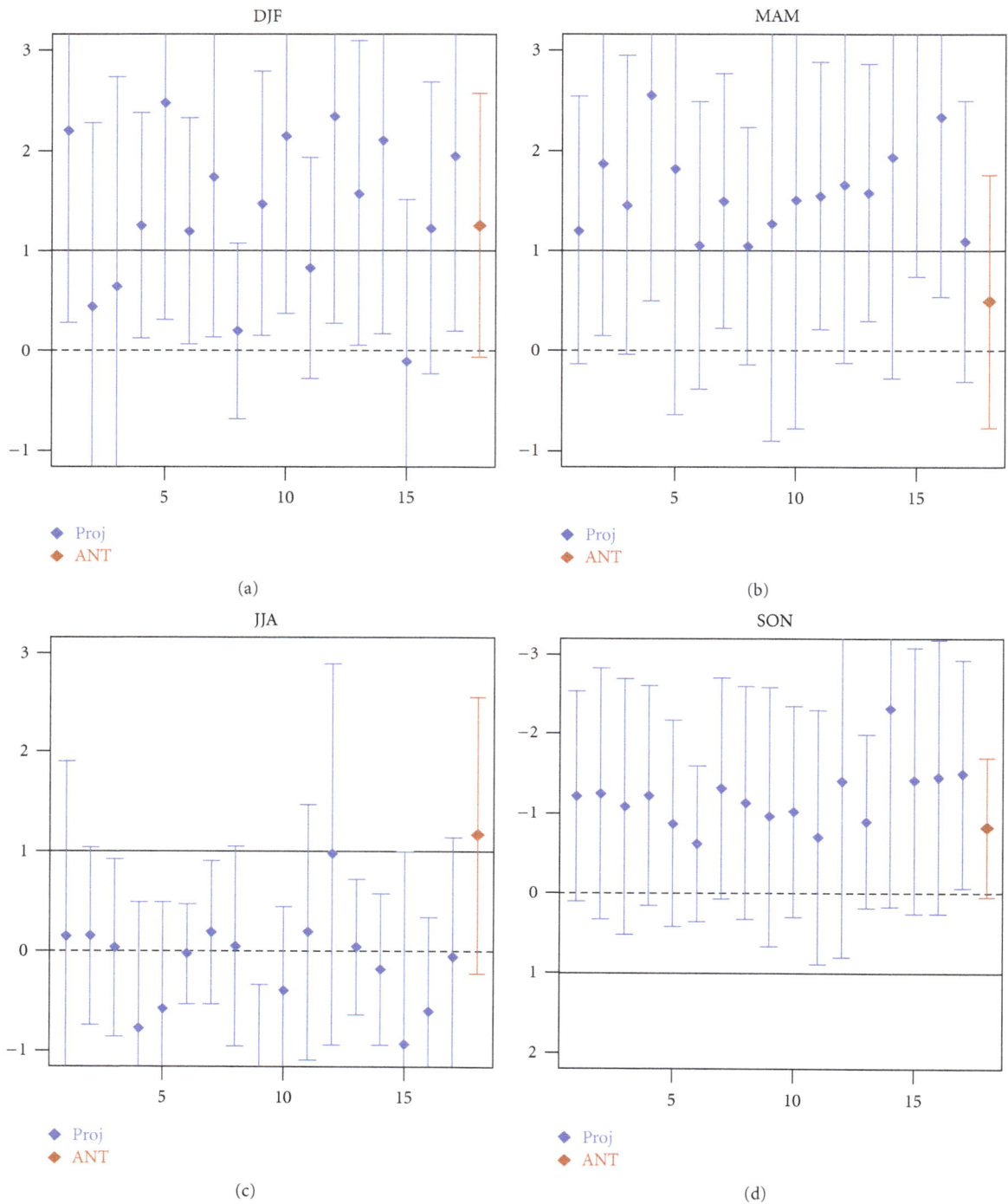

FIGURE 4: Results from consistency of anomaly patterns. Blue bars: seasonal centred regression coefficients (y-axes) of observed SLP changes onto the GS signal patterns estimated from 17 climate change projections (x-axes) based on the SRES A1B scenario. Red bars: regression coefficient of observed SLP onto the ensemble mean of 17 ANT simulations (conducted with 9 models). The bars show the 90th percentile uncertainty ranges of regression coefficients derived from model-based estimates of natural (internal) variability (206 nonoverlapping 30-year segments derived from 6200-year control runs). The solid lines mark regression indices equal to unity indicating consistency with GS forcing.

Investigating the Influence of Anthropogenic Forcing on Observed Mean and Extreme Sea Level Pressure
Trends over the Mediterranean Region

107

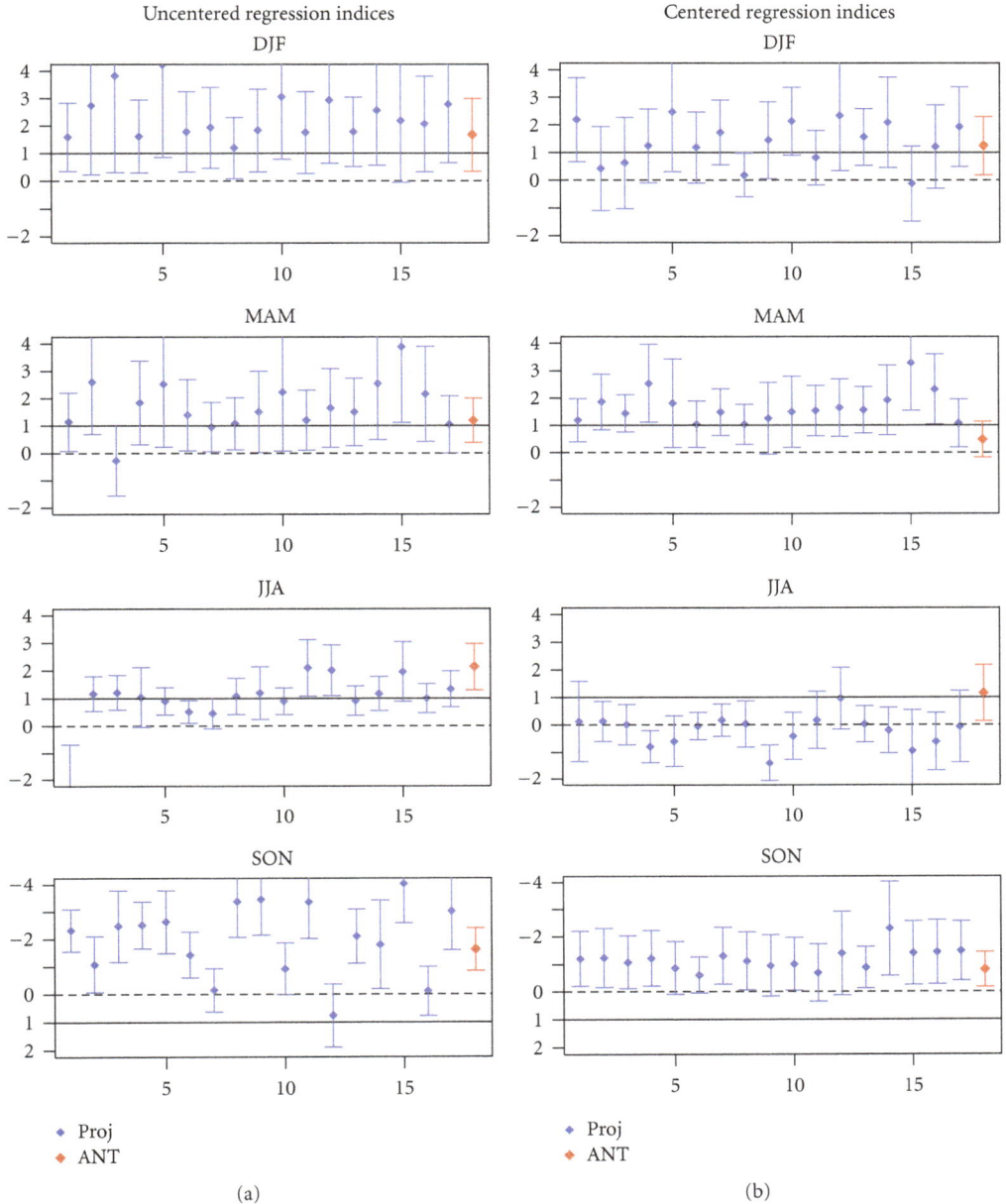

FIGURE 5: Seasonal uncentred (a) and centered (b) regression indices (y-axes) of observed SLP changes onto the changes in response to GS forcing derived from 17 climate change projections (x-axes) based on the SRES A1B scenario (blue bars) and regression coefficient of observed SLP changes onto the simulated GS signal estimated from ensemble mean of 17 ANT simulations (red bars). Whiskers show the 90% uncertainty ranges of regression coefficients derived from observationally based internal variability, estimated from block bootstrapping. The solid lines mark regression indices equal to unity indicating consistency with GS forcing.

natural variability of this atmospheric circulation. Therefore, in this section we explore the consequences of subtracting in the observations that part of the sea level pressure variability that can be attributable to the NAO.

For the NAO we use a station-based NAO index [25]. The fingerprint of the NAO is defined as the fraction of the variability in SLP time series, which covaries with the NAO index. Thus we define the fingerprint of the NAO as the slope of the regression of SLP time series on the NAO index for each grid box separately. The NAO signal is removed from

observed trends by subtracting the trend in the NAO index times the NAO signal from the trend in the observations.

The obtained results after removing the fingerprint of NAO is shown are Figure 6. Results show that the NAO hardly affects the SLP trends, and the removal of the NAO signal does not considerably change the observed trends. Therefore the detection of the influence of anthropogenic (GS) forcing in winter and summer is robust against the removal of the

NAO fingerprint.

7. Analysis of the Frequency of Extreme SLP Days

In this section we investigate the role of GS forcing in the frequencies of extreme low and high mean SLP days over the Mediterranean area. The frequencies in the extremes of low/high mean SLP days were computed using daily mean sea level pressure reconstructions for the period 1974–2003 [13]. The extreme values of mean SLP were established considering as threshold values those of the 10th percentile, for the lowest values and of the 90th percentile for the highest one. Days with SLP less than the 10th percentile value are defined as low SLP days and those greater than the 90th percentile are defined as high SLP days. The remaining days are considered as normal mean SLP days [10, 11].

The same procedure as that used for mean SLP is applied for estimating the frequency of extreme SLP days in response to GS forcing. One time-slice experiment, conducted with the INGV climate model, is used to estimate anthropogenic climate change signal. The scenario run with the INGV model covers the period until 2070, and therefore, the GS signal is defined as the difference between 2041–2070 and the reference climatology (1961–1990). The resulting signal is scaled to change per decade. Control simulations (1,200 year) are used to estimate the variability of the extreme mean SLP values in a stationary climate.

Figure 7 shows the temporal evolution of the frequencies of low and high SLP days in each season over the period 1974–2003. The trend lines included in the plots point to the existence of trends in all seasons during the study period. Figure 8 displays seasonal area mean changes of low SLP days (days with SLP less than the 10th percentile) and high SLP days (days with SLP higher than the 90th percentile) over the period from 1974 to 2003 in comparison with GS signal pattern derived from INGN model. The red whiskers denote the 90% uncertainty range of observed trends estimated from 1,200-year control simulations (40 nonoverlapping 30-year segments).

In winter (DJF) the trends are found to be positive for the frequency of the high SLP days and negative for those of low SLP days (Figures 7 and 8). In spring also the observed trends in the frequency of high SLP days are positive indicating, consequently, a reduction in observed winter and springtime precipitation amount over the Mediterranean region. In contrast to winter and spring, in summer and autumn the observed trends are positive for the case of low SLP days, whereas in the case of high SLP days the trends are negative, indicating an increase in cyclonic activities and a decrease in anticyclonic activities. After 1980, in summer and autumn the trends in the frequency of low (high) SLP days are seen to increase (decrease) with +5 (−0.3) and +4 (−2.5) days per decade, respectively.

As shown by red whiskers in Figure 8 observed trends in winter and autumn are inconsistent with simulated natural (internal) variability, and thus externally forced changes are detectable at 5% level. Therefor, we estimate a less than 5%

probability that the changes in the frequency of low/high SLP days as strong as that observed in winter and autumn could occur in unforced coupled models.

In the following we assess the probability that GS forcing is a plausible explanation for the observed changes in the frequency of low/high SLP days over the Mediterranean region, which can not be explained by natural (internal) variability. For this purpose we compare the observed seasonal trends (1974–2003) with those estimated by the INGV model as the response of extreme SLP to GS forcing.

In winter the pattern of observed extreme low (high) SLP days shows a high correlation with the pattern of GS signal correlation 0.88 (0.80) and significantly greater than zero (with less than 5% risk). Figure 9 displays uncentred regression coefficients of observed changes in the frequency of low/high SLP days against changes projected by INGV model in response to GS forcing. As shown in Figure 9, in winter the large-scale component of the GS signal is detectable in the observed pattern of the frequency of low and high SLP days. However observed SLP changes are inconsistent with projected response to GS forcing, and INGV model strongly underestimates the increasing (decreasing) tendency in the frequency of observed high (low) SLP days (the centered regression indices are greater than 3). In autumn, the negative regression indices (see Figure 9) point to the present of an external forcing, which is not GS forcing. Therefore, we conclude that the increasing (decreasing) tendency of the frequency of the low (high) SLP days in autumn, which goes along with more precipitation over the Mediterranean region, might have rather a natural than anthropogenic (GS) origin.

8. Summary and Conclusions

In this study, we try to assess the influence of GS forcing in observed mean SLP trends, using a large set of climate change projections. We investigate whether the observed mean SLP trends over the period 1975–2004 are similar to what models projected and simulated as response of SLP to anthropogenic forcing. We separate the consistency process in two parts. The first is searching for projected-observed data correspondence at the spatial-mean level by using uncentered pattern similarity indices. The second is considering this also at the pattern level, by employing centered pattern similarity indices. The statistical significance of this similarity is estimated in two ways: by using 6,200-year control runs (control simulations with all forcings held constant) and by estimating the internal variations of the SLP trends by resampling the observations. We further examine the role of GS forcing in the frequencies of the extremes of low/high mean SLP days. To determine whether the days is a low or high mean SLP day, the statistical percentile thresholds are computed based on the daily data for the period 1974–2003 during the seasons. Control simulations (1,200-year) are used to estimate the variability of the extreme mean SLP values in a stationary climate.

Between 1975 and 2004 period observed mean SLP has increased in winter and spring, while observed record

Investigating the Influence of Anthropogenic Forcing on Observed Mean and Extreme Sea Level Pressure
Trends over the Mediterranean Region

109

Uncentered regression indices
NAO signal removed

Centered regression indices
NAO signal removed

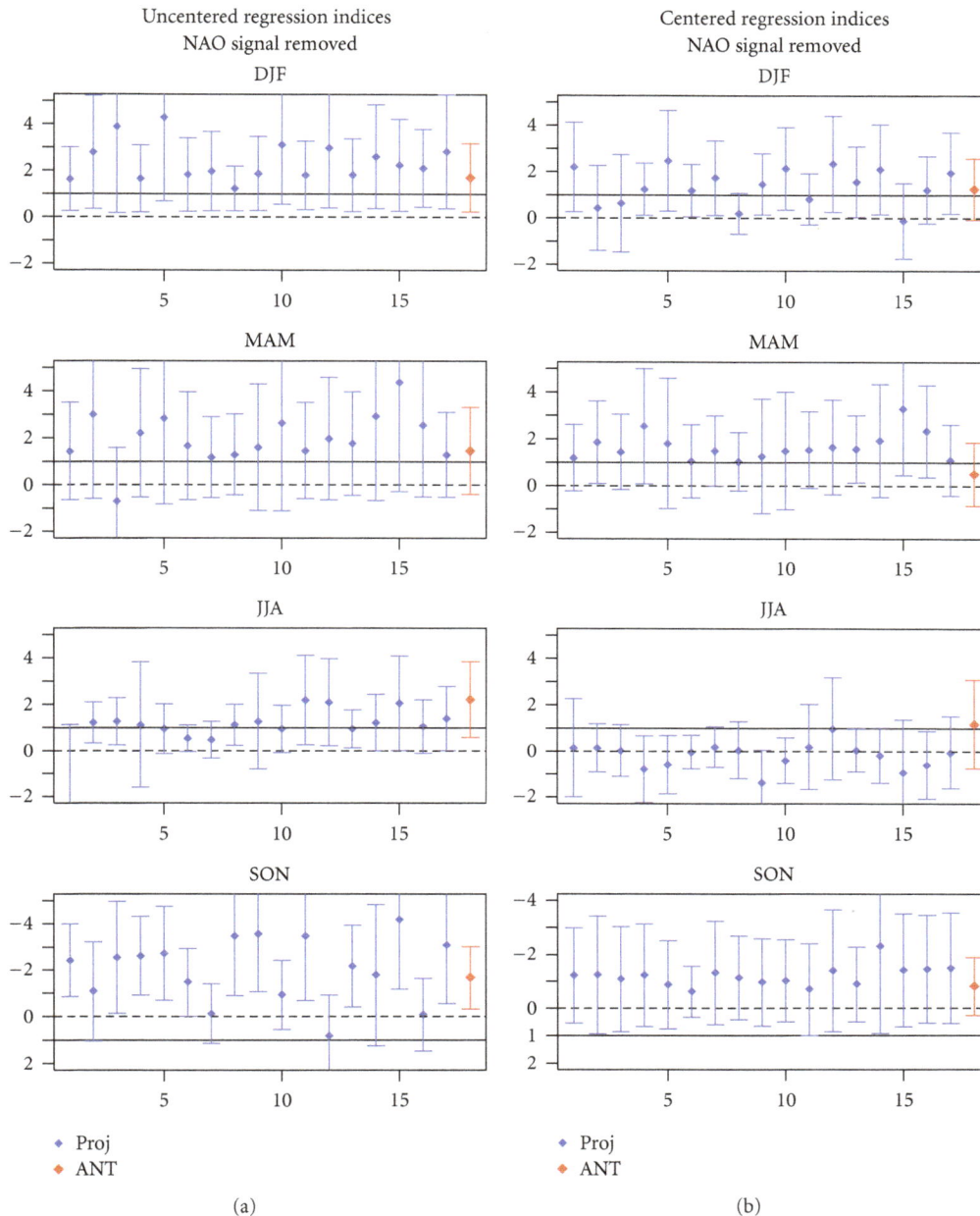

FIGURE 6: Same as Figures 3 and 4, but after removing the NAO signal.

suggests a decrease in mean SLP in summer and autumn over the Mediterranean region. Results indicate that no single sample of 206 segments, derived from control variability, yields a trend of SLP as strong as that observed during the period 1975–2004, in all seasons except spring (with probability of error of less than 5%). Therefor we conclude that observed SLP trends can not be explained by natural(internal) variability alone and externally forced changes are significantly (at 5% level) detectable in all seasons, except spring.

The significantly greater than zero high uncentred correlation coefficients, along with significant uncentred regression indices, clearly indicate that the large-scale component

(spatial mean) of the GS signal is detectable in the observed mean SLP trends in winter and summer and that the observed trends are significantly consistent with all 17 GS signal patterns in winter and in 12 of the 17 GS signal patterns in summer (at 5% level). Our consistency analysis of the anomaly patterns reveals that the smaller-scale (spatial anomalies about the spatial mean) component of the GS signal is only detectable within 9 of the 17 models in winter. However, in summer we failed to find consistency between the observed and projected spatial anomaly patterns, indicating that in summer the spatial mean of GS signal is the dominant component. These results are robust to using simulated GS signal, derived from ensemble mean of 17

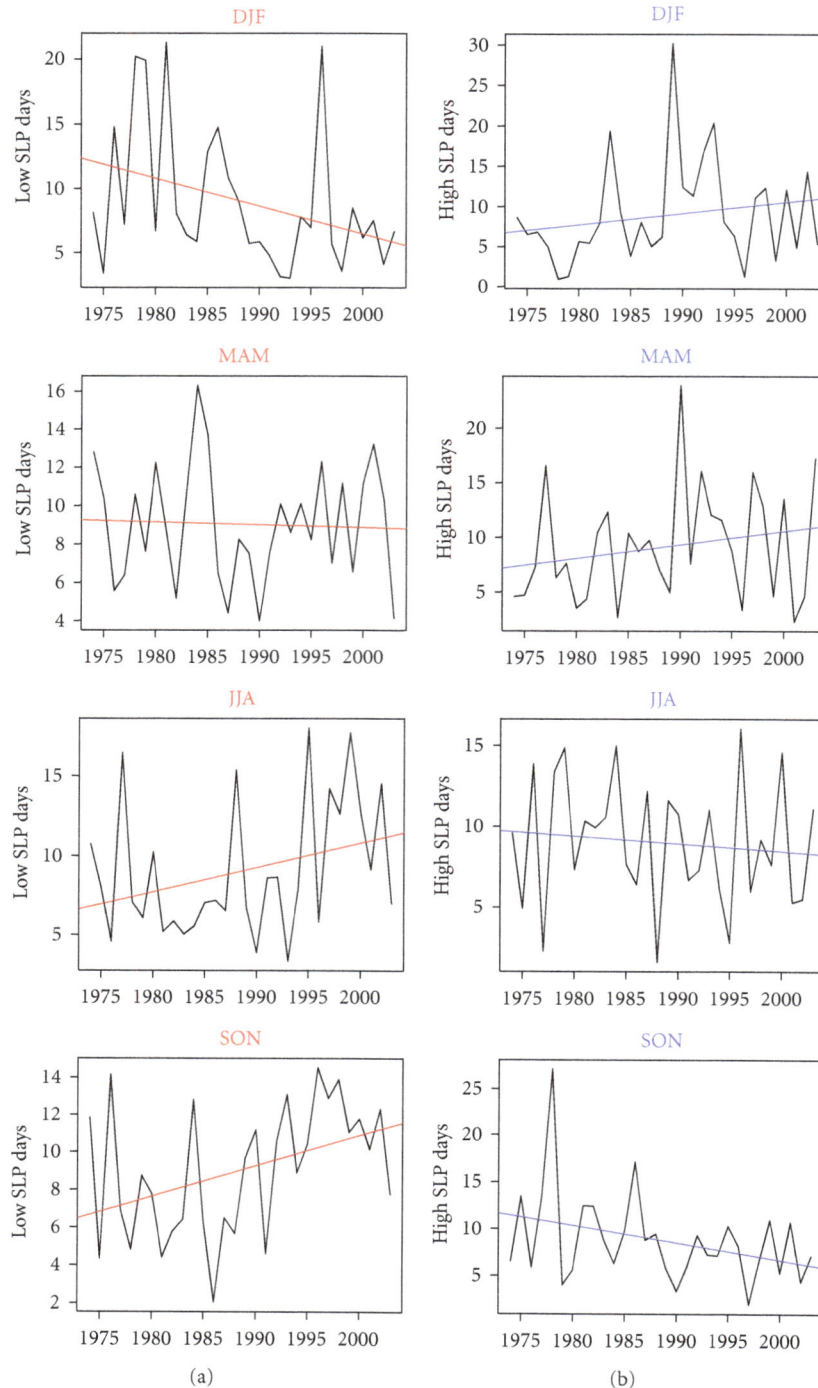

FIGURE 7: (b) Temporal evolutions of the frequencies of HIGH SLP days. (a) Temporal evolutions of the frequencies of low SLP days, in DJF (winter), MAM (spring), JJA (summer) and SON (autumn) over the time period from 1974 to 2003. The corresponding best-fit linear trend lines are included.

ANT simulations (conducted with 9 models). Our results are also robust against using observationally based internal variability. We further investigate the influence of the NAO and find that obtained results in this study are robust against the removal of the NAO fingerprint.

In addition, we find that in winter the observed trends (1974–2003) in the frequency of low/high SLP days are not consistent with internal variability, and the response to external forcing is detected (at 5% level). Obtained results show that the large-scale component of GS signal is detectable (at 5% level) in winter. However, there are discrepancies in

Investigating the Influence of Anthropogenic Forcing on Observed Mean and Extreme Sea Level Pressure
Trends over the Mediterranean Region

111

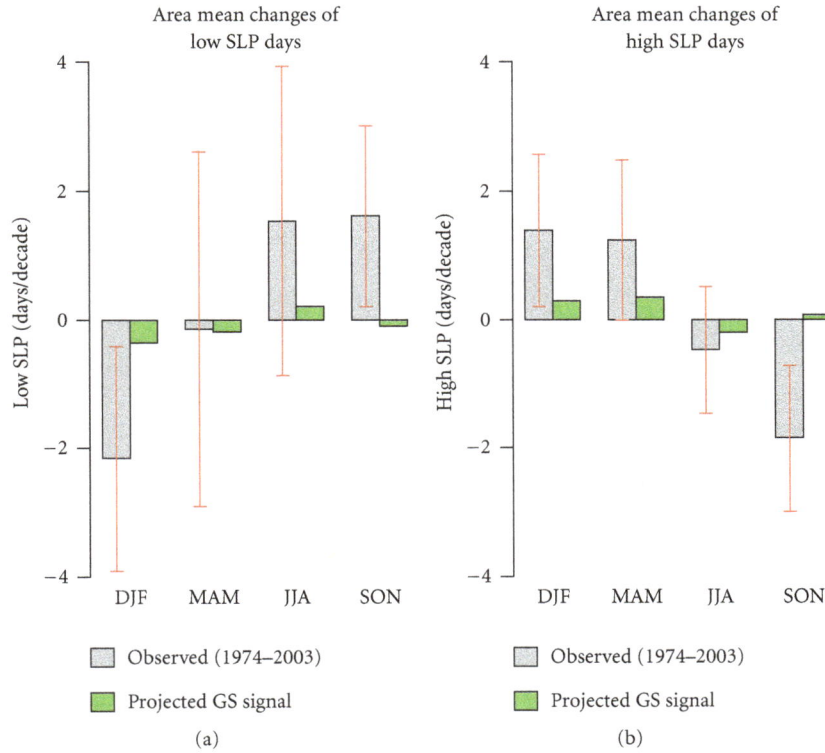

FIGURE 8: (a) Observed seasonal area mean change (grey bars) in the frequency of low SLP days (days with mean SLP less than the 10th percentile). (b) Observed seasonal area mean change (grey bars) in the frequency of high SLP days (days with mean SLP greater than the 90th percentile) over the period from 1974 to 2003 in comparison with GS signal pattern (GS) according to A1B scenario derived from INGV model (green bars). The vertical axes show trends in the number of low (high) days per decade. The red whiskers indicate the 90% uncertainty range of observed trends derived from natural (internal) variability estimated from 1,200-year control simulations (40 nonoverlapping 30 year segments).

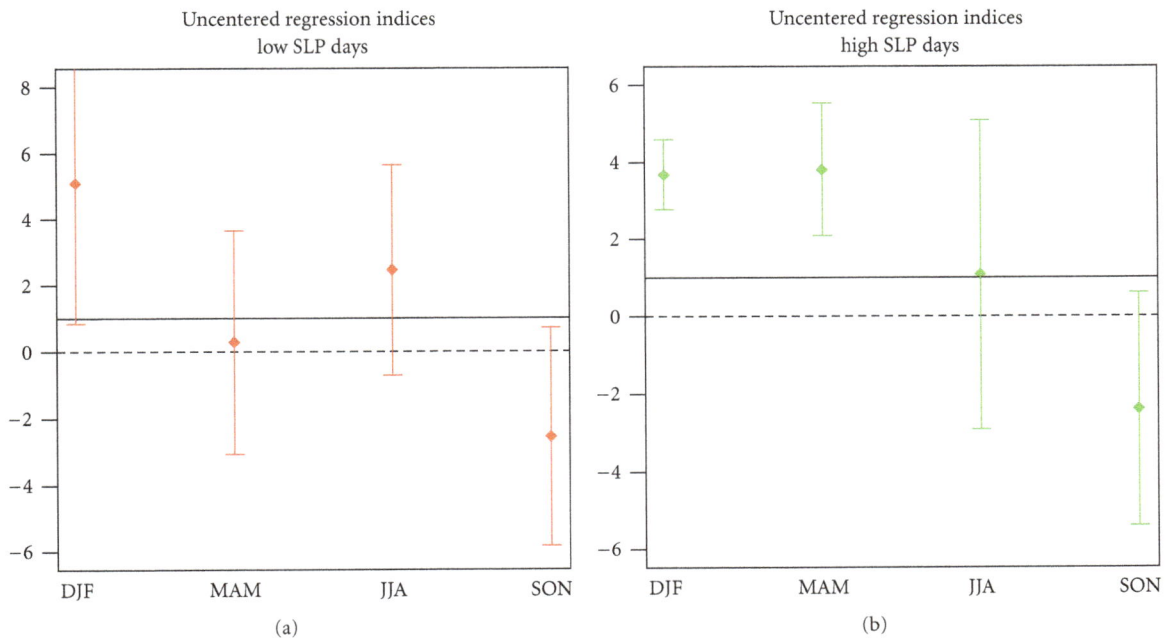

FIGURE 9: Uncentred regression coefficients of observed changes in the frequency of low/high SLP days against changes projected by INGV model in response to GS forcing. The whiskers indicate the 90% uncertainty range of regression indices derived from control variability (1,200-year control simulations). Solid lines indicate regression indices equal to unity, indicating consistency with GS forcing.

the magnitude of changes. The projected GS signal strongly underestimates the observed negative (positive) trends in the number of low (high) SLP days in winter and thus consequently may underestimate the anticyclonic activities in winter, which may have already had a significant effect on agriculture in the Mediterranean region. These results support the results of our analysis of precipitation trends [8], which demonstrates that whereas the precipitation response to GS is detectable in wintertime, the observed negative trends in precipitation are several times larger than expected changes due to GS forcing.

In winter (DJF), the decreasing (increasing) tendency of the frequency of the low (high) SLP days suggests a consistency between the extreme SLP and its associated rainfall activities. In 1983, 1989, 1992, and 1993 Mediterranean region has experienced extreme high pressure systems during the 21, 32, 18, and 22 days per 90 days, respectively, and strong decrease in the amount of precipitation [8]. Interestingly, two major volcanic eruptions in the last two centuries occurred during this time period: one in 1982 and 1991. These eruptions injected large amount of ashes and gases in the form of aerosols into the atmosphere and could impact the global climate (about 15 million tons of sulfur dioxide). Therefore, the large volcanic eruptions might have triggered, to some extent, the extreme high frequency of high SLP days in 1992, 1993, and 1983 in winter and ultimately reducing the rainfall activity over the region. Thus, the observed extreme daily high pressure systems and consequently reduction in wintertime precipitation amount over the Mediterranean region might be partly natural.

While the detection of an anthropogenic influence on observed mean and extreme SLP trends is robust in winter, there are inconsistencies in autumn, where analysis points to the presence of an external forcing, which is not GS forcing. In autumn, observed changes show a large negative trend in the mean SLP, as well as increase (decrease) in low (high) SLP days, these trends are inconsistent with internal variability and also inconsistent with GS signal patterns. Therefore, we conclude that the increasing (decreasing) tendency of the frequency of the low (high) SLP days in autumn, which goes along with more precipitation over the Mediterranean region [8], might have a natural rather than an anthropogenic (GS) origin. This kind of inconsistency must be reconciled if we want to employ the scenarios as realistic anticipations of possible future regional climate change.

Acknowledgments

The author would like to thank Hans von Storch and Eduardo Zorita. She acknowledges the modelling groups, the Program for Climate Model Diagnosis and Intercomparison (PCMI), and the WCRPs Working Group on Coupled Modeling (WGCM) for their roles in making available the WCRP CMIP3 multi-model dataset. The office of Sciences, U.S. Department of Energy, provides support of this dataset.

References

[1] N. P. Gillett, F. W. Zwiers, A. J. Weaver, and P. A. Stott, "Detection of human influence on sea-level pressure," Nature, vol. 422, no. 6929, pp. 292–294, 2003.

[2] N. P. Gillett, R. J. Allan, and T. J. Ansell, "Detection of external influence on sea level pressure with a multi-model ensemble," Geophysical Research Letters, vol. 32, no. 19, Article ID L19714, pp. 1–4, 2005.

[3] D. T. Shindell, G. A. Schmidt, R. L. Miller, and D. Rind, "Northern Hemisphere winter climate response to greenhouse gas, ozone, solar, and volcanic forcing," Journal of Geophysical Research D, vol. 106, no. 7, pp. 7193–7210, 2001.

[4] G. J. Marshall, P. A. Stott, J. Turner, W. M. Connolley, J. C. King, and T. A. Lachlan-Cope, "Causes of exceptional atmospheric circulation changes in the Southern Hemisphere," Geophysical Research Letters, vol. 31, no. 14, pp. L14205–4, 2004.

[5] X. L. Wang, V. R. Swail, F. W. Zwiers, X. Zhang, and Y. Feng, "Detection of external influence on trends of atmospheric storminess and northern oceans wave heights," Climate Dynamics, vol. 32, no. 2-3, pp. 189–203, 2009.

[6] N. P. Gillett and P. A. Stott, "Attribution of anthropogenic influence on seasonal sea level pressure," Geophysical Research Letters, vol. 36, no. 23, Article ID L23709, 2009.

[7] A. Barkhordarian, J. Bhend, and H. von Storch, "Consistency of observed near surface temperature trends with climate change projections over the Mediterranean region," Climate Dynamics. In press.

[8] A. Barkhordarian, H. von Storch, and J. Bhend, "The expectation of future precipitation change over the Mediterranean region is different from what we observe,". In press.

[9] G. A. Meehl, C. Covey, T. Delworth et al., "The WCRP CMIP3 multimodel dataset—a new era in climatic change research," Bulletin of the American Meteorological Society, vol. 88, no. 9, pp. 1383–1390, 2007.

[10] J. L. Labajo, Q. Martín, A. L. Labajo, A. Piorno, M. Ortega, and C. Morales, "Recent trends in the frequencies of extreme values of daily maximum atmospheric pressure at ground level in the central zone of the Iberian Peninsula," International Journal of Climatology, vol. 28, no. 9, pp. 1227–1238, 2008.

[11] J. L. Labajo, A. L. Labajo, A. Piorno, Q. Martín, M. T. Ortega, and C. Morales, "Analysis of the behavior of the extreme values of minimum daily atmospheric pressure at ground level over the Spanish Central Plateau," Atmosfera, vol. 22, no. 2, pp. 125–139, 2009.

[12] R. Allan and T. Ansell, "A new globally complete monthly historical gridded mean sea level pressure dataset (HadSLP2): 1850–2004," Journal of Climate, vol. 19, no. 22, pp. 5816–5842, 2006.

[13] T. J. Ansell, P. D. Jones, R. J. Allan et al., "Daily mean sea level pressure reconstructions for the European-North Atlantic region for the period 1850–2003," Journal of Climate, vol. 19, no. 12, pp. 2717–2742, 2006.

[14] S. Gualdi, E. Scoccimarro, and A. Navarra, "Changes in tropical cyclone activity due to global warming: results from a high-resolution coupled general circulation model," Journal of Climate, vol. 21, no. 20, pp. 5204–5228, 2008.

[15] J. Bhend and H. von Storch, "Consistency of observed winter precipitation trends in northern Europe with regional climate change projections," Climate Dynamics, vol. 31, no. 1, pp. 17–28, 2008.

[16] J. Bhend and H. von Storch, "Is greenhouse gas forcing a

plausible explanation for the observed warming in the Baltic Sea catchment area?" *Boreal Environment Research*, vol. 14, no. 1, pp. 81–88, 2009.

[17] N. P. Gillett, F. W. Zwiers, A. J. Weaver, G. C. Hegerl, M. R. Allen, and P. A. Stott, "Detecting anthropogenic influence with a multi-model ensemble," *Geophysical Research Letters*, vol. 29, no. 20, p. 31, 2002.

[18] G. Hegerl and F. Zwiers, "Use of models in detection and attribution of climate change," *Wiley Interdisciplinary Reviews: Climate Change*, vol. 2, no. 4, pp. 570–591, 2011.

[19] B. D. Santer, T. M. L. Wigley, and P. D. Jones, "Correlation methods in fingerprint detection studies," *Climate Dynamics*, vol. 8, no. 6, pp. 265–276, 1993.

[20] B. D. Santer, K. E. Taylor, T. M. L. Wigley, J. E. Penner, P. D. Jones, and U. Cubasch, "Towards the detection and attribution of an anthropogenic effect on climate," *Climate Dynamics*, vol. 12, no. 2, pp. 77–100, 1995.

[21] D. S. Wilks, "Resampling hypothesis tests for autocorrelated fields," *Journal of Climate*, vol. 10, no. 1, pp. 65–82, 1997.

[22] E. Xoplaki, J. F. Gonzalez-Rouco, J. Luterbacher, and H. Wanner, "Wet season Mediterranean precipitation variability: influence of large-scale dynamics and trends," *Climate Dynamics*, vol. 23, pp. 63–78, 2004.

[23] J. C. Rogers, "North Atlantic storm track variability and its association to the North Atlantic oscillation and climate variability of Northern Europe," *Journal of Climate*, vol. 10, no. 7, pp. 1635–1647, 1997.

[24] P. Ribera, R. Garcia, H. F. Diaz, L. Gimeno, and E. Hernandez, "Trends and interannual oscillations in the main sea-level surface pressure patterns over the Mediterranean, 1955–1990," *Geophysical Research Letters*, vol. 27, no. 8, pp. 1143–1146, 2000.

[25] T. J. Osborn, "Recent variations in the winter North Atlantice Oscillation," *Weather*, vol. 61, no. 12, 2006.

Modeling Pan Evaporation for Kuwait by Multiple Linear Regression

Jaber Almedeij

Civil Engineering Department, Kuwait University, P.O. Box 5969, 13060 Safat, Kuwait

Correspondence should be addressed to Jaber Almedeij, j.almedeij@ku.edu.kw

Academic Editors: S. Beguería, J. Corte-Real, and N. Hatzianastassiou

Evaporation is an important parameter for many projects related to hydrology and water resources systems. This paper constitutes the first study conducted in Kuwait to obtain empirical relations for the estimation of daily and monthly pan evaporation as functions of available meteorological data of temperature, relative humidity, and wind speed. The data used here for the modeling are daily measurements of substantial continuity coverage, within a period of 17 years between January 1993 and December 2009, which can be considered representative of the desert climate of the urban zone of the country. Multiple linear regression technique is used with a procedure of variable selection for fitting the best model forms. The correlations of evaporation with temperature and relative humidity are also transformed in order to linearize the existing curvilinear patterns of the data by using power and exponential functions, respectively. The evaporation models suggested with the best variable combinations were shown to produce results that are in a reasonable agreement with observation values.

1. Introduction

Estimation of the water loss by evaporation is important for modeling, survey, and management of many projects related to hydrology and water resources systems [1, 2]. For example, in many arid and semiarid countries in the Middle East, domestic and industrial water are mostly obtained by desalination. This is an expensive process and other water sources such as harvested rainwater and treated wastewater are being considered for landscape irrigation and other purposes. These projects, however, may require large volumes of water to be stored in artificial lakes and reservoirs open to the atmosphere. In the arid regions with flat terrains, suitable reservoir sites will be limited, and the reservoirs may be shallow with large surface areas. In these cases, significant amounts of water may be lost due to evaporation [3, 4]. Consequently, determination of evaporation losses will be critical in the assessment of the design and operation of these reservoirs.

Evaporation depends on the supply of heat energy and vapor pressure gradient, which in turn depend on meteorological factors such as temperature, relative humidity, wind speed, atmospheric pressure, and solar radiation [5]. In hydrological practice, actual evaporation may be measured by using pan evaporimeters. However, in arid areas of sufficiently low humidity, it is widely acknowledged the existence of a complementary relationship that can influence the reliability of the measurements of pan evaporation [6, 7]. When water availability is limited, actual (regional) evaporation falls below its potential, and a certain amount of energy becomes available. This energy excess increases the temperature and humidity gradients of the overpassing air and leads to an increase in potential evaporation equal in magnitude to the decrease in actual evaporation. If water availability is increased, actual evaporation increases while potential evaporation decreases. This process continues until the actual and potential evaporations become equivalent upon the prevailing conditions of moisture availability. The latter condition is likely present in arid coastal countries where moisture availability is high enough by which pan evaporation measurements become nearly representative to actual evaporations.

Models to estimate evaporation based on meteorological data have also been used by many researchers [8–10]. The well-known Penman's equation may be recommended as a standard method [11]. Penman derived constants for water bodies based on research in lakes. However, to solve this equation, information is needed not only on the external meteorological conditions, but also on the heat storage within the water body, which requires temperature profile measurements within the water [7, 12]. Accordingly, Penman equation cannot be applied if one or more of its parameters are not available from meteorological weather station measurements.

For this reason, several simpler analytical and empirical equations have been developed [2, 5, 13–17]. For example, Cahoon et al. [14] and Fennessey and Vogel [18] employed regression methods to develop models for regional monthly average evaporation in USA as a function of readily available variables such as temperature and site's longitude and elevation. The empirical models were shown to be an improvement over other temperature-based models such as that of Linacre [19] and that of Hargreaves and Samani [20]. Tabari et al. [17] estimated evaporation in semiarid region of Iran using both techniques of artificial neural network and of multivariate nonlinear regression. The techniques comprise various combinations of meteorological variables of temperature, relative humidity, wind speed, solar radiation, and precipitation. They showed that both techniques provide acceptable results. Other similar attempts were also made successfully in the arid region of Saudi Arabia [21–23].

The aim here is to develop simple empirical relations for the estimation of daily and monthly evaporation in Kuwait in terms of available meteorological data. Pan evaporation will be used as the dependent variable. Multiple linear regression techniques will be employed to fit the possible model forms. The performance of the models chosen with the best independent variable combinations will then be compared and discussed.

2. Meteorological Data

Kuwait, which is about $18,000\,\mathrm{km}^2$, is a desert country characterized by long, hot, and dry summers and short winters. The average depth of annual evaporation is high approaching a value of 4000 mm, while that of precipitation is low varying from 50 mm to 250 mm. Temperature during summer (winter) reaches an average daily high of 43°C (15°C), with the average daily low falling to 23°C (5°C). Winter temperatures can be classified as mild, but occasionally become cold when northerly or north-westerly winds bring cold air from the north. Summers are uniformly hot, and temperatures can be very high when hot winds blow from the desert. Owing to the coastal location of the country, the heat is often rendered even more uncomfortable by high humidity approaching a maximum of 90 percent or more.

The meteorological data used in this study are daily average measurements of pan evaporation (mm day^{-1}), temperature at 2 m height (°C), relative humidity (%), and wind speed at 2 m height (m s^{-1}). The effect of precipitation on evaporation rates will not be considered in this study due to the rare rainfall events in the country [24]. The climatological data adopted are readily available from the

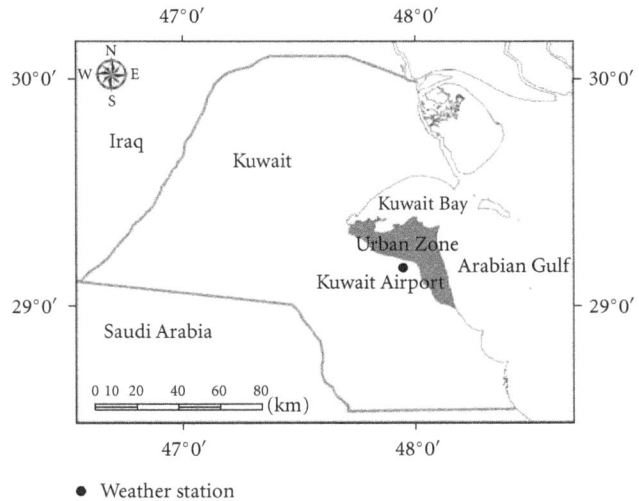

FIGURE 1: Location of the weather station in Kuwait.

Meteorological Department of the Directorate of Civil Aviation collected at a local weather station near Kuwait Airport (Figure 1). These data are of substantial continuity coverage, within a period of 17 years between January 1993 and December 2009 and considered representative of the climate within the urban zone. The reason for such a meteorological point estimate to be considered representative is that the urban zone of the country spans a small area, within latitudes from 29°20′N to 29°03′N and longitudes from 47°37′E to 48°10′E, and is characterized by nearly flat surface elevations. It should also be mentioned that the data for relative humidity provided by this weather station includes only the daily maximum and minimum measurements. Accordingly, the average daily values considered here are obtained by calculating the arithmetic mean of the maximum and minimum measurements. Table 1 presents statistical description for the daily average measurements of the climatological parameters. It can be seen that within the specified time duration, the measurements for each parameter are highly variable reflecting the typical arid climate of the country.

The relation between evaporation and the climatological parameters of temperature, relative humidity, and wind speed can be described by considering a correlation analysis. Table 2 shows that the temperature has the highest correlation coefficient with evaporation of +0.86. This is not a surprising result for a desert country since the physical mechanisms responsible for evaporation are directly proportional to the heat provided by high temperature conditions. The relative humidity has the second highest correlation coefficient with evaporation equaling −0.81. This high correlation reflects the coastal location of the country along the Arabian Gulf. Relative humidity refers to the amount of water in the air, as a fraction of the total amount a saturated air can hold. The more humidity is in the air, the less space will be available for evaporation. Once the air reaches an upper limit of 100 percent relative humidity, it is no longer able to hold additional water molecules. The wind speed has the lowest correlation coefficient with evaporation

TABLE 1: Descriptive statistics of daily average meteorological data for Kuwait from 1993 to 2009.

	Evaporation (mm/day)	Temperature (°C)	Relative humidity (%)	Wind speed (m/s)
Mean	11.66	26.82	38.54	4.03
Standard deviation	7.85	9.68	20.48	1.86
Sample variance	61.64	93.65	419.37	3.44
Kurtosis	−0.38	−1.35	−0.88	−0.15
Skewness	0.68	−0.16	0.42	0.62
Range	39.90	38.04	92.00	11.36
Minimum	0.10	5.14	5.50	0.10
Maximum	40.00	43.18	97.50	11.46

TABLE 2: Correlation coefficients of daily and monthly average meteorological data for Kuwait from 1993 to 2009.

			Evaporation	Temperature	Humidity	Wind speed
Untransformed T and H	Monthly	Evaporation	1.00			
		Temperature	0.94	1.00		
		Humidity	−0.92	−0.92	1.00	
		Wind speed	0.74	0.59	−0.67	1.00
	Daily	Evaporation	1.00			
		Temperature	0.86	1.00		
		Humidity	−0.81	−0.78	1.00	
		Wind speed	0.53	0.32	−0.34	1.00
Transformed T and H	Monthly	Evaporation	1.00			
		Temperature	0.96	1.00		
		Humidity	0.95	0.93	1.00	
		Wind speed	0.74	0.60	0.72	1.00
	Daily	Evaporation	1.00			
		Temperature	0.87	1.00		
		Humidity	0.84	0.82	1.00	
		Wind speed	0.53	0.32	0.39	1.00

of +0.53. The speed at which wind flows over the water surface affects the rate of evaporation by sweeping away water particles that are in the air, allowing more particles to evaporate in the space above the water surface.

3. General Model Form

Multiple linear regression techniques can be used to model the pan evaporation data for Kuwait in terms of the local climatological parameters of temperature, relative humidity, and wind speed. For a multiple linear regression model, the dependent variable y is assumed to be a function of k independent variables $x_1, x_2, x_3, \ldots, x_k$. The model is expressed in the form

$$y_i = b_0 + b_1 x_{1,i} + \cdots + b_k x_{k,i} + e_i, \quad (1)$$

where $b_0, b_1, \ldots,$ and b_k are fitting constants; $y_i, x_{1,i}, \ldots, x_{k,i}$ represent the ith observations of each of the variables $y, x_1, \ldots, x_k$, respectively; e_i is a random error term representing the remaining effects on y of variables not explicitly included in the model. For simple regression models, e_i can be assumed to be an uncorrelated variable with zero mean.

The most common procedure for estimating the values of $b_0, b_1, \ldots,$ and b_k is to employ the least squares criterion with the minimum sum of squares of error terms (S); that is to find $b_0, b_1, \ldots,$ and b_k to minimize

$$S = \sum_{i=1}^{n} \left(y_{i_{\text{observed}}} - b_0 - b_1 x_{1,i} - \cdots - b_k x_{k,i} \right)^2$$

$$= \sum_{i=1}^{n} \left(y_{i_{\text{observed}}} - y_{i_{\text{calculated}}} \right)^2 \quad (2)$$

$$= \sum_{i=1}^{n} e_i^2.$$

As a result, $b_0, b_1, \ldots,$ and b_k must satisfys

$$\frac{\partial S}{\partial b_j} = 2 \sum_{i=1}^{n} e_i \frac{\partial e_i}{\partial b_j} = 0, \quad j = 0, 1, \ldots, k \quad (3)$$

and since $e_i = y_{i_{\text{observed}}} - y_{i_{\text{calculated}}}$, the above equation becomes

$$\frac{\partial S}{\partial b_j} = -2 \sum_{i=1}^{n} e_i \frac{\partial y_{i_{\text{calculated}}}}{\partial b_j} = 0, \quad j = 0, 1, \ldots, k. \quad (4)$$

The meteorological data of Kuwait can be examined for the suitability of fitting this type of regression model. Figures 2(a), 3(a), and 4(a) present the relation between evaporation and the other climatological parameters of temperature, relative humidity, and wind speed on a daily basis. Obviously, this time scale presentation shows high scatter in the pattern of the data by which the average individual relations expressing evaporation in terms of the other parameters become difficult to estimate accurately. In order to smooth the pattern and recognize the most suitable model form, the data can possibly be plotted on a monthly average basis. This is shown in Figures 2(b), 3(b), and 4(b) with substantial reduction in scatter compared to that of the daily average basis.

A general multiple regression model expressing the evaporation in terms of those climatological parameters can initially be assumed as

$$E_i = b_1 T_i + b_2 H_i + b_3 u_i, \tag{5}$$

where E is pan evaporation (mm day^{-1}); T is temperature (°C); H is relative humidity (%); u is wind speed (m s^{-1}). Based on the classical assumptions of multiple regression modeling, the above equation suggests linear correlations between the evaporation and the independent variables. However, Figures 2 and 3 show a definite curvilinear appearance for the relations of evaporation with temperature and relative humidity. It can be shown that those relations are best expressed correspondingly as power and exponential functions such that

$$E = bT^\alpha = 0.04T^{1.69}, \tag{6}$$

$$E = be^{\alpha H} = 40.5e^{-0.038H}, \tag{7}$$

and the above multiple regression equation can thus be linearized by transforming the independent variables of temperature and relative humidity as

$$T^* = T^{1.69}, \tag{8}$$

$$H^* = e^{-0.038H}. \tag{9}$$

Accordingly, the general regression model considering all climatological parameters examined becomes

$$E_i = b_1 T_i^{1.69} + b_2 e^{-0.038H_i} + b_3 u_i \tag{10}$$

and is in the linearized form

$$E_i = b_1 T_i^* + b_2 H_i^* + b_3 u_i. \tag{11}$$

A measure of the strength of a linear association between two variables can be presented by examining the correlation coefficients before and after the transformation. Table 2 presents this analysis, showing improved correlations for both temperature and relative humidity after applying the transformation. This is evident for both data patterns of daily and monthly bases.

It is worth mentioning that (11) neglects the influence of other meteorological parameters on evaporation rates, as it

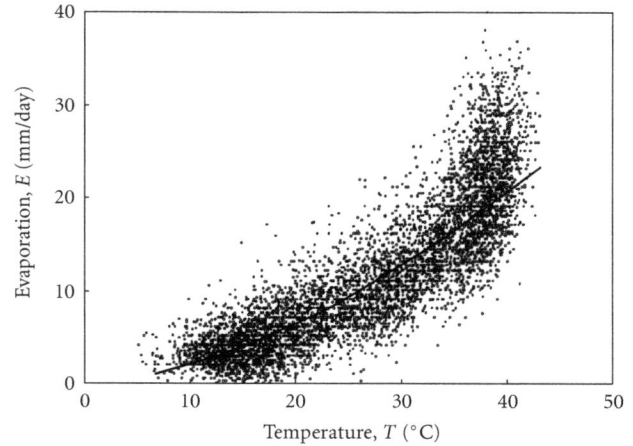

(a)

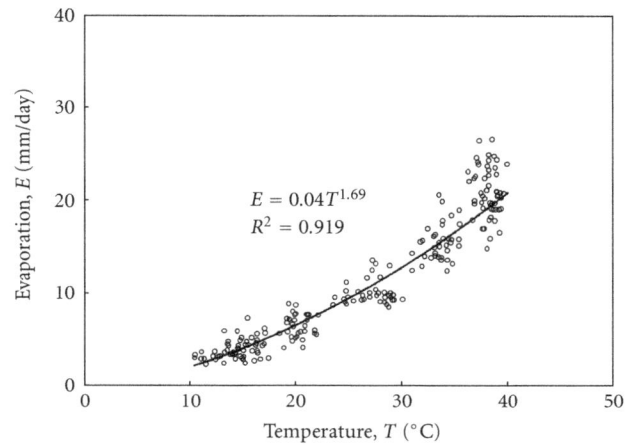

(b)

Figure 2: Evaporation as a function of temperature for the data of Kuwait from 1993 to 2009. The solid curves represent suggested relations fitted using regression analysis. (a) Daily average data; (b) monthly average data.

considers an intercept equal to zero, that is, $b_0 = 0$. Regarding the effect of wind speed, although Figure 4(b) suggests that a model with a nonzero intercept can account for a linear correlation with evaporation of the form

$$E = b_0 + bu, \tag{12}$$

regression might not result with a significant fit for the coefficients because of the considerable data scatter, regardless of the time scale chosen whether of a daily or monthly average basis. It is possible though to assume that the intercept of this correlation is equal to zero

$$E = bu. \tag{13}$$

Although this assumption results with a fitting accuracy less significant for the average data presented on a monthly basis, Figure 4(a) shows that the nonzero intercept trend constitutes a possible relation fitted by eye for the data on daily basis.

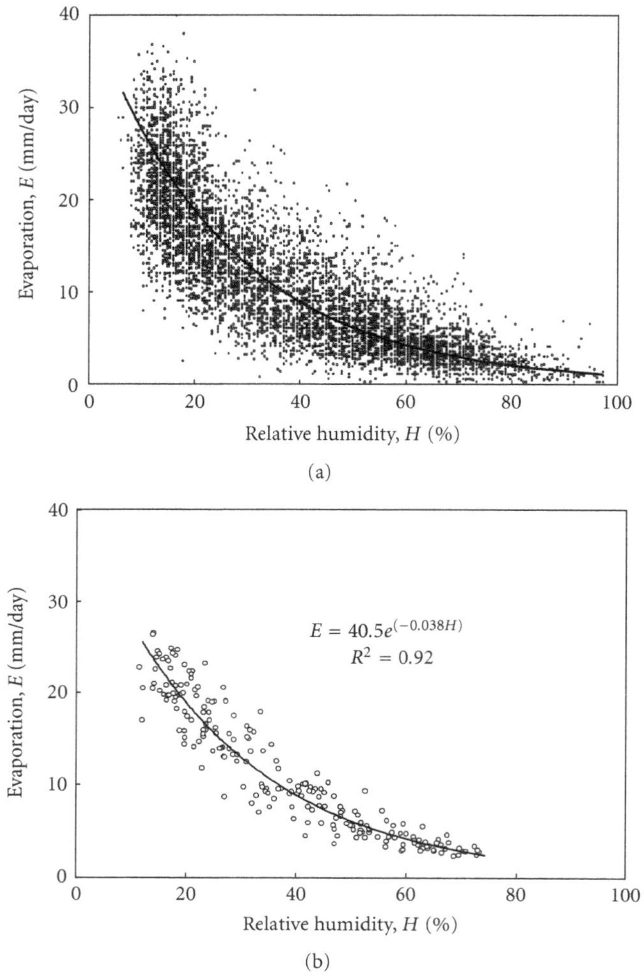

(a)

(b)

FIGURE 3: Evaporation as a function of relative humidity for the data of Kuwait from 1993 to 2009. The solid curves represent suggested relations fitted using regression analysis. (a) Daily average data; (b) monthly average data.

(a)

(b)

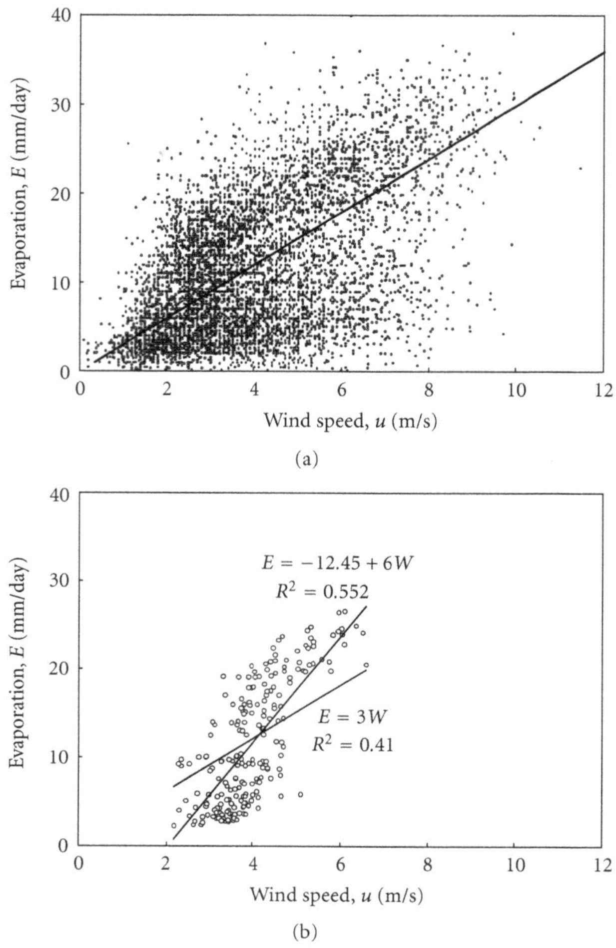

FIGURE 4: Evaporation as a function of wind speed for the data of Kuwait from 1993 to 2009. The solid curves represent suggested relations fitted using regression analysis. (a) Daily average data; (b) monthly average data.

4. Model Selection

For a multiple linear regression, model building by a variable selection procedure involves many steps reported in the literature. Typically, there are two aspects here to consider: selecting the number of variables for the models and evaluating each model selected. Including more variables in the model is not necessarily better, and it may result in overfitting. Such a model will perform poorly when applied to a new sample drawn from the same population. There is no specific test to determine the best number of variables included in the model. A possible strategy that can be applied here is to first enter the independent variables in the model one by one. Then the best of all one-variable models can be chosen as Model 1. Next all combinations of two-variable models are considered, and the best pair is chosen as Model 2. The full model with the three variables can be selected as Model 3. The three selected models can then be evaluated and discussed. The range of monthly data from January 1993

TABLE 3: Fitting accuracy of regression models using monthly average meteorological data for Kuwait from 1993 to 2006 (calibration range).

Variable	Accuracy measure			
	R_a^2	MAPE	RMSE (mm)	NSE
One-variable models				
T	0.972	16.27	2.05	0.9121
H	0.971	16.15	2.24	0.8945
u	0.843	72.65	5.30	0.4100
Two-variable models				
T, H	0.980	12.74	1.68	0.9405
T, u	0.977	17.38	2.01	0.9150
H, u	0.972	17.68	2.22	0.8968
Three-variable model				
T, H, u	0.979	13.23	1.69	0.9402

until December 2006 with 179 measurements can be used to fit all the models, and the remaining data up to December

TABLE 4: ANOVA results for the suggested models fitted using monthly average meteorological data for Kuwait from 1993 to 2006 (calibration range).

Model number		df	SS	MS	F	Significance F
1	Regression	3	32722.44	10907.48	4718.01	0.000
	Residual	176	434.52	2.47		
	Total	179	33156.96			
2	Regression	2	32721.38	16360.69	6648.19	0.000
	Residual	177	435.58	2.46		
	Total	179	33156.96			
3	Regression	1	32375.68	32375.68	7376.18	0.000
	Residual	178	781.28	4.39		
	Total	179	33156.96			

df: degrees of freedom, SS: sum of squares, and MS: mean square.

TABLE 5: Regression results for the suggested models fitted using monthly average meteorological data for Kuwait from 1993 to 2006 (calibration range).

Model number		Coefficients	Standard error	t stat	P value
1	Intercept	0			
	Temperature	0.024	0.002	9.5010	0.000
	Humidity	18	1.875	11.450	0.000
	Wind speed	0.033	0.074	0.6560	0.513
2	Intercept	0			
	Temperature	0.028	0.002	9.9750	0.000
	Humidity	15	1.799	11.852	0.000
3	Intercept	0			
	Temperature	0.0422	0.002	85.885	0.000

2009 with 37 measurements may be employed to verify the performance of the final model chosen.

Suitable quantitative criteria to evaluate the accuracy of the models will be needed. A possible one is the coefficient of determination R^2, which describes the percentage of total variation explained by the model. A high R^2 value close to 1.0 indicates a good model fit with observed data. However, R^2 also increases as the number of explanatory variables used in the model increases even if these variables are not significant in explaining the variability of the dependent variable. Hence, it is preferable to use the adjusted R_a^2 defined as

$$R_a^2 = 1 - \frac{n-1}{n-p} \times (1 - R^2), \qquad (14)$$

where n is sample size; p is number of explanatory variables. In addition to the coefficient of determination, the performance of the models can be evaluated using statistical error tests such the Mean Absolute Percentage Error (MAPE), Root Mean Square Error (RMSE), and Nash-Sutcliffe equation (NSE). These indicators can be calculated as follows:

$$\text{MAPE} = \frac{1}{n} \sum_{i=1}^{n} \left| \frac{y_{i\text{calculated}} - y_{i\text{observed}}}{y_{i\text{observed}}} \times 100 \right|,$$

$$\text{RMSE} = \sqrt{\frac{1}{n} \sum_{i=1}^{n} \left(y_{i\text{calculated}} - y_{i\text{observed}} \right)^2}, \qquad (15)$$

$$\text{NSE} = 1 - \frac{\sum_{i=1}^{n} \left(y_{i\text{observed}} - y_{i\text{calculated}} \right)^2}{\sum_{i=1}^{n} \left(y_{i\text{observed}} - \overline{y}_{i\text{observed}} \right)^2},$$

where $\overline{y}$ is the mean value of the dependent variable. For better data modeling, MAPE and RMSE statistics should be closer to zero. The NSE criterion compares the model performance to the use of the mean value of the dependent variable as an estimate. Given a perfect fit, the NSE criterion is equal to 1.0; whereas if the model is worse than the mean value of the dependent variable, the NSE statistic will be negative. In general, one can expect satisfactory results from a model with NSE criterion higher than 0.8.

Table 3 presents the accuracy measures for all possible models. In all one-variable models, the best variable based on all evaluation criteria is temperature, termed Model 1. For all combinations of two-variable models, the best two parameters based on all evaluation criteria are temperature and relative humidity, termed Model 2. Now Model 3 considers all parameters of temperature, relative humidity, and wind speed. The three models have the forms:

$$\text{Model 1,} \quad E = 0.0422T^{1.69},$$

$$\text{Model 2,} \quad E = 0.028T^{1.69} + 15e^{-0.038H}, \quad (16)$$

$$\text{Model 3,} \quad E = 0.024T^{1.69} + 18e^{-0.038H} + 0.033u.$$

Among the three models selected, Model 2 is considered the best one based on all the evaluation criteria shown in Table 3, while Model 1 has the lowest performance. This finding is emphasized further by recognizing that the tests of MAPE and RMSE have corresponding percentages of difference between Model 2 and Model 1 equal to 28% and 22%.

While Model 1 has the lowest performance among the three models, it has the advantage of simplicity as it can estimate evaporation from only one climatological parameter of temperature. From a practical point of view, this model can be considered suitable to serve as a tool to estimate evaporation when input meteorological variables are insufficient. Model 3 has the advantage of including all the three parameters examined in this study; however, the accuracy of this model can be influenced by the high scatter found in the data of wind speed. Model 2 excludes wind speed by taking the advantage of the better performance of the two parameters of temperature and relative humidity. The ANOVA analysis and the regression results are shown for the three models in Tables 4 and 5, respectively. For the three models, the F test statistic provides a strong evidence of the presence of a satisfactorily significant linear trend between the evaporation and the other variables. However, as it was expected for Model 3, the P value for the wind speed variable is large, suggesting that the null hypothesis of the slope being equal to zero is true. Accordingly, Model 2 can be chosen here as the best evaporation model form.

Owing to the reason that the accuracy of results presented previously are based on monthly average data, it is crucial to test the performance of Model 2 using the daily average data. For this time scale, the accuracy results of Model 2 based on R_a^2, MAPE, RMSE, and NSE are equal to 0.84, 44.4, 3.4 mm, and 0.807, respectively. Although these results are satisfactory, the performance of the model for the daily data is lower than that for the monthly data. This lower performance is due to the higher scatter observed in the daily data pattern as shown in Figure 5, plotting the observed versus calculated evaporation values for both daily and monthly average data. Here, the solid line represents the condition of perfect agreement, and the other lines represent discrepancies of ±2 mm, ±3 mm, and ±4 mm. It is seen that the higher scatter in the daily evaporation data is evident. The

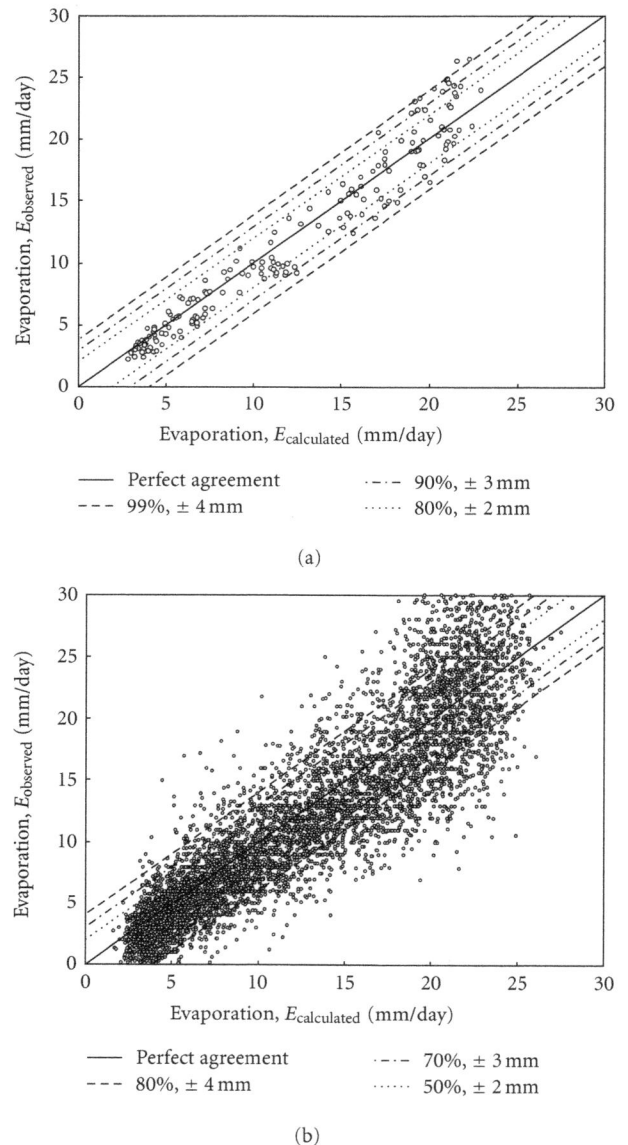

Figure 5: Observed versus calculated evaporation values based on Model 2 for the data of Kuwait from 1993 to 2006 (calibration range). (a) Monthly average data; and (b) daily average data.

percentage of the monthly (daily) average evaporation values calculated with a discrepancy up to ±2 mm is 80% (50%), up to ±3 mm is 90% (70%), and up to ±4 mm is 99% (80%).

Model 2 can be used to provide forecasts. Initially, the model accuracy will be verified by using the range of evaporation data from January 2007 to December 2009, with 37 measurements, which has not been used for model calibration. The monthly average meteorological data can be adopted for this analysis. Figure 6 shows a reasonable agreement between the observed and calculated evaporation values for the entire range of data from January 1993 to December 2009. The same figure also shows that the model is successful in representing most of the seasonal variation-patterns during the years. Regarding the verified range, the

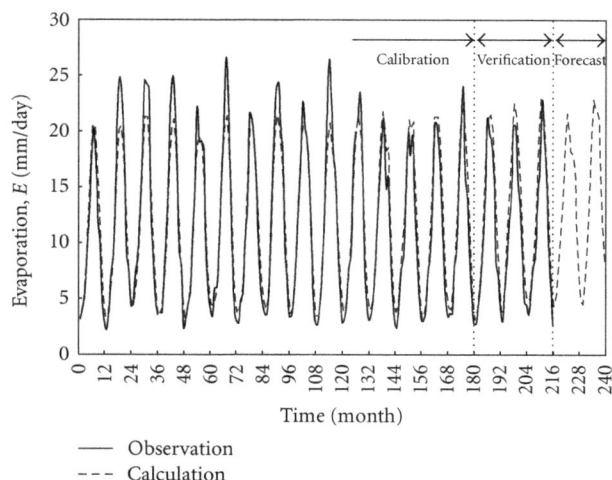

FIGURE 6: Evaporation as a function of time in months for Model 2. Calibration is performed for the duration from January 1993 (corresponding to month number 1) to December 2006 (month number 179); verification is performed for the duration from January 2007 (month 180) to December 2009 (month 216); forecast is performed for the duration from January 2010 (month 217) to December 2011 (month 240).

model produces accuracy values of R_a^2, MAPE, RMSE, and NSE equal to 0.973, 13.36, 1.84 mm, and 0.921, respectively. These values are acceptable compared to the corresponding ones obtained for the range of model calibration shown in Table 3. It is also interesting to examine the model accuracy within this range in terms of the percentage of discrepancy criterion. Here, for the verified range, the percentage with the model discrepancy up to ±2 mm is 57%, up to ±3 mm is 82%, and up to ±4 mm is 98%.

Since the verification results are acceptable, forecasts can be produced by Model 2, given that the conditions used to derive the model remain the same. Based on the previous verification range, the mean forecasting error is assumed to be nearly within accuracy of R_a^2, MAPE, RMSE, and NSE equal to 0.973, 13.36, 1.84 mm, and 0.921, respectively. Figure 6 presents two years ahead of monthly average evaporations for the period from 2010 to 2011, with 24 data points. As can be seen, the evaporation rates follow nearly the same seasonal pattern recognized for the monthly average data.

5. Conclusions

This study derived empirical relations for modeling pan evaporation suitable for application in Kuwait. Plotting the daily data on a monthly average basis has smoothened the high scatter in the pattern by which the relations between evaporation and the independent variables of temperature, relative humidity and wind speed became possible to estimate more accurately. The transformations considered for both variables of temperature and relative humidity by using

the corresponding power and exponential functions have improved the correlation results. While these transformations are applicable locally for the meteorological data of Kuwait, it is hoped that this will bring the attention of others to examine whether such correlations exist universally in data collected from other locations. In general, the three models chosen here provided acceptable results with reasonable accuracy. Model 2 has been considered as the best one based on the evaluation criteria employed, while Model 1, although it produced the lowest performance values, has the advantage of simplicity as it requires only one climatological parameter of temperature to estimate evaporation. It is also shown that Model 2 is capable of simulating the seasonal variation pattern typically observed with monthly average evaporation data.

Acknowledgments

The author is grateful to the Meteorological Department of the Directorate General of Civil Aviation of Kuwait, Climatological Division, for providing the relevant data of pan evaporation, temperature, relative humidity, and wind speed.

References

[1] J. M. Molina Martínez, V. Martínez Alvarez, M. M. González-Real, and A. Baille, "A simulation model for predicting hourly pan evaporation from meteorological data," *Journal of Hydrology*, vol. 318, no. 1–4, pp. 250–261, 2006.

[2] P. B. Shirsath and A. K. Singh, "A comparative study of daily pan evaporation estimation using ANN, regression and climate based models," *Water Resources Management*, vol. 24, no. 8, pp. 1571–1581, 2010.

[3] B. Gallego-Elvira, A. Baille, B. Martín-Górriz, and V. Martínez-Álvarez, "Energy balance and evaporation loss of an agricultural reservoir in a semi-arid climate (south-eastern Spain)," *Hydrological Processes*, vol. 24, no. 6, pp. 758–766, 2010.

[4] I. H. Abou El-Magd and E. M. Ali, "Estimation of the evaporative losses from lake Nasser, Egypt using optical satellite imagery," *International Journal of Digital Earth*, vol. 5, no. 2, pp. 133–146, 2012.

[5] C. Y. Xu and V. P. Singh, "Dependence of evaporation on meteorological variables at different time-scales and intercomparison of estimation methods," *Hydrological Processes*, vol. 12, no. 3, pp. 429–442, 1998.

[6] R. J. Bouchet, *Evapotranspiration Reelle Evapotranspiration Potentielle, Signification Climatique*, vol. 2 of *International Association of Scientific Hydrology, General assembly of Berkeley, Transactions*, Evaporation, Berkeley, Calif, USA, 1963.

[7] W. Brutsaert, *Hydrology: An Introduction*, Cambridge University Press, New York, NY, USA, 2005.

[8] C. Vallet-Coulomb, D. Legesse, F. Gasse, Y. Travi, and T. Chernet, "Lake evaporation estimates in tropical Africa (Lake Ziway, Ethiopia)," *Journal of Hydrology*, vol. 245, no. 1–4, pp. 1–18, 2001.

[9] H. Gavin and C. A. Agnew, "Modelling actual, reference and equilibrium evaporation from a temperate wet grassland," *Hydrological Processes*, vol. 18, no. 2, pp. 229–246, 2004.

[10] F. J. Chang, L. C. Chang, H. S. Kao, and G. R. Wu, "Assessing the effort of meteorological variables for evaporation estimation by self-organizing map neural network," *Journal of Hydrology*, vol. 384, no. 1-2, pp. 118–129, 2010.

[11] H. L. Penman, "Natural evaporation from open water, bare soil and grass," *Proceedings of the Royal Society A*, vol. 193, pp. 120–145, 1948.

[12] G. Stanhill, "Changes in the rate of evaporation from the Dead Sea," *International Journal of Climatology*, vol. 14, no. 4, pp. 465–471, 1994.

[13] E. A. Fitzpatrick, "Estimates of pan evaporation from mean maximum temperature and vapor pressure," *Journal of Applied Meteorology*, vol. 2, pp. 780–792, 1963.

[14] J. E. Cahoon, T. A. Costello, and J. A. Ferguson, "Estimating pan evaporation using limited meteorological observations," *Agricultural and Forest Meteorology*, vol. 55, no. 3-4, pp. 181–190, 1991.

[15] R. D. Crago and W. Brutsaert, "A comparison of several evaporation equations," *Water Resources Research*, vol. 28, no. 3, pp. 951–954, 1992.

[16] L. D. Rotstayn, M. L. Roderick, and G. D. Farquar, "A simple pan-evaporation model for analysis of climate simulations: evaluation over Australia," *Geophysical Research Letters*, vol. 33, Article ID L17715, 5 pages, 2006.

[17] H. Tabari, S. Marofi, and A. A. Sabziparvar, "Estimation of daily pan evaporation using artificial neural network and multivariate non-linear regression," *Irrigation Science*, vol. 28, no. 5, pp. 399–406, 2010.

[18] N. M. Fennessey and R. M. Vogel, "Regional models of potential evaporation and reference evapotranspiration for the northeast USA," *Journal of Hydrology*, vol. 184, no. 3-4, pp. 337–354, 1996.

[19] E. T. Linacre, "A simple formula for estimating evaporation rates in various climates, using temperature data alone," *Agricultural Meteorology*, vol. 18, no. 6, pp. 409–424, 1977.

[20] G. H. Hargreaves and Z. A. Samani, "Estimating potential evapotranspiration," *Journal of the Irrigation & Drainage Division*, vol. 108, no. 3, pp. 225–230, 1982.

[21] H. M. Abo-Ghobar, "Evaporation and drift losses from sprinkler irrigation systems under hot and dry conditions," *Journal of King Saud University*, vol. 5, no. 2, pp. 153–164, 1993.

[22] H. M. Al-Ghobari, "Estimation of reference evapotranspiration for southern region of Saudi Arabia," *Irrigation Science*, vol. 19, no. 2, pp. 81–86, 2000.

[23] M. I. Al-Saud, "Reduction of evaporation from water surfaces—preliminary assessment for Riyadh Region, Kingdom of Saudi Arabia," *Journal of Engineering and Applied Sciences*, vol. 4, no. 3, pp. 205–209, 2009.

[24] J. Almedeij, "Modeling rainfall variability over urban areas: a case study for Kuwait," *The Scientific World Journal*, vol. 2012, Article ID 980738, 8 pages, 2012.

13

Climatological Aspects of Aerosol Physical Characteristics in Tunisia Deduced from Sun Photometric Measurements

Mabrouk Chaâbane,[1] Chafai Azri,[1] and Khaled Medhioub[2]

[1]*Faculté des Sciences, Université de Sfax, B.P. 1171, Sfax 3000, Tunisia*
[2]*Institut Préparatoire aux Etudes d'Ingénieurs de Sfax, Université de Sfax, BP 805, Sfax 3018, Tunisia*

Correspondence should be addressed to Mabrouk Chaâbane, chaabane_mabrouk@yahoo.fr

Academic Editor: Yaping Shao

Atmospheric and climatic data measured at Thala site (Tunisia) for a long-time period (1977–2001) are used to analyse the monthly, seasonal, and annual variations of the aerosol optical depth at 1 μm wavelength. We have shown that aerosol and microphysical properties and the dominating aerosol types depend on seasons. A comparison of the seasonal cycle of aerosol optical characteristics at Thala site showed that the contribution of long-range transported particles is expected to be larger in summer as a consequence of the weather stability typical of this season. Also, the winter decrease in atmospheric turbidity may result from increases in relative humidity and decreases in temperature, leading to increased particle size and mass and increased fall and deposition velocities. The spring and autumn weather patterns usually carry fine dust and sand particles for the desert area to Thala region. The annual behaviour of the aerosol optical depth recorded a period of stead increase started in 1986 until 2001. Trends in atmospheric turbidity after 1988 could be explained other ways by the contribution of the eruption of Mount Pinatubo in 1991 and by local or regional changes in climate or in aerosol emissions.

1. Introduction

Atmospheric aerosols play a key role in the functioning of the earth system. Their influence on radiation passing through the atmosphere cannot be neglected, especially in urban or industrialized areas. About three billion tons of particles are injected annually into the atmosphere by natural processes (soil erosion, volcanoes, ocean spray) or by human activities (industrial, traffic). Typical residence time of these particles is around a week in the troposphere. During this period, they absorb or scatter a portion of the telluric and solar radiation (direct effect), they are involved in cloud formation and influence their lifetime and their optical properties (indirect effect). For these two effects, aerosols significantly affect the radiation balance on earth. Because of their small size, these particles are subjected to a long-range atmospheric transport (several thousand kilometres). This ability to transport is such that for some ecosystems and for some elements, aerosols are the major driver of biogeochemical cycling.

To receive comprehensive information about the atmospheric aerosols' optical and microphysical properties, ground-based stations joined together into globally distributed networks and equipped with radiometric instrumentation are now paid special attention. Aerosol optical depth (AOD) data provide a unique opportunity for a comprehensive analysis of diurnal and daily aerosol variations according to aerosol types and landscape.

One of the major components of the aerosol loading is dust which is found over the Pacific ocean and the Mediterranean sea with highest concentrations found over the equatorial and tropical north Atlantic [1]. Mineral dust lifted from arid regions of our planet exerts a large influence on radiative transfer processes [2]. Model studies suggest that the direct radiative forcing of dust on regional as well as on a global scale may be comparable to or even exceed the forcing by anthropogenic aerosols [3, 4]. Nevertheless, net radiative forcing and climate impact of dust particles are still largely uncertain. This fact is due to a still limited knowledge of production sources, transport patterns, particle properties, and evolution and changes during the particles' lifetime.

On a global scale, the dominant sources of mineral dust are all located in the Northern Hemisphere, mainly in North Africa, the Middle East, Central Asia, and the Indian subcontinent [5]. North-West Africa is the most important source of mineral aerosols over the Mediterranean basin [1, 6, 7].

Aerosol and water vapour are somewhat interrelated because aerosols serve as condensation nuclei in the formation of water droplets from water vapour. Next to clouds, aerosols and water vapour are quantitatively the two main parameters affecting transmission of solar radiation through the atmosphere but which are also the least well known. This is mainly due to their variability with location and time and difficulties associated with their measurement.

Attenuation of solar irradiance is strongly dependent on conditions of the sky, cleanliness of the atmosphere, and composition of gaseous constituents. In a clean and dry atmospheric condition, solar irradiance is attenuated by permanent atmospheric constituents of air molecules, gases and ozone, whose contents are nearly invariable. Two additional attenuation processes, which are the absorption by water vapor and scattering by aerosol particles, take place in a real atmosphere. The additional attenuation caused by these two processes is known as being due to the turbidity of the atmosphere.

Atmospheric turbidity is an important parameter for assessing the air pollution in local areas, as well as being the main parameter controlling the attenuation of solar radiation reaching the earth's surface under cloudless sky conditions. Among the different turbidity indices, the Angstrom turbidity coefficient β is frequently used [8].

The establishment of the Aerosol Robotic Network (AERONET) which is a federated international network funded in 1993 and coordinated by the NASA that maintains more than 200 permanent automatic sun/sky radiometers worldwide [9, 10] has significantly contributed in the last years to obtain a global coverage and good sampling of aerosol properties.

In this study, two series of sun photometer measurements at Thala site (Tunisia) are used to analyse monthly, seasonal, and annual variations in Tunisia of aerosol optical depth AOD at 1 μm wavelength which is known as Angstrom atmospheric turbidity β. The first photometric data are provided by WMO (World Meteorological Organisation) for a long-series experiment (1981–1988) and used to characterize dust particles advected from North-West Africa (Background Air Pollution Monitoring Network BAPMoN) [11]. The results are then compared to those found at the same site with AERONET photometer instrument operated during the period of March–October 2001 by the EMAGPOT Franco-Tunisian project [12–14] and installed by L.O.A (Laboratoire d'Optique Atmosphérique, Lille, France).

Monthly and annual variations of the mean AOD are investigated by examining the origin of the air masses. The temporal extent of Thala region's historical climate record provided the opportunity to examine both seasonal and long-term variability in turbidity in that region. A statistical study of climate data allowed investigating and analyzing the correlation of the atmospheric turbidity with measured meteorological parameters.

2. Measured Data Set

2.1. Optical Properties of Atmospheric Aerosols. Mie theory characterizes the optical properties of a particle similar in size to the wavelength of the incident light. The theory defines the parameter $\alpha_{Mie} = 2\pi r/\lambda$ where r is the radius of the particle and λ is the wavelength of the incident radiation. The aerosol properties also depend on their refractive index m, which expresses the radiative effect of aerosols according to the nature of the components. Incident radiation on a particle can be either scattered or absorbed. These two phenomena combined lead to the extinction of the transmitted radiation. The efficiency coefficients of extinction and scattering Q_{ext} (α_{Mie}, m) and Q_{scat} (α_{Mie}, m) define, respectively, the variation of extinction and the scattering in function of the Mie parameter and the refraction index. The additivity of optical properties of spherical particles contained in the atmospheric column allows easily defining the characteristic quantities of particles in the atmosphere. The aerosol optical depth due to extinction $\tau_{aer}^{ext}(\lambda, m)$ describes the extinction of direct solar radiation by its scattering and absorption by aerosols. This value is directly proportional to the amount of aerosols and depends on the efficiency factor of extinction Q_{ext}:

$$\tau_{aer}^{ext}(\lambda, m) = \int_{r_{min}}^{r_{max}} \pi r^2 \cdot Q_{ext}(\alpha_{Mie}, m) \cdot n(r) dr, \quad (1)$$

where $n(r)$ is the aerosol size distribution.

The spectral dependence of extinction is expressed for two the wavelengths λ_1 and λ_2 by the Angstrom coefficient α:

$$\alpha = -\frac{\ln \tau_{aer}^{ext}(\lambda_1, m)/\tau_{aer}^{ext}(\lambda_2, m)}{\ln \lambda_1/\lambda_2}. \quad (2)$$

This coefficient is an indicator of particle size. It is often used as an indicator of the proportion between the number of large and small particles. The α fluctuations reflect changes in size distributions of the aerosol. The Angstrom coefficient increases when particle size decreases, the maximum value of α equal to 4 corresponds to molecules. For Saharan dust, this coefficient is low and even negative.

2.2. Photometric Measurements from the Ground. The measurement of the attenuation of solar radiation by the atmosphere follows a simple observation technique under the direct sun. The analytical expression is written

$$I(\lambda) = I_0(\lambda) \cdot \exp(-m_{air} \cdot \tau_{tot}^{ext}(\lambda)). \quad (3)$$

The irradiance $I_0(\lambda)$ is the measurement of solar radiation outside the atmosphere, $I(\lambda)$ is the radiation attenuated by the atmospheric components, and m_{air} is the air mass which could be approximated by the relationship:

$$m_{air} = \frac{1}{\cos \theta_s}, \quad (4)$$

for solar angles less than 75°.

The total optical depth $\tau_{\text{tot}}^{\text{ext}}(\lambda)$ is composed of two terms outside the absorption bands of gaseous. One is due to molecular scattering $\tau_{\text{Ray}}^{\text{dif}}(\lambda)$ and the other is due to the scattering and absorption by aerosols $\tau_{\text{aer}}^{\text{ext}}(\lambda)$:

$$\tau_{\text{tot}}^{\text{ext}}(\lambda) = \tau_{\text{aer}}^{\text{ext}}(\lambda) + \tau_{\text{Ray}}^{\text{scat}}(\lambda). \tag{5}$$

Under these conditions, the aerosol optical depth determined from photometric measurements is given by

$$\tau_{\text{aer}}^{\text{ext}}(\lambda) = \frac{1}{m_{\text{air}}} \cdot \ln\left[\frac{I_0(\lambda)}{I(\lambda)}\right] - \tau_{\text{Ray}}^{\text{scat}}(\lambda). \tag{6}$$

The measured optical depth includes all aerosols in the atmospheric column, but it is usually largely due to tropospheric aerosols concentrated in the lower layers. This photometric method can be considered the reference technique for measuring the aerosol optical depth in the solar spectrum. Photometric measurements are used to determine the atmospheric turbidity factor β and Angstrom coefficient α [15] from data of aerosol optical depth. These coefficients are related by the Angstrom formula:

$$\tau_{\text{aer}}^{\text{ext}}(\lambda) = \beta \cdot \lambda^{-\alpha}. \tag{7}$$

β expresses the amount of aerosols in the atmosphere and characterizes the degree of air pollution. α is an indicator of the size of atmospheric particles.

The experimental determination of the Angstrom's turbidity coefficient β requires measurements of spectral direct normal irradiance at two wavelengths where absorption is negligible. The method consists in making a direct fit of Angstrom equation (7) to the experimental data. This fit produces a single value for each of the coefficients α and β valid for the whole band. Values of the Angstrom turbidity coefficient have been determined using spectral sun photometer data.

Data reported in this study are related to two different periods. The first spreads from 1981 to 1988 and measurements were made with a sun-sky radiometer installed at the meteorological Thala station and operated by WMO (World Meteorological Organization) in the framework of the global atmospheric background monitoring for selected environmental parameters (WMO, BAPMoN data for atmospheric aerosol depth) held on several sites on the globe at a rate of three measurements per day [16]. Volz introduced a hand-held Sun photometer 50 years ago [17]. Improved versions of this instrument [18] were used in various networks, including 95 stations of BAPMoN.

A Sun photometer should detect a relatively narrow band of optical wavelengths since the optical depth of the clear sky is strongly wavelength dependent. Optical interference filters have long been used to meet this requirement, but such filters are expensive and subject to unpredictable and very significant degradation. Filter degradation is such a serious problem that it caused the international BAPMoN network to be closed.

The second period is only for 2001 and which data are provided by CIMEL sun/sky radiometer within AERONET. This is an automatic, robotically operated instrument. Two detectors are used for the measurements of direct sun and sky radiance. Spectral observations of sun radiance are generally made at seven spectral channels: 340, 380, 440, 500, 675, 870, and 1020 nm, while measurements of sky radiance are made at 440, 675, 870, and 1020 nm every 15 minutes [9, 10]. A flexible inversion algorithm developed by Dubovik et al. [19] is used to retrieve columnar aerosol volume size distribution, real and imaginary refractive indices, and single scattering albedo from direct sun and diffuse sky radiance measurements.

It is generally accepted that the calibration is the paramount problem of sun photometry. The experience shows that good primary calibrations are only possible from high-altitude stations with exceptionally clear sky conditions. Experiments from rockets demonstrate the validity of the high mountain calibrations by the Langley method but only if the calibration days are carefully selected [20].

3. Climatic Characteristics of the Measurement Site

Thala station occupies a strategic position in point of view of meteorology (the only mountain resort in Tunisia, at an altitude of around 1090 m). It was chosen in nearly the majority of projects using the national meteorological network. It was place of a draft measure of pollution (chemical precipitation, aerosols and gases) in the late seventies with equipment provided by the World Meteorological Organization (sampling of rainfall using an automatic rain gauge). Thala measurement site (35° 33′ North; 08° 41′ East; 1091 m asl) is located in the mountains, the vegetation consists mainly of wild plants such as cactus. Olives are also cultivated but rarely. The lands around the station are operated as a breeding ground.

Since its opening, the meteorological station operates with conventional measuring instruments normed by WMO. The measured data are daily or monthly submitted on technical documents to regional center of the subdivision of Sfax where they will be corrected and then sent to the national center of Tunis to capture. They are stored in a database after undergoing digital control operations (application of several statistical procedures on data sets, error identification and correction or elimination). The dataset presented in this section is a result of a statistical analysis to a measurement series between 1977 and 2001.

Figure 1 shows the geographical location of the site selected for this study. Thala is a small town situated in the Middle West of the Tunisian territory at about 250 km south of the capital of Tunisia and close to the Algerian border. The measurement site is located in the meteorological station (about 6 km from Thala city) which is a part of the World Meteorological Organization network. Thanks to its geographical position, the instrument is particularly appropriate to detect both frequent dust outbreaks from the African Sahara and marine particles from the Mediterranean Sea. A minor impact of anthropogenic pollution is expected at this site.

3.1. Air Temperature Variations. Thala is the coldest region of Tunisia. Mean monthly values of temperature vary between

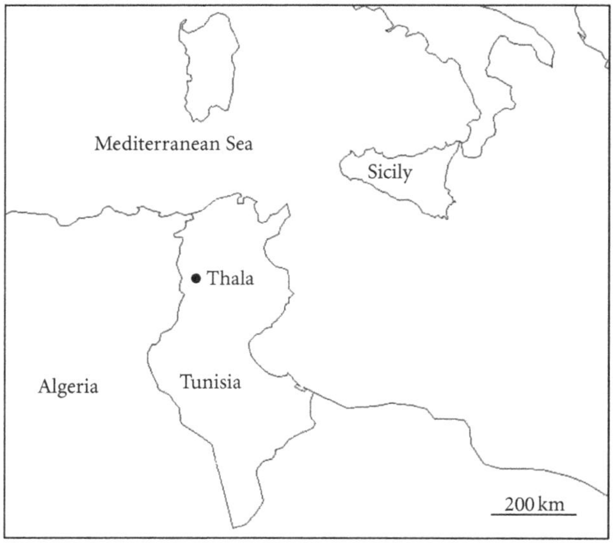

Figure 1: Geographical location of Thala Tunisian site.

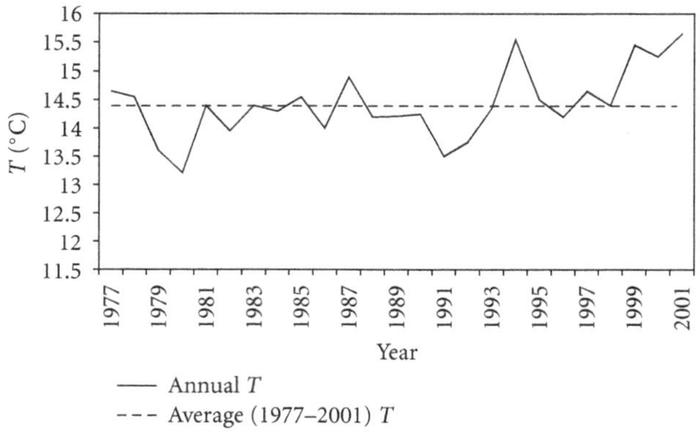

Figure 2: Annual averaged temperature at Thala site between 1997 and 2001.

5.7°C in January and 24.6°C in July (Table 1). Annual averaged temperature is 14.4°C computed for the time period 1977–2001. It varies between 13.2°C in 1980 and 15.7°C in 2001. Apart a notable peak recorded in 1994, the period spanning from 1977 to 1995, has been characterized by mean temperatures generally below the annual average. However, we note between 1996 and 2001 a considerable growth exceeding 1°C at the annual scale (Figure 2).

At the seasonal scale, we note that winter in Thala is characterized by low temperature with mean value of 6.5°C. The coldest winter from the studied period has been observed in 1981 with averaged value around 5°C and a number of 28 days where temperature is less than 0°C. Spring and autumn are characterized by temperature varying between 12.1°C and 15.7°C. The highest seasonal temperatures for spring and autumn are, respectively, 14.6°C (1981) and 17.2°C (1978) and the lowest are 9.5°C (1980) and 13.9°C (1979).

The diurnal thermal amplitude (difference between maximum and minimum temperatures) varies from one month to another. It peaked in summer months, with a value equal to 11.2°C. The minimum is observed in January and December, with a value not exceeding 4.2°C.

3.2. Wind Speed and Direction.
Although Thala site is located at altitude (1091 meters), we find that the wind is relatively moderate to fairly strong, as showed by the weakness of calm winds with an average frequency of 1.8% and strong winds with a frequency not exceeding 1.9%. We note also that at annual basis, half of the wind speeds did not exceed 5 m/s.

On the other hand, the wind blows from all directions and this in proportions is quite different in both frequencies and speeds. Winds from NNW, N, and NW sectors are still the most frequent and also the strongest. We can conclude that the prevailing wind at Thala station is the North to North-West, with a total frequency exceeding 47%. Wind

sectors ESE, E, and SE are rare and blow with relatively low speeds (Table 2). South blowing winds are relatively frequent in the period between April and October.

Moreover, we note that the most common values of wind speed are those between 3 and 8 m/s (Table 3). Strong winds with average frequency of not exceeding 1.9% are rare. They blow from NNW direction, South and North, with frequencies equal, respectively, to 27, 14, and 13%. It is noteworthy that despite the low frequency of winds from a southerly direction (only 9.3% of total sales), the majority of these winds are strong (14% of the winds).

Considering the geographical position of Thala station, the sand storm is a meteorological phenomenon very uncommon. During the period 1977–2001, this phenomenon has shown only in 1977, 1978, 1998, and 2001.

The sirocco, known in Tunisia as the "Chehili", is a wind of dynamic origin, blowing from South and Southwest areas. It is usually accompanied by a very significant rise in temperature and a remarkable drop in relative humidity. A total of 263 days of sirocco are registered at Thala station with the majority during the summer season (June, July and August) (Table 4). This shows that our station, despite being a mountainous region near the north-west, is not sheltered from the sirocco blows.

3.3. Relative Humidity.
Thala station has a fairly high relative humidity during the months of September to May. It ranges from 75% in January and 60% in May and September. It suffered a significant decline during the three summer months, due to the continental aspect of the station (only 50%). In the beginning of the autumn, it climbs rapidly above 60% (Figure 3). A wide daily variation of humidity is much more significant. Maximum values are observed between midnight and early morning hours, then declines rapidly to reach the minimum around 15 h.

3.4. Amount of Rainfall.
With an annual rainfall averaged over the period 1977–1997 equal to 440 mm, the Thala station belongs to a region ranked as moderated rainfall in Tunisia. Between 1981 and 1988 annual amount of rain varies

TABLE 1: Monthly averaged temperature at Thala site.

	January	February	March	April	May	June	July	August	September	October	November	December	Average
Temperature (°C)	5.7	6.6	8.8	11.3	16.3	21.3	24.6	24.4	20.2	16.5	10.3	7.1	14.4

TABLE 2: Monthly frequencies of mean wind directions at Thala site.

Wind direction	January	February	March	April	May	June	July	August	September	October	November	December
N	12	11	15	13	15	17	17	18	16	12	13	12
NNE	2	1	2	1	2	3	3	3	3	2	2	1
NE	1	2	2	2	3	3	3	3	3	2	1	1
ENE	1	1	2	1	2	2	3	3	3	2	1	1
E	1	1	2	1	2	2	2	2	2	2	1	1
ESE	1	1	1	1	1	2	1	1	1	1	1	0
SE	2	2	2	3	4	6	5	5	5	6	3	1
SSE	3	4	4	8	9	11	9	7	6	6	3	1
S	6	5	7	10	14	11	13	11	11	13	6	4
SSW	3	3	3	4	4	3	4	4	4	5	3	2
SW	5	4	4	4	4	3	4	3	5	5	5	5
WSW	6	5	4	4	4	4	3	4	5	5	6	7
W	13	13	10	11	8	7	7	8	8	8	12	17
WNW	10	9	7	7	5	4	4	4	5	6	9	10
NW	16	17	17	12	8	8	8	8	9	11	16	17
NNW	18	21	20	18	14	15	15	15	14	13	17	18

TABLE 3: Monthly frequencies of mean wind speeds at Thala site.

Wind speed (m/s)	January	February	March	April	May	June	July	August	September	October	November	December
0	2	2	1	1	1	1	2	2	2	2	3	1
1	4	3	3	3	4	5	5	6	6	6	5	4
2	9	8	7	7	9	11	12	12	13	12	9	8
3	10	8	10	10	11	14	14	15	15	13	11	10
4	9	9	10	9	12	14	14	14	12	12	10	9
5	9	10	11	10	12	13	12	14	13	12	11	10
6	8	10	11	11	12	11	11	11	11	11	10	10
7	10	10	10	11	10	10	10	8	9	8	9	10
8	8	8	8	7	7	6	6	6	5	6	7	8
9	7	7	6	7	6	5	5	4	4	5	7	7
10	6	6	6	7	5	4	4	3	3	4	6	7
11	7	5	4	5	4	3	3	2	2	3	4	5
12	4	4	3	3	2	2	2	1	2	2	3	3
13	3	3	2	2	2	1	1	1	1	2	3	3
14	2	2	2	2	1	1	0	1	1	1	1	2
15	2	2	1	1	1	0	0	0	1	1	1	1
>16	4	4	4	3	1	1	0	0	1	1	2	4

TABLE 4: Number of sirocco days recorded in Thala site.

	January	February	March	April	May	June	July	August	September	October	November	December
Number of sirocco days	0	0	0	3	26	59	111	53	11	0	0	0

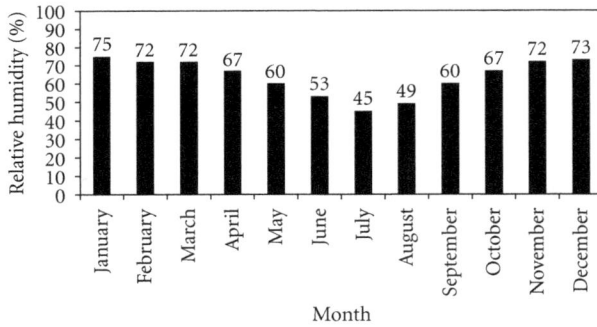

FIGURE 3: Monthly means of relative humidity at Thala station.

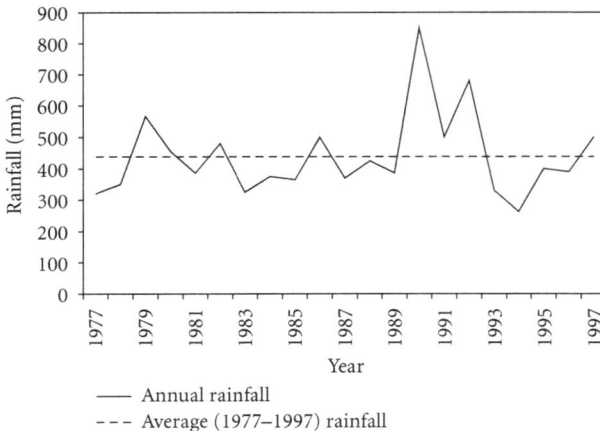

FIGURE 4: Annual rainfall in Thala station for (1977–1997) period.

between around 330 and 500 mm (Figure 4). In 2001, the recorded annual value falls to 313 mm.

4. Monthly and Seasonal Variations of Aerosol Optical Properties

4.1. Aerosol Optical Depth AOD. The mean monthly values of AOD measured for the two studied periods are deduced from individual measurements performed at Thala site with sun photometer instrument. The number of observations for each month is illustrated in Table 5.

The AOD ($\lambda = 1 \mu$m) monthly variability is clearly revealed by Figure 5 showing different results for the two studied periods. First, for the period (Mars–October 2001), data are determined by high monthly variability which will be explained latter, with two notable peaks in July and October. In the contrary, the second period between 1981 and 1988 is characterized by small monthly variability showing that AOD ($\lambda = 1 \mu$m) takes averaged values larger than 0.1 from June to August (summer) and lower or equal to 0.1 from September to May with a peak in July and a minimum in December to January. These results are quite comparable to those found for 1999 at Sidi Bou Saïd station (Northern Tunisia) in spite of the difference between the geographical characteristics of the two sites. We showed [21] that mean monthly

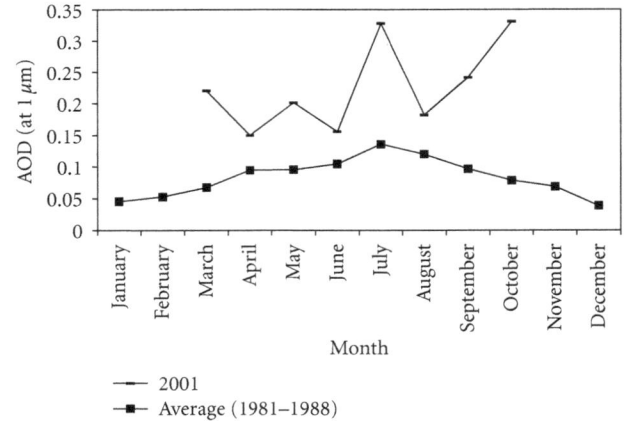

FIGURE 5: Mean monthly variations of AOD ($\lambda = 1 \mu$m) at Thala site for the two series of measurements.

values of Linke turbidity factor T_L present two levels. The first, with highest turbidity, is observed between April and August. The second, characterized by lowest values, is related to rest of the year (September to March). We note moreover that, Linke's T_L factor refers to the whole solar spectrum and represents the turbidity caused by aerosols and water vapour, which affects solar radiation by absorption in the visible and near infrared regions. On the contrary, β coefficient has been accepted as an index of the turbidity caused just by aerosols, as it represents the aerosol load in the air and it is obtained from spectral measurements.

The larger Angstrom turbidity coefficient values observed in summer months may be due to a larger concentration of aerosol size distribution characterized by higher scattering and/or absorption coefficients. In fact, turbidity depends on the vertical profile of the aerosol extinction coefficient that is made of two parts: the scattering and the absorption component [22].

Probably, turbidity seasonal variability revealed by Figure 5 is mainly determined by the seasonal dependence of aerosol removal processes. The weather stability typical of summer regimes favours the accumulation of atmospheric particles over North Africa. As a consequence, the contribution of long-range transported particles is expected to be larger in summer regimes. Moreover, the aerosol wet removal is practically absent over the east Mediterranean basin in summer and the larger amount of solar radiation reaching the earth's surface may favour photochemical reactions that affect optical and microphysical properties of the accumulated atmospheric aerosols. Also, the winter decrease in atmospheric turbidity may result from increases in relative humidity and decreases in temperature (as it is presented in Section 3), leading to increased water uptake in the aerosols, hence increased particle size and mass and increased fall and deposition velocities. The two main causes of the low aerosol optical characteristics in the winter season are washout by rainfall and high relative humidity leading to aerosol size increases and subsequent deposition.

It is worth observing from Figure 5 that some months from the first period (1981–1988) are characterized by mean

TABLE 5: Number of observations per month for the two series of measurements.

Number of observations	January	February	March	April	May	June	July	August	September	October	November	December
from1981 to 1988	111	126	148	149	173	259	385	312	319	224	116	104
At 2001	—	—	242	629	574	1086	1011	959	468	517	—	—

AOD ($\lambda = 1\,\mu$m) values larger than 0.15. Moreover, for 2001 studied period, some averaged monthly values exceed 0.2 with rapid change from month to month. This may indicates long-range transport of aerosols in Thala. The high-aerosol-load days are attributed to Saharan dust. In fact, we have already mentioned (Section 3) that during the two studied periods, sandstorms have only been recorded in 2001. We can also note that the nonhomogeneity between the monthly turbidity trends for the two series of measurements (Figures 5 and 6) could be explained, among others, by climate change effects. This is also well proved by the rise of more than 1°C in annual averaged temperature between 1981 and 2001 and on the other hand by a decrease in rainfall amount in 2001 compared to the first period.

From Figure 6, we could also observe that particularly for 1987 and 1988, turbidity at Thala presents two levels. The first, with averaged β values greater than 0.1, is related to the spring-summer period (from April to September) reaching a maximum of around 0.2 in July. The second period spreads from October to March and for which monthly averaged turbidity is lower than 0.1.

The spring season (March to May) over Thala is characterized by moderately high temperature and wind speed and also by low relative humidity as shown in Section 3. These spring weather patterns usually carry fine dust and sand particles from the desert area to Thala region.

The summer season (from June to August) is characterized by high temperature, low relative humidity, and relatively low wind speeds over the studied site. This climate situation leads to increased photochemical processing in the atmosphere and hence greater anthropogenic aerosol production. The aerosols formed through primary emissions and secondary reactions are thus potential causes for the observed turbidity values during this season. In addition, strong vertical temperature gradients in the Thala area in late summer lead to enhanced vertical convection, carrying aerosols aloft and enhancing turbidity. Similar results have been found by Perrone et al. [7] for one year (March 2003 to March 2004) of sun-photometer measurements over South-East Italy. They have shown that aerosol optical and microphysical properties and the dominating aerosol types depend on seasons.

In conclusion, it is difficult to attribute seasonal trends in turbidity to anyone causal factor. Meteorological parameters are only one of many factors that can influence turbidity at any given location. Other climatic factors include convective activity, seasonal differences in air mass origin, and the synoptic situation, as well as dispersion, transformation, and removal processes.

These preliminary results that based on a yearly comparison of measurements reveal that concentration, size distribution, and chemical composition of aerosol particles depend

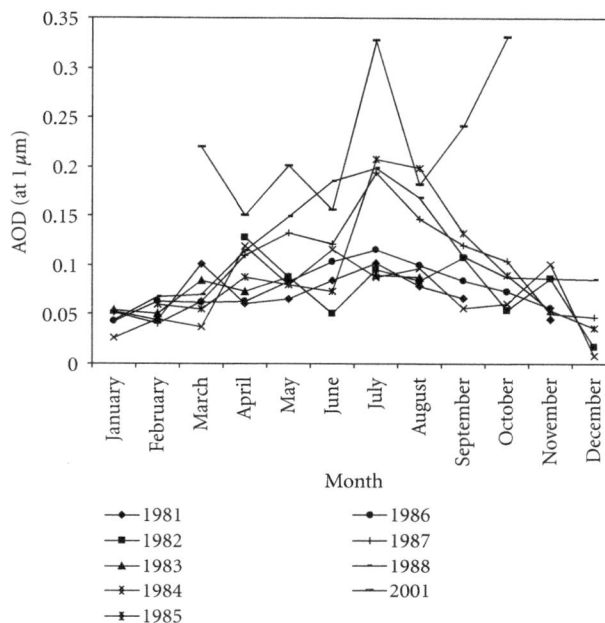

FIGURE 6: Monthly variations of AOD ($\lambda = 1\,\mu$m) at Thala site for each studied year.

on seasons and lead at first of the assumption that the higher aerosol loads are characterized by a predominant contribution of larger, scattering particles over west regions of Tunisia in summer.

4.2. Angstrom Coefficient and Volume Size Distribution. Recent studies have shown many useful applications of measures of Angstrom coefficient to characterize the radiative properties of aerosols [23, 24]. These measurements are also used to determine the size distribution of aerosols which is the most important parameter to characterize a population of aerosols. The size distribution of atmospheric aerosols can be described in terms of modes that characterize their size range. The submicron particles are classified into accumulation mode ($0.05 < r < 0.5$ micron), Aitken mode ($0.01 < r < 0.05$ microns), and the nucleation mode ($r < 0.01$ micron). Coarse particles are defined for a range of radius $r > 1$ micron [25]. In the accumulation mode, we found organic compounds and sulfates from industrial emissions of sulfur dioxide (SO_2).

The electromagnetic scattering theory indicates that small particles with radius comparable to the wavelength λ have large α values of order of 1, while large particles have small α values close to 0. Typical α values that range $0.5\sim1$ for accumulation mode aerosols in the urban type air mass were

TABLE 6: Number of inversions for the computation of aerosol size distribution at Thala in 2001.

	March	April	May	June	July	August	September	October
Number of inversions	21	50	36	84	94	85	31	50

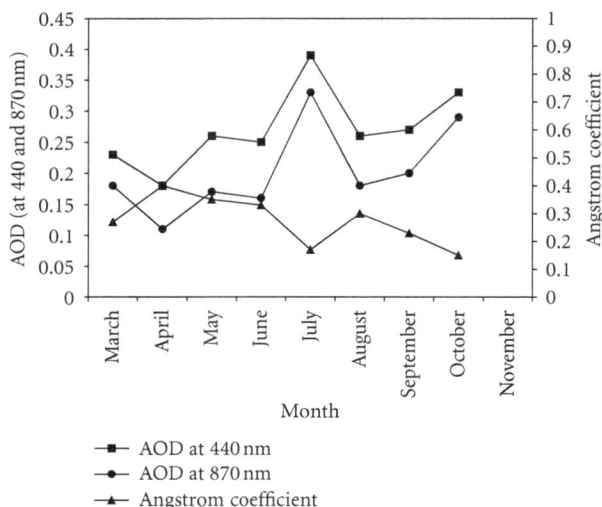

FIGURE 7: Monthly evolution of mean AOD (at 440 and 870 nm) and Angstrom exponent α at Thala site in 2001.

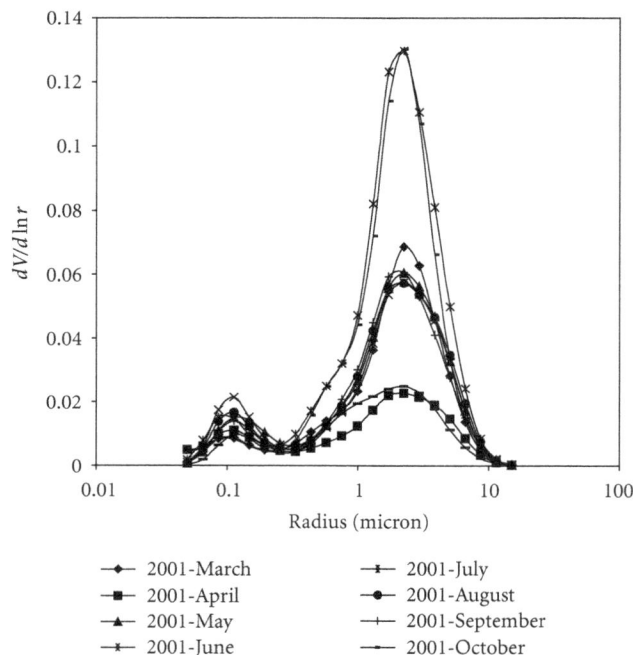

FIGURE 8: Mean volume size distribution at Thala site for 2001 months.

observed by ground-based measurements. Small α values (less than 0.5) originate from areas of prevailing soil-derived particles such the area near the Saharan desert. High α values (greater than 1) are caused by anthropogenic aerosols around industrial areas or by biomass burning aerosols [19, 23].

The Angstrom exponent α is defined by the spectral dependence of the aerosol optical depth. The temporal evolution of the Angstrom exponent (computed from AOD values at 440 and 870 nm) is also reported in Figure 7. We clearly observe that AODs and Angstrom exponents are following opposite trends at Thala site for 2001 period.

It is worth observing that July and October 2001 are characterized by mean monthly AOD larger than 0.25 (Figure 7). Relatively low averaged Angstrom coefficient α computed from 440 and 870 μm wavelengths (less than 0.2) is also observed. The high turbidity linked to the low values of α recorded in these two months suggests that the influence of the southwesterly winds blowing in from the Sahara is maximal [13].

Figure 8 shows the mean volume size distribution obtained at Thala site by averaging different volume size distribution profiles retrieved during the 2001 months. The number of inversions per month is illustrated on Table 6. The domination of coarse mode particles in desert aerosols is clearly pointed out: particles radius peak values span the 1.7–3 μm range. Aerosols with a dominant contribution of moderately absorbing particles are advected from Europe and West Africa. Air masses from these regions have the common property to travel across large cities and industrialized areas before reaching Thala region and as a consequence these air

masses can be affected by urban-industrialized aerosols [6] as well as by seasonal biomass burning [26].

Monthly volume size distributions plotted on Figure 8 reveal that the coarse mode is predominant and as a consequence, Angstrom coefficient α takes averaged values lower than 0.5, in accordance with typical desert properties. In fact, AERONET sun photometer measurements performed from 1993 to 2000 at Cape Verde (Africa) during dust outbreaks have shown that the median radius of the coarse mode was 1.9 ± 0.03 μm [27]. Also, the scanning electron microscopy analysis of several dust samples from rainfall residues collected at Leece (Italy) during dust storm occurring from April to June 2002 have provided size distributions with median radius values between 0.85 and 1.2 μm [28].

Previous studies [7, 29, 30] have shown that the optical and chemical properties of dust particles are quite less dependent on wavelength. Then, lower AOD dependence on wavelength can be used as marker for the dominant contribution of desert type particles. As a consequence, the Angstrom coefficient α takes values ranging from 0.1 to 1 along dust events. The AOD wavelength dependence is determined by size distribution, shape, and chemical composition (real and imaginary refractive indices) of aerosol particles. Characterizations of water insoluble components of Sahara dust samples from rainfall residues have shown that dust particles

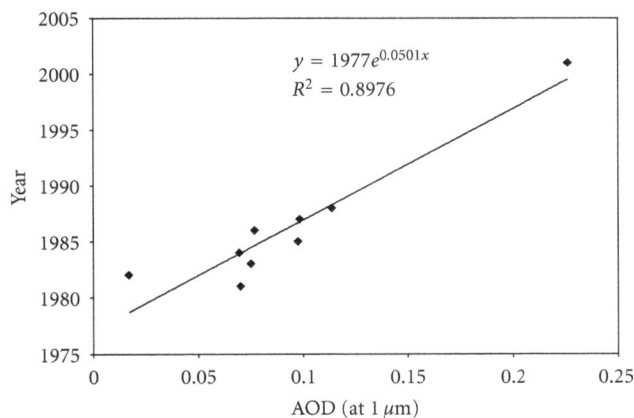

FIGURE 9: Annual variation of mean AOD ($\lambda = 1 \, \mu$m) at Thala site for the two series of measurements.

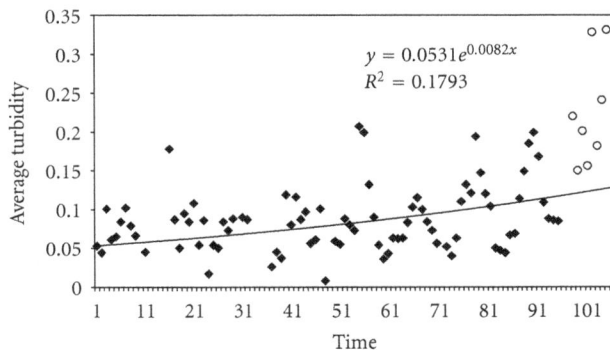

FIGURE 10: Linear trend fit to monthly AOD ($\lambda = 1 \, \mu$m) at Thala site from January 1981 to December 1988 (series 1) and from March to October 2001: the last eight points (series 2).

with a high content of illite are mainly advected over the Mediterranean basin during dust storms [28]. These results are also in accordance with those obtained by Avila et al. [31] by analyzing 11-year records of Africa dust rain in the Montseny Mountain (north eastern Spain). They have observed that illite was the most abundant mineral identified in all dust samples by X-ray diffraction and that the illite concentration was 41% when the source region was western Sahara, and 34.5% for the air masses comes from central Algeria.

5. Interannual Variability of Atmospheric Turbidity

In order to examine long-term trends in turbidity over central west Tunisian region, moving averages were computed at Thala station. Although interannual variability in turbidity is evident, the first data series (1981–1988) is generally characterized by stable turbidity. However, a linear model fitted to the Thala time series shows long-term trends in turbidity. Long-term averaged turbidity is around 0.1, β has, on average, increased by approximately 0.044 between 1981 and 1988 and by 0.156 between 1981 and 2001. Figure 9 presents the annual variation of averaged β Angstrom turbidity coefficient for the two studied periods: 1981 to 1988 and 2001. Relatively low values (between 0.05 and 0.12) are observed during the first period (1981–1988); a rapid increase appears in 2001 with an averaged annual value around 0.23. A linear increasing trend is observed for the variation of β coefficient between 1981 and 2001. Minimum averaged value of the turbidity coefficient ($\beta = 0.017$) occurred in 1982 was accompanied by low air temperature (averaged yearly temperature near 14°C) and high rainfall (around 480 mm per year) in the period under study.

In Thala site, a period of steadily increasing turbidity commenced in 1986 and continued until 1988 (Figures 9 and 10) and then to 2001. In the contrary, turbidity is generally stable ($0.05 < \beta < 0.1$) through the rest time series (1981–1985). Between 1988 and 2001, there are insufficient turbidity data to put in evidence the increase in aerosol load in Thala region and to enable the assessment of any possible effects on the aerosol climatology. However, short-term trends after 1988 can be explained by the contribution of the eruption of Mount Pinatubo. Power et al. [32] have shown a rise in Angstrom turbidity starting in 1991 to 1997 over south African sites (mainly Cape Town, Pretoria and Upington). The authors explain this trend by aerosols ejected from the June 1991 eruption of Mount Pinatubo. Anomalously high aerosol optical depth (AOD) values and low global direct radiation were observed in mid 1991 over Cairo and early 1992 over Aswan [33]. Compared to El-Chichon's eruption, that of Pinatubo was distinguished by a large magnitude, leading to a 50% increase in stratospheric extinction as demonstrated by Michalsky et al. [34]. Some relevant Pinatubo AOD measurements can be also found in Schmid et al. [35]. These aerosols first produced a strong depleting effect on direct radiation, as documented elsewhere but subsequently decayed exponentially over time. The increase of particle load in the atmosphere site may probably result on the possible effects of this fact on the aerosol climatology. In fact, volcanic eruptions eject gas and dust into the earth's upper troposphere, where dispersion processes lead to worldwide transport of the ejected material. Sulfur dioxide gas emitted by the volcanoes is oxidized over a period of weeks to form sulfuric acid gas. This gas nucleates and/or condenses to form aerosols that impact the direct solar radiation.

Furthermore, the long-term trend in turbidity that does exist in Thala should also be interpreted by local or regional changes in climate or in aerosol emissions (natural or anthropogenic). As with the seasonal trends identified above, meteorological parameters are one of many factors that can influence turbidity at any given location. As is shown in Section 3, 2001 has been characterized by particular meteorological conditions, compared to the first period of measurement (1981–1988): mean annual temperature, rainfall amount, and sandstorm phenomenon.

6. Conclusion

This research provides a preliminary climatology of aerosols previously unavailable for Thala Tunisian site. A long-time period of data has been used to analyse the monthly, seasonal and annual variations for the atmospheric Angstrom turbidity coefficient β (AOD at $\lambda = 1\,\mu$m) computed for two different periods (the first: 1981–1988; the second: March–October 2001). It was found that in spring and summer seasons, the aerosol optical characteristics recorded significant high values, due to the effects of sand and dust particles emitted from the African Sahara. In summer, photochemical processes became the main origin of the aerosol and this leads to slight increases in the values of aerosol optical characteristics, despite lower wind speeds relative to other seasons. In the autumn and winter seasons, the lowest aerosol optical characteristics were measured over Thala. This was due to the washout by rain and high relative humidity.

The interannual variability of the Angstrom turbidity coefficient β is also investigated. β coefficient recorded a period of steadily increasing turbidity commenced in 1986 until 2001. Short-term trends after 1988 are explained among others, by the contribution of the eruption of Mount Pinatubo. The long-term trend in turbidity is interpreted by local or regional changes in climate or in aerosol emissions. 2001 has been characterized by particular meteorological conditions compared to the first period (1981–1988).

It is shown that aerosols of different origin and type can be advected at this region. However, depending on travel distance and residence time over source regions and monitoring site, the particle number concentrations, the physical and chemical state, and thus the optical properties of aerosols change. As a consequence, there is no straightforward one-to-one relationship between enhanced aerosol load and long-range transport from Europe and/or Africa.

Acknowledgments

The authors gratefully thank Professors Philippe Goloub, the leader of the Centre of Environmental Research PHOTON-AERONET and Didier Tanré from Atmospheric Optics Laboratory (LOA, Lille-France) for their help and their assistance during the development of this paper. They also thank Mr Fayçal Elleuch from National Institute of Meteorology of Tunis (Sfax section), for providing them climatological data for Thala meteorological station.

References

[1] D. Tanré, Y. J. Kaufman, B. N. Holben et al., "Climatology of dust aerosol size distribution and optical properties derived from remotely sensed data in the solar spectrum," *Journal of Geophysical Research D*, vol. 106, no. 16, pp. 18205–18217, 2001.

[2] M. O. Andreae, "The dark side of aerosols," *Nature*, vol. 409, no. 6821, pp. 671–672, 2001.

[3] I. Tegen, "Contribution to the atmospheric mineral aerosol load from land surface modification," *Journal of Geophysical Research*, vol. 100, no. D9, pp. 18707–18726, 1995.

[4] I. N. Sokolik and O. B. Toon, "Direct radiative forcing by anthropogenic airborne mineral aerosols," *Nature*, vol. 381, no. 6584, pp. 681–683, 1996.

[5] J. Prospero, P. Ginoux, O. Torres, and S. E. Nicholson, "Environmental Characterization of Global sources of atmospheric soil dust derived from the NIMBUS-7 TOMS absorbing aerosol product," *Reviews of Geophysics*, vol. 40, no. 1, p. 1002, 2002.

[6] F. De Tomasi and M. R. Perrone, "Lidar measurements of tropospheric water vapor and aerosol profiles over southeastern Italy," *Journal of Geophysical Research D*, vol. 108, no. 9, pp. AAC14-1–AAC14-12, 2003.

[7] M. R. Perrone, M. Santese, A. M. Tafuro, B. Holben, and A. Smirnov, "Aerosol load characterization over South-East Italy for one year of AERONET sun-photometer measurements," *Atmospheric Research*, vol. 75, no. 1-2, pp. 111–133, 2005.

[8] A. Angstrom, "Techniques of determining the turbidity of the atmosphere," *Tellus*, vol. 13, pp. 214–223, 1961.

[9] B. N. Holben, T. F. Eck, I. Slutsker et al., "AERONET—a federated instrument network and data archive for aerosol characterization," *Remote Sensing of Environment*, vol. 66, no. 1, pp. 1–16, 1998.

[10] B. N. Holben, D. Tanré, A. Smirnov et al., "An emerging ground-based aerosol climatology: aerosol optical depth from AERONET," *Journal of Geophysical Research D*, vol. 106, no. 11, pp. 12067–12097, 2001.

[11] WMO, "Report on the measurements of atmospheric turbidity in BAPMoN," World Meteorological Organization Report GAW 94, 1994.

[12] M. Masmoudi, M. Chaabane, D. Tanré, P. Gouloup, L. Blarel, and F. Elleuch, "Spatial and temporal variability of aerosol: size distribution and optical properties," *Atmospheric Research*, vol. 66, no. 1-2, pp. 1–19, 2003.

[13] M. Masmoudi, M. Chaabane, K. Medhioub, and F. Elleuch, "Variability of aerosol optical thickness and atmospheric turbidity in Tunisia," *Atmospheric Research*, vol. 66, no. 3, pp. 175–188, 2003.

[14] M. Chaâbane, M. Masmoudi, K. Medhioub, and F. Elleuch, "Daily and monthly averaged aerosol optical properties and diurnal variability deduced from AERONET sun-photometric measurements at Thala site (Tunisia)," *Meteorology and Atmospheric Physics*, vol. 92, no. 1-2, pp. 103–114, 2006.

[15] A. Angstrom, "On the atmospheric transmission of sun radiation and on the dust in the air," *Geografiska Annaler 2*, vol. 2, pp. 156–166, 1929.

[16] WMO, A joint publication of the World Meteorological Organization. Global Atmospheric Background Monitoring for Selected Environmental Parameters. BAPMoN data (Aerosol Optical Depth), 1984–1989.

[17] F. E. Volz, "Photometer mit selen-photoelement zur spektralen Messung der sonnenstrahlung und zur bestimmung der wallenlangenabhangigkeit der dunsttrubung," *Archiv für Meteorologie, Geophysik und Bioklimatologie*, vol. 10, pp. 100–131, 1959.

[18] F. E. Volz, "Economical multispectral sun photometer for measurements of aerosol extinction from 0.44 μm to 1.6 μm and precipitable water," *Applied Optics*, vol. 13, no. 8, pp. 1732–1733, 1974.

[19] O. Dubovik, A. Smirnov, B. N. Holben et al., "Accuracy assessments of aerosol optical properties retrieved from Aerosol Robotic Network (AERONET) Sun and sky radiance measurements," *Journal of Geophysical Research D*, vol. 105, no. 8, pp. 9791–9806, 2000.

[20] B. Schmid and C. Wehrli, "Comparison of sun photometer calibration by Langley technique and standard lump," *Applied Optics*, vol. 34, no. 21, pp. 4500–4512, 1995.

[21] M. Chaâbane, "Analysis of the atmospheric turbidity levels at two Tunisian sites," *Atmospheric Research*, vol. 87, no. 2, pp. 136–146, 2008.

[22] V. Matthias, D. Balis, J. Bösenberg et al., "Vertical aerosol distribution over Europe: statistical analysis of Raman lidar data from 10 European Aerosol Research Lidar Network (EARLINET) stations," *Journal of Geophysical Research D*, vol. 109, no. 18, article D18201, 12 pages, 2004.

[23] T. Nakajima and A. Higurashi, "A use of two-channel radiances for an aerosol characterization from space," *Geophysical Research Letters*, vol. 25, no. 20, pp. 3815–3818, 1998.

[24] J. S. Reid, T. F. Eck, S. A. Christopher, P. V. Hobbs, and B. Holben, "Use of the Ångstrom exponent to estimate the variability of optical and physical properties of aging smoke particles in Brazil," *Journal of Geophysical Research D*, vol. 104, no. 22, pp. 27473–27489, 1999.

[25] E. Weingartner, S. Nyeki, and U. Baltensperger, "Seasonal and diurnal variation of aerosol size distributions ($10 < D < 750$ nm) at a high-alpine site (Jungfraujoch 3580 m asl)," *Journal of Geophysical Research D*, vol. 104, no. 21, pp. 26809–26820, 1999.

[26] D. S. Balis, V. Amiridis, C. Zerefos et al., "Raman lidar and sunphotometric measurements of aerosol optical properties over Thessaloniki, Greece during a biomass burning episode," *Atmospheric Environment*, vol. 37, no. 32, pp. 4529–4538, 2003.

[27] O. Dubovik, B. N. Holben, T. Lapyonok et al., "Non-spherical aerosol retrieval method employing light scattering by spheroids," *Geophysical Research Letters*, vol. 29, no. 10, 4 pages, 2002.

[28] A. Blanco, F. De Tomasi, E. Filippo et al., "Characterization of African dust over southern Italy," *Atmospheric Chemistry and Physics*, vol. 3, no. 6, pp. 2147–2159, 2003.

[29] G. A. D'Almeida, P. Koepke, and E. P. Shettle, *Atmospheric Aerosol-Global Climatology and Radiative Characteristics*, Deepak Hampton, 1991.

[30] J. Ackermann, "The extinction-to-backscatter ratio of tropospheric aerosol: a numerical study," *Journal of Atmospheric and Oceanic Technology*, vol. 15, no. 4, pp. 1043–1050, 1998.

[31] A. Avila, I. Queralt-Mitjans, and M. Alarcón, "Mineralogical composition of African dust delivered by red rains over northeastern Spain," *Journal of Geophysical Research D*, vol. 102, no. 18, pp. 21977–21996, 1997.

[32] H. C. Power and C. J. Willmott, "Seasonal and interannual variability in atmospheric turbidity over South Africa," *International Journal of Climatology*, vol. 21, no. 5, pp. 579–591, 2001.

[33] A. S. Zakey, M. M. Abdelwahab, and P. A. Makar, "Atmospheric turbidity over Egypt," *Atmospheric Environment*, vol. 38, no. 11, pp. 1579–1591, 2004.

[34] J. J. Michalsky, R. Perez, R. Seals, and P. Ineichen, "Degradation of solar concentrator performance in the aftermath of Mount Pinatubo," *Solar Energy*, vol. 52, no. 2, pp. 205–213, 1994.

[35] B. Schmid, C. Mätzler, A. Heimo, and N. Kämpfer, "Retrieval of optical depth and particle size distribution of tropospheric and stratospheric aerosols by means of sun photometry," *IEEE Transactions on Geoscience and Remote Sensing*, vol. 35, no. 1, pp. 172–182, 1997.

14

Trends of Rural Tropospheric Ozone at the Northwest of the Iberian Peninsula

S. Saavedra,[1] A. Rodríguez,[1] J. A. Souto,[1] J. J. Casares,[1] J. L. Bermúdez,[2] and B. Soto[2]

[1] Department of Chemical Engineering, University of Santiago de Compostela, 15782 Santiago de Compostela, Spain
[2] Department of Environment, As Pontes Power Plant, Endesa Generación S.A., 15320 La Coruna, Spain

Correspondence should be addressed to J. A. Souto, ja.souto@usc.es

Academic Editors: L. Gimeno and D. T. L. Shek

Tropospheric ozone levels around urban and suburban areas at Europe and North America had increased during 80's–90's, until the application of NO_x reduction strategies. However, as it was expected, this ozone depletion was not proportional to the emissions reduction. On the other hand, rural ozone levels show different trends, with peaks reduction and average increments; this different evolution could be explained by either emission changes or climate variability in a region. In this work, trends of tropospheric ozone episodes at rural sites in the northwest of the Iberian Peninsula were analyzed and compared to others observed in different regions of the Atlantic European coast. Special interest was focused on the air quality sites characterization, in order to guarantee their rural character in terms of air quality. Both episodic local meteorological and air quality measurements along five years were considered, in order to study possible meteorological influences in ozone levels, different to other European Atlantic regions.

1. Introduction

Apart from the stratospheric ozone, around 10–15% of the natural produced atmospheric ozone is located at the troposphere [1]. In addition, global ozone data show an increment of the ozone concentration with height, as an indicative of higher stratospheric-tropospheric exchange, and a more efficient ozone production in the upper troposphere.

At lower troposphere, the analysis of ozone ground level concentration measurements at rural and suburban areas in North America and Europe shows a typical diurnal cycle in rural sites, with minimum levels at sunrise and maximum between afternoon and evening. This pattern comes from the photochemical activity along the day, which is more favorable when low thermal inversions and soft winds keep the pollutants close to the ground level during hours.

On the other hand, sites close to large NO sources produce significant nocturnal low ozone, because of its fast reaction with NO; this can be extended to urban areas with spare but significant NO nocturnal emissions, producing strong variations of the ozone levels along every day. A different situation is usually observed downwind the city emissions, because NO has time enough to produce NO_2 and, after that, NO_2 is photolized by solar radiation to promote ozone

production. In fact, the highest ozone peaks are usually observed in suburban and rural areas around the big cities, with hourly averages that can achieve 400–800 $\mu g/m^3$; on the other hand, isolated rural sites usually achieve lower ozone levels around 40–80 $\mu g/m^3$.

Long-term trends in tropospheric ozone are not only affected by emissions but they can also be masked by meteorological conditions. However, an annual increment of 1-2% at Europe in the period 1958–1988 was reported [2], which is coincident to similar trends in Asia [3] and Pacific [2]. At Northern Hemisphere, these changes increased the ozone levels over the Northern Atlantic [4] producing persistent levels of ozone at rural and urban areas of the European Atlantic Coast: Mace Head rural site (Ireland) increased its ozone levels in 0.14 ppb in the period 1987–1997, keeping constants after 2000 [5]. Similar trends can be observed at the US Pacific Coast [6].

At Europe, peak levels of ozone at UK decreased around 30% in the period 1986–1999 [7], but annual concentration showed a small increment. Derwent et al. [8] explained the peaks decrease because of the trends of VOCs levels during 90's. Similar results were obtained by Solberg et al. [9] at Nordic countries during 90's, probably due to the reduction of the continental emissions. However, the EMEP rural sites data are not consistent with these regional trends in the period 1990–2006, due to three different factors that affect the rural ozone levels: (a) reduction of ozone peaks, (b) reduction of NO emissions from transport and other sources, with less ozone destruction, and (c) increase of the background ozone levels at Hemispheric and global scale. As an additional factor, not yet quantified, climate change can move these trends to either positive or negative [10].

On the other hand, at cities where NO_x road traffic emissions decreased, winter ozone levels increased (less ozone destruction by NO), but summer peaks decreased. Therefore, ozone urban cycles at these cities are more similar to suburban and rural patterns.

At the Iberian Peninsula, tropospheric ozone is detected in all regions and, particularly, in mediterranean and other coastal areas. This is related to the industrial development, road traffic increment, and the high isolation in central, east, and south regions. However, ozone levels distribution is very heterogeneous between regions, because of their strong differences in climate patterns, emissions profiles, and, even, possible external apportioning.

From the Iberian regions, Mediterranean has been the most widely studied, because emissions and high isolation are usually high, especially during summertime. Works of Millán et al. [11–14], Sanz and Millán [15], Alonso et al. [16], Gangoiti et al. [17, 18], and Sanz et al. [19] explained the influence of both synoptic and local coastal patterns in the ozone levels at this region. Other authors supported their studies in this region with photochemical modeling, at urban [20–22] and regional scales [23–25]. Finally, several studies were focused on topographic and land use effects [23, 26–30]. Some of their results were extended to the Cantabric [16] and Atlantic Coasts [21].

In other Iberian regions close to the northwest, some studies about tropospheric ozone were published. Different studies over rural and urban coastal areas of the Portuguese Atlantic Coast show the influence of sea breezes on the transport of ozone precursors and the production of ozone at the coastal line [31, 32], including topographic effects [33]. At rural areas, the sensitivity of ozone to BVOCs levels was also considered. Other works were focused on the Lisbon urban area [21, 34].

Cantabric Coast is also a region with significant ozone levels, mainly because the transboundary transport of ozone from the East (Central Europe). In fact, most of the studies were centered in the Cantabric Eastern Coast. Long-range transport studies [16, 18] based in field experiments were completed with the application of dispersion models [35, 36], establishing the ozone transport along stable nocturnal layers from Central Europe and, even, from Central Iberia. Finally, some studies were focused on the meteorological patterns associated to ozone peaks [37, 38].

On the other hand, at the northwest of the Iberian Peninsula, just a few of local studies were done, over industrial [41, 42] and urban areas [43]. However, due to its geographical location and climate variability, rural ozone levels at this region show a complex variability (Table 1), as they are affected by coastal meteorology, complex terrain, transboundary transport (including Atlantic background

ozone transport), and, even, local ozone production at summertime. In this work, a study of the rural ozone trends along 6 years were done, based on the analysis of meteorological and air quality time series from suburban and rural sites. The problem was focused on five sites located at the northwest of this region, although measurements from surrounding areas were also considered. Particularly, a specific classification of the sites was done, in order to guarantee a representativeness of their measurements.

2. Area under Study

For the analysis of ozone trends in the extreme northwestern corner of the Iberian Peninsula, five air quality monitoring stations were selected. We analyzed hourly O_3, NO_2, and NO_x data recorded from those sites between 2003 and 2008. This area is centered on As Pontes Valley (Figure 1); it comprises the roughly E-W-oriented lowlands around the River Eume, together with the surrounding geographic features: to the east and north mountain ranges, which reach an altitude of 1000 m; to the north, a series of hill ranges running roughly N-S from the coast, with maximum altitudes of 550–750 m; to the west, low coastal hills (<200 m) bordering the Atlantic; to the south maximum altitudes of 750–850 m; to the south east, communicating with the river Eume, the high plain of Terra Chá. Therefore, it is a complex terrain, with several granitic mountains and valleys mixed in the same environment.

The northwestern part of Iberian Peninsula has a wide-range climate because it is a transitional zone between Oceanic and Mediterranean regimes, but the area under study is a coastal Atlantic zone with mainly Atlantic climate. The Atlantic climate is characterized by mild temperatures, with small annual temperature oscillations and abundant rainfall [44]. This region is characterized by rains distributed along the year, with an annual precipitation rate between 1000 and 1600 mm, more usual during Autumn and Spring and sporadic during summer, but not unusual, as isolated storms in the afternoon. Summers are mild, as the sea breeze refreshes the coastal areas and the altitude regulates the temperature in the inner areas. Summer days are usually sunny, with low moisture and maximum temperatures from 20 to 30°C. On the other hand, heat waves are not usual in summer, and they only spend a few days, as the proximity of the coast keeps the average temperature at 20–25°C, with higher temperatures (up to 30°C) at the inner valleys. Main winds come from the SW and NW during Winter and Autumn, with low-pressure conditions; on the other hand, high-pressure conditions typical during summertime usually correspond to NE winds. The atmosphere-topography interactions are very important in this environment, specially the mountains, with their effects in rain production, and the complex coastal topography that increases the sea breeze [45].

3. Classification of Air Quality Sites

We selected seven monitoring stations (Figure 2, Table 2) as representative of the main types of stations (rural and

TABLE 1: Summary of exceedances of the ozone thresholds stablished by the Ambient Air Quality and Cleaner Air for Europe Directive [39] in five stations at the Northwestern Iberian Peninsula during the period 2003–2008. Source [40].

| Year | Daily maximum of 8-h mean > 120 $\mu g/m^3$ | | Hourly mean > 180 $\mu g/m^3$ | | |
	Number of days	3 years average	No. of days	No. of hours	No. of stations
2003	No data	—	5	11	4
2004	No data	—	3	8	3
2005	No data	—	0	0	0
2006	No data	—	0	0	0
2007	26	—	0	0	0
2008	6	—	0	0	0

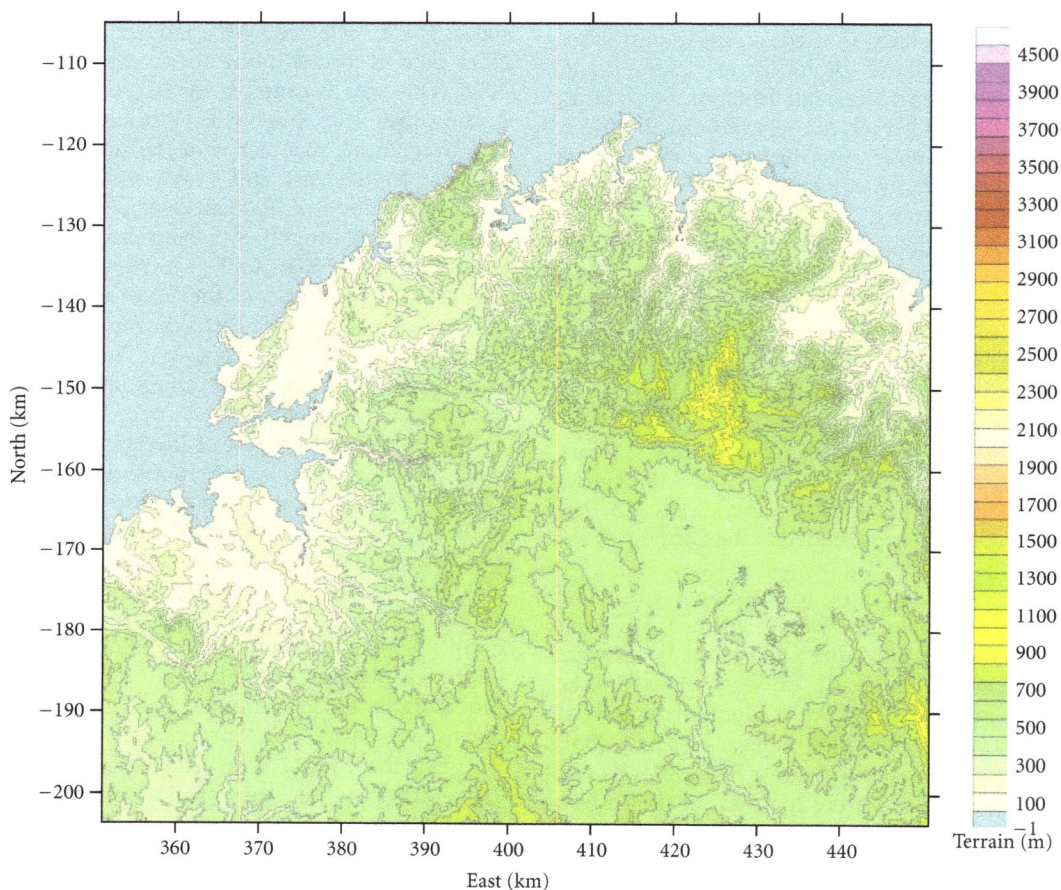

FIGURE 1: Topography of the area under study.

suburban). They cover a variety of environmental conditions ranging from near-sea level to 1000 m of altitude and from sea shore to 240 km inland.

All data analysis are based upon the hourly averaged data. Table 3 shows the percentage of ozone data coverage. The data capture is optimal in most of the stations, higher than 95% overall, except in 2007 when the percentage of valid data covers only January–August (eight months); therefore 2007 data were not applied in all the statistics.

The analysis of air quality time series has to take into account all the information about the location of the air quality measurements, in order to get representative values of the pollutants concentrations and to understand possible local phenomena that affect the ozone levels.

An air quality site can be classified depending on the pollutants levels to be measured, its main goal (environmental and health protection, research), and its surrounding environment [46]. For the purposes of this work, the last criterion is the most appropriate, as measurement should be representative of a rural or, sometimes, a suburban environment where local sources influence is well characterized. However, different classifications can be considered.

FIGURE 2: Map of the monitoring sites in NW Iberian Peninsula.

TABLE 2: Air quality stations whose data have been used for comparative purposes in this study. The coordinates are Universal Transverse Mercator (UTM) and grid zone 29T, except Peñausende and Niembro stations, in grid zone 30T.

Station	UTMx (km)	UTMy (km)	Distance to the sea (km)	Altitude (m)	Monitoring network
ES16-O Saviñao	606.1	4721.2	88.1	506	EMEP
ES08-Niembro	350.3	4811.7	0.0	134	EMEP
ES13-Peñausende	259.9	4574.3	238.0	985	EMEP
B1-A Magdalena	593.3	4811.4	28.0	363	Regional
B2-Louseiras	601.8	4821.1	18.0	540	Regional
C9-Mourence	606.0	4796.4	41.0	465	Regional
F2-Fraga Redonda	581.9	4806.3	16.0	480	Regional
G2-Vilanova	578.5	4822.9	11.0	290	Regional

(a) Decision 97/101/CE [47] proposes nine site types, as a combination of the land use (urban, suburban, and rural) and the main local sources (transport, industry, or none of them, as a background site).

(b) In addition, the topic center ETC-AQ [48] distinguished different subclasses for the background sites: urban/suburban, near-city, regional, and isolated stations.

(c) For ozone, Directive 92/72/CE [49] established three different sites: street, urban, and rural; and Directive 2008/50/EC [39] distinguishes background rural, rural, suburban, and urban sites.

For the five northwestern sites used to identify ozone peaks, Directive 2008/50/EC [39] was applied, because of our interest in identifying the rural stations following the most recent European methodology. First, a short description of every site location is introduced; after that, sites classification based on their ozone and precursors time series is shown.

3.1. Surrounding Sites Environment. B1 site (Figure 3(a)) is located at 363 asl-m, in the small town of As Pontes, surrounded by cultived and built lands; therefore, it can be considered as a suburban site. This town is located in a valley, with an SSW drainage following the River Eume, and surrounded by small elevations (500–600 asl-m).

B2 site is located in a complex terrain area (Figure 3(b)), at 540 asl-m, in the western side of a hill close to its top (601 asl-m). A narrow valley close to the site started in the SE-NW direction, changing to SSW-NNE, creating a complex terrain environment with multiple soft hills and valleys.

C9 site (Figure 3(c)) is located at the NW of Vilalba small town, close to it (1.5 km), so it is classified as suburban; in addition, several roads are near this site. However, the surroundings are mainly grass and cultivated land. It is located in a plateau (465 asl-m) with elevations (1000 asl-m) at 11 km to the northeast; this feature is significant, as typical northeastern dry winds in the region are softer at this site.

G2 site (Figure 3(d)) seems to be located in a rural area, with grass and cultivated land, and wood production. However, this area is surrounded by coastal towns and industrial areas, with a significant road network; so, emissions from these sources affect this site, changing its typical rural air quality pattern. About its topography, the site is located in

TABLE 3: Percentage of valid hourly O_3 measurements per year and station. Data of the the year 2007 are refered to its first eight months.

Station	Data capture (%)					
	2002	2003	2004	2005	2006	2007[*]
B1-A Magdalena	47.9	99.1	95.0	92.7	98.1	97.3
B2-Louseiras	94.7	96.5	95.1	96.7	98.5	99.5
C9-Mourence	96.4	98.8	99.4	98.3	98.8	99.0
F2-Fraga Redonda	95.7	98.7	94.6	96.1	98.7	96.6
G2-Vilanova	50.2	99.4	98.3	99.2	98.7	97.2
O Saviñao	97.7	95.9	97.6	97.9	96.6	96.3
Niembro	97.5	92.2	97.5	97.6	98.5	99.0
Peñausende	93.3	92.4	97.7	96.7	97.1	98.1

[*] Year 2007 only includes the first eight months.

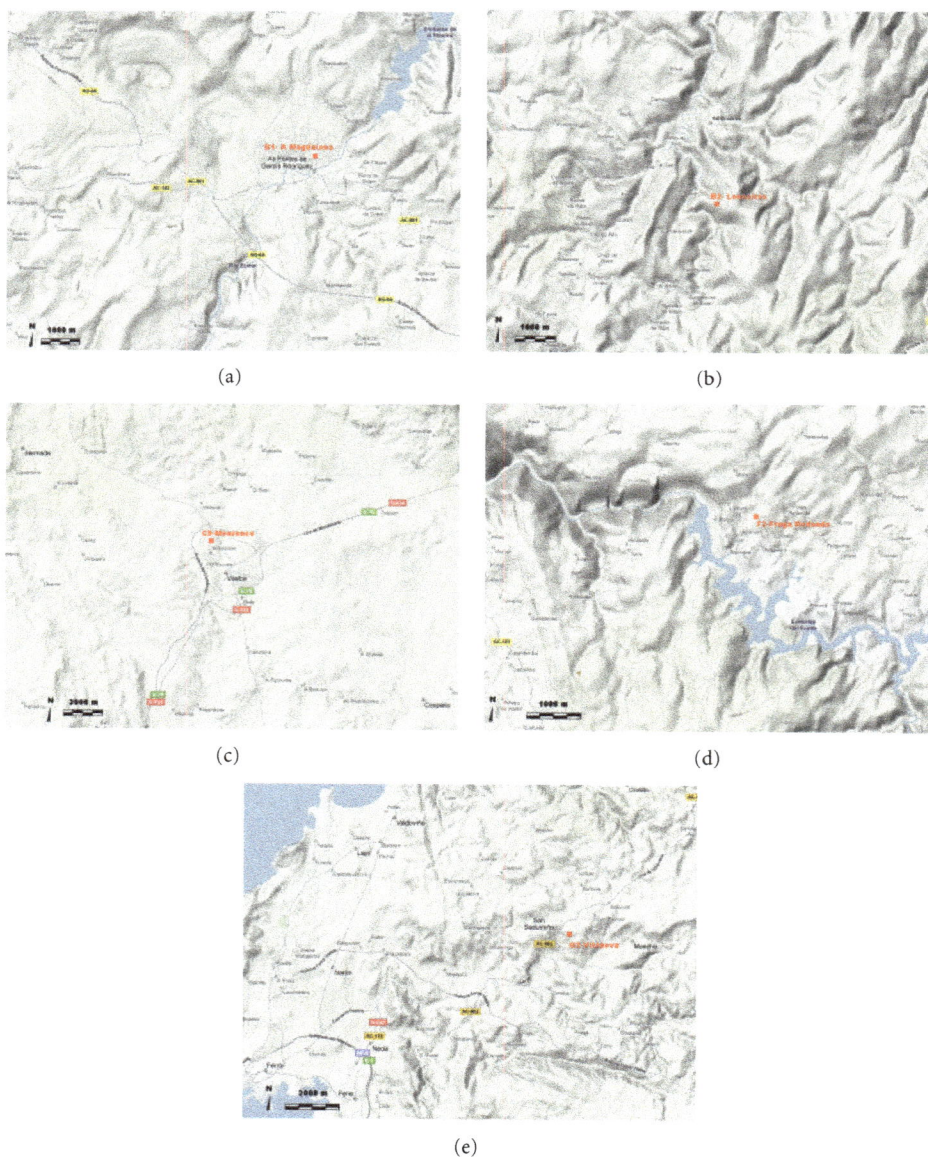

(a)

(b)

(c)

(d)

(e)

FIGURE 3: Geographic location and topography of the analyzed Galician air quality stations.

a valley (290 asl-m) with soft hills around it; the closest tops are located to the south at 7 km (600 asl-m) and to the east at 14 km (800 asl-m). However, it is the proximity to the coast the main factor that affect the wind on this site, with sea breezes that mix to the soft valley breezes.

F2 site (Figure 3(e)) is located in an Atlantic forest natural park, a typical rural area, close to the Eume river dam. This zone is dominated by the complex terrain, with the site located in the east side of the dam valley (480 asl-m), that crosses from East to West and, following the river, to the coast. Winds are dominated by valley and sea breezes that usually come from the coast, 15 km far, following the river path; however, they are strongly affected by the local topography, as it will be shown.

3.2. Sites Classification.

3.2. *Sites Classification.* In this work, the standard EU air quality sites level 1 classification [49] was applied, which considers the human activities, land use, and local emissions surrounding the site. Apart from this subjective classification based on the direct analysis of the local environment around the site, an objective analysis of air quality data can help to establish the site type [46]. For O_3 sites, Fromage [50] used the O_3/O_x ratio, with $O_x = O_3 + NO_2$, that is,

(i) urban sites: low O_3/O_x ratio along the year, from 0.10 (winter) to 0.50 (summer), due to the strong influence of the local NO_x emissions;

(ii) rural-regional sites: high variability in the O_3/O_x ratio, from 0.20 (winter) to 0.95 (summer), that is, O_3 is significant in summer, but in winter the urban NO_x emissions near the ratio;

(iii) background rural sites: O_3/O_x is close to 1.00 along the year.

Figure 4 shows the results of this analysis applied to the five sites described above, for the period January 2002 to August 2007. At the same time, classification based in the sites environment [51] shows different results, as it is shown on Table 4.

The objective classification can facilitate the accurately setting of each station to a given category, only based on the analysis of the data series of O_3 and NO_2, which is very useful when working with sites in unknown environments. However, the use of the subjective classification requires a thorough knowledge of the surroundings of the station, which is not always available. Moreover, the types of stations considered in the objective classification (urban, rural-regional, and rural background) are too generic to characterize a site, especially for rural stations-regional type: the criterion of high variability in the O_3/O_x ratio (0.20 to 0.95) seems too loose and bring together some sites affected by different factors (proximity to a city, a highway, or industrial sources) that should be taken into account in the analysis of ozone concentrations. This is an advantage of the subjective classification which directly considers the influence of these factors.

In the analyzed sites (Table 4), both classifications show significant discrepancies in four of the eight stations (highlights O Saviñao and G2-Vilanova stations). According to

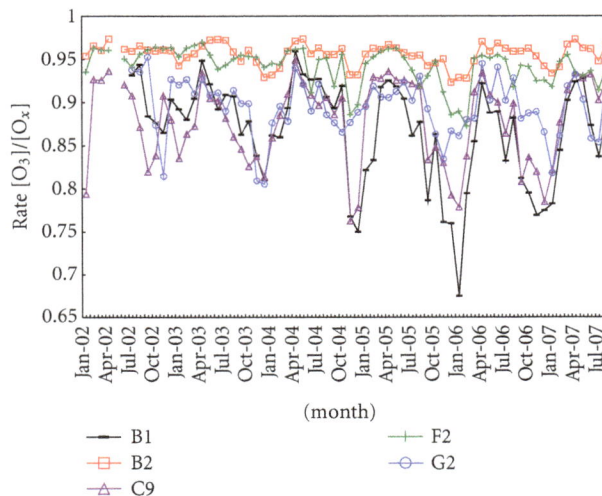

FIGURE 4: Monthly rate $[O_3]/[O_x]$, with $O_x = O_3 + NO_2$, at five monitoring stations (B1-A Magdalena, B2-Louseiras, C9-Mourence, F2-Fraga Redond, and G2-Vilanova) during the period from January 2002 to August 2007.

the subjective classification, O Saviñao is an EMEP site so it is a "rural background site" by definition, but objective classification assigns it to the "rural-regional" type. In this case, we have doubts about the rightness of the subjective classification applied to this EMEP station because O Saviñao often shows a marked diurnal cycle in ozone ground level concentration. Therefore, it is likely this station can be affected by a nearby unknown anthropogenic emission source, despite being listed as EMEP site. G2-Vilanova is considered "rural-background" according to the objective classification despite that site is affected by emissions from a nearby wood board factory. Other two stations that could be discussed are B1-A Magdalena and C9-Mourence, characterized as "rural-regional" following the objective classification, despite being suburban stations affected by traffic emissions, as evidenced by their daily ozone cycles. Therefore, even though the disadvantages of application of the subjective classification, in this work, it was preferred to the objective classification, because of the accurate knowledge of the sites environment and the limited influence on the surrounding sources in this region. For the complementary air quality sites applied in this work, all of them are EMEP sites; so they are classified as background rural sites.

Because the complex terrain in the area under study can affect the ozone dynamic, as it was studied in other regions [14], a level 2 classification dependant on the topography (Table 5) was also done. Finally, following the Decission 97/101/CE [47], a classification by influence, based on the main emission sources around the sites, was done. Results of the three classifications (level 1, level 2, and influence) are summarized in Table 6. Again, for other sites different than the reference sites, classifications provided by the network managers were applied.

TABLE 4: Classification of air quality monitoring stations following the *subjective* [51] and *objective* level 1 classification [46].

Station	Subjective classification [51]		Objective classification [46]		
	Type	Influence	Type	$[O_3]/[O_x]$ ratio	
				Summer (max.)	Winter (min.)
B1-AMagdalena	Suburban	Traffic	Rural-regional	0.95	0.67
B2-Louseiras	Rural	Background	Background rural	0.97	0.92
C9-Mourence	Suburban	Traffic	Rural-regional	0.95	0.76
F2-Fraga Redonda	Rural	Background	Background rural	0.97	0.87
G2-Vilanova	Rural	Industrial	Background rural	0.95	0.81
O Saviñao (EMEP)	Rural	Background	Rural-regional	0.96	0.75
Niembro (EMEP)	Rural	Background	Background rural	0.97	0.80
Peñausende (EMEP)	Rural	Background	Background rural	0.98	0.80

TABLE 5: Classification of air quality monitoring stations, namely, Level 2, according to their topographic environment.

Level 2	Orographic location
EI	Elevated inland
IC	Intermediate elevated mountain slope between coast and inland
CL	Coastal low altitude
IM	Intermediate elevated mountain slope at inland

TABLE 6: Final classification of air quality monitoring stations; measurements from 2002 to 2007 were used for statistical ozone analysis.

Station	Level 1	Level 2	Influence
B1-AMagdalena	Suburban	IM	Traffic
B2-Louseiras	Rural	IC	Background
C9-Mourence	Suburban	IM	Traffic
F2-Fraga Redonda	Rural	IC	Background
G2-Vilanova	Rural	CL	Industrial
O Saviñao	Rural	IM	Background
Niembro	Rural	CL	Background
Peñausende	Rural	EI	Background

4. Results

For the analysis of ozone trends in this region, first possible episodic relationships between the different reference sites are obtained. After that, general trends (annual, monthly, and daily) are derived from the sites measurements.

4.1. Sites Relationships. The geographical location and characteristics of the five sites considered to identify ozone episodes in this region can drive to establish relationships between the ozone levels observed at all of them, which can affect the study of ozone trends. This dependence between sites was estimated by the Pearson correlation coefficient (r) of the hourly ozone measurements along ozone episodes from 2002 to 2007 [52]. Results are shown in Table 7.

TABLE 7: Correlations of ozone levels between the five air monitoring stations during ozone episodes, using the Pearson correlation coefficient r. Periods containing less than 80% of the total data were withdrawn from the analysis.

Pair of stations	Pearson coefficient (r)			
	Mean	Standard deviation	Min.	Max.
B1-B2	0.48	0.18	0.18	0.85
B1-C9	0.87	0.05	0.68	0.93
B1-F2	0.63	0.17	0.19	0.90
B1-G2	0.68	0.12	0.33	0.87
B2-C9	0.49	0.16	0.09	0.77
B2-F2	0.78	0.15	0.17	0.92
B2-G2	0.68	0.16	0.29	0.89
C9-F2	0.61	0.14	0.20	0.82
C9-G2	0.63	0.11	0.32	0.81
F2-G2	0.72	0.13	0.44	0.92

The highest correlation (Table 7) is obtained between B1 and C9 site measurements, located in suburban environments, with similar altitude and traffic emissions influence. Then, their measurements will be less useful for the study of rural ozone trends.

Lower but significant correlation (Table 7) is obtained between B2 and F2 site measurements, as rural sites with similar surroundings; however, their specific locations are different, with F2 on a valley side, and B2 at the top of a hill. Even though their measurements follow similar patterns, local phenomena and different sources influence could affect their ozone levels. Therefore, their measurements will be analyzed separately.

F2 and G2 site measurements show lower but significant correlation (Table 7), as both can be considered as background stations. However, their different topography and some influence of traffic emissions at G2 site produce those differences.

Other sites show poor correlations along all the episodes. However, in some of them B1, B2, and C9 sites showed high correlations, which can indicate a general rise of ozone levels in the region, not related to local effects site by site. This feature could be considered for episodic analysis in this region.

4.2. Seasonal and Interannual Trends of Tropospheric Ozone.

For this analysis, ozone measurements in the period 2002–2006 from the five reference sites and the three EMEP at the northwest of the Iberian Peninsula were considered.

For interannual trend, annual averages, 50 and 80 percentiles, were considered. Results are shown in Figure 5. Mean value along the 5 years moves between 50 and 73 $\mu g/m^3$ for the five reference sites, which are lower to the Peñausende site mean value (77 $\mu g/m^3$). On the other hand, F2, G2, Niembro, and O Saviñao sites show mean values around 63 $\mu g/m^3$, which are higher than B1 and C9 suburban sites mean values (58 $\mu g/m^3$), as expected.

Annual means of the maximum hourly data (Figure 6) show a range of 82–88 $\mu g/m^3$ at the five reference sites; as for the mean value, this range is lower to the mean of maximum at Peñausende (96 $\mu g/m^3$).

For the reference sites, both mean and maximum ozone trends are similar, keeping small variations at rural stations; except on the year 2003, due to the effects of the forest fires in the Iberian Peninsula and the summer heat wave over Western Europe [53, 54]. Previous annual trends in this region [41, 55] show similar results.

About EMEP sites, Peñausende shows the highest levels, due to its altitude (985 m) that reduces the ozone destruction by deposition and NO surface reaction, and increases its production because of the higher solar radiation [26, 56–58]. In fact, relatively higher means at B2 site should be related to its higher altitude, respect to the other reference sites, and the lower deposition over coastal areas. This effect is also observed at Niembro site, with a high annual mean; however, Niembro shows a low maximum daily annual mean, because of its low NO_x level. O Saviñao shows an opposite behavior, because of the higher deposition in this inland location.

G2 site shows a behavior similar to Niembro, because of its proximity to the coast; in the middle, between Niembro and O Saviñao, we can find B2 and F2 sites. Finally, B1 and C9, as inland suburban sites with traffic NO emissions and lower deposition close to the sea [59–61], show annual means lower than in rural sites, while maximum annual means are higher than F2 and G2 values; only B2 gets both higher annual and maximum annual means higher, because of its higher altitude.

Interannual trends were studied following a nonparametric test of Mann-Kendall [62], with 50th and 98th percentiles. The first percentile shows lower sensitivity to changes in emissions than the last one. 50th percentiles (Figure 7) show an increment at Niembro and a small reduction at Peñausende, considering the maximum achieved in 2003. More significant is the decrease at B1 site.

Again, 98th percentiles (Figure 8) show a significant decrease at B1 site. These results are in agreement to the small variations of annual means but show a small trend to the ozone levels reduction. However, other studies covering global European trends from 1990 to 2004 [63, 64] show a 50th percentile stable or a bit rised (especially, during winter), and significant reductions in 98th percentile. Both European trends are related to the general NO_x emissions reduction, that reduces both the ozone peaks [8, 9], and the available NO to destruct ozone during winter [65–67],

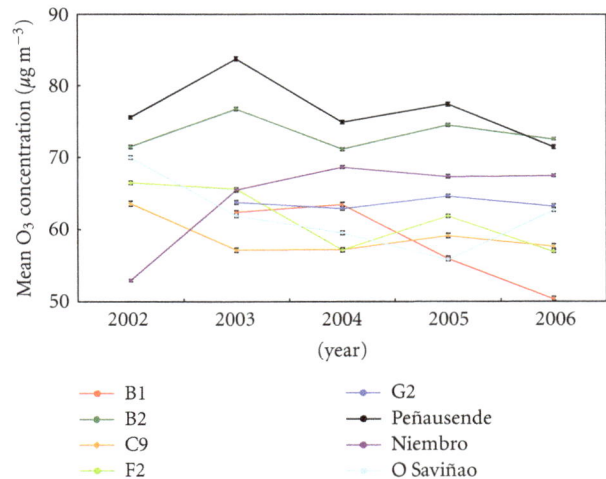

FIGURE 5: Series of annual mean 24-h ozone concentrations throughout the 5-year-study period (2002–2006) for each site.

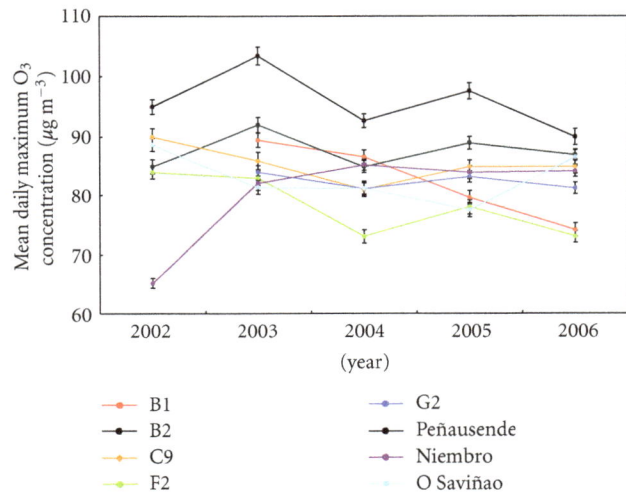

FIGURE 6: Series of annual mean of daily maximum ozone concentrations throughout the 5-year-study period (2002–2006) for each site.

increasing the background ozone, although this trend seems to be changed after the year 2000 [5].

These differences in the ozone trends between the northwest of the Iberian Peninsula and the rest of Europe can be explained because of the increment of NO_x emissions at the Iberian Peninsula: from 1990 to 2006 NO_x, emissions have increased 19% at Spain and 5% at Portugal [68, 69]. This can explain the stability of annual means and percentile 50 at this region, because of the levels stability.

The slight decline in the 98th percentile between 2002 and 2006 is in agreement with the small but steady reduction of the peak ozone concentrations recorded in events analyzed between 2002 and 2007. Three quarters of these high-ozone events in Northern Galicia were associated with synoptic atmospheric conditions characterized by a blocking anticyclone to the north of the Iberian Peninsula and a

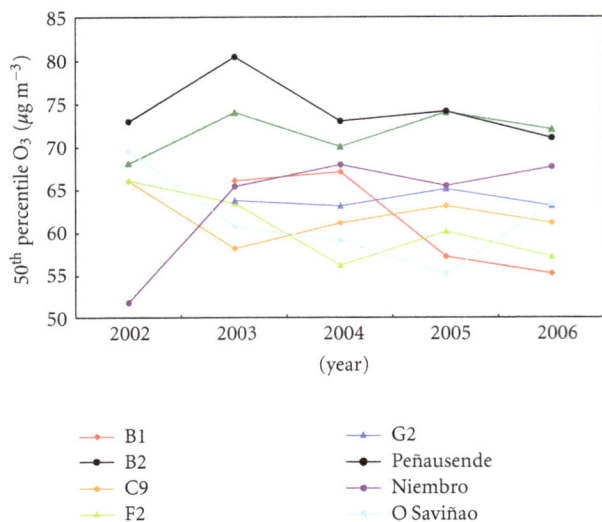

FIGURE 7: Internannual variability of tropospheric ozone concentration represented by the 50th percentile throughout the 5-year-study period (2002–2006) for each site.

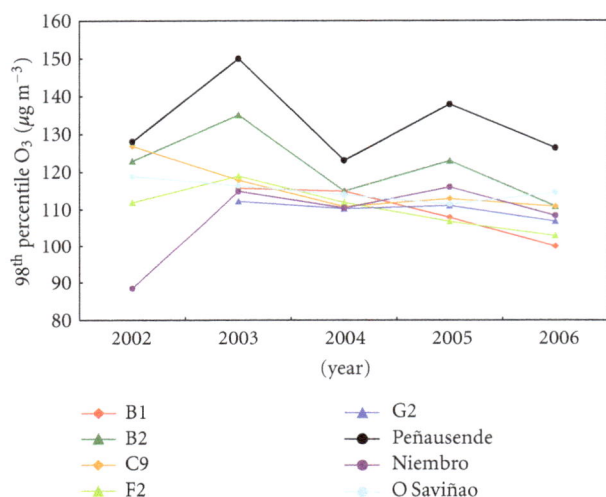

FIGURE 8: Internannual variability of tropospheric ozone concentration represented by the 98th percentile throughout the 5-year-study period (2002–2006) for each site.

high-altitude ridge of high pressure that extended across the Peninsula from Africa, following Saavedra et al. [52]. This synoptic situation causes a prevailing eastern-southeastern synoptic circulation, characterized by high temperatures, low humidity, and light wind speed. Keeping in mind that the distribution of pollutants is not only dependent on the spread of its emissions but is also affected by various weather/climatic drivers cannot be excluded that the decrease in 50th and 98th percentiles is also affected by changes in the large-scale circulation patterns over the NW Iberian Peninsula [70, 71], in turn related to the real-physical local circulations and weather types. Changes in the frequency of synoptic patterns favorable for high-ozone events could

contribute significantly to the alteration of ozone trends [72]. However, the time series analyzed in this work are not long enough to establish a relationship between changes in tropospheric ozone concentrations with alterations in global synoptic circulation patterns.

The annual cycle of tropospheric ozone at Northern Hemisphere has been widely studied, with seasonal variations and patterns related to the latitude and altitude [60, 73, 74]. At Europe, maximum levels are usually achieved in summer at Mediterranean regions and Central Europe because of photochemical production [75]. Otherwise, spring peaks are more usual at the Atlantic coast and the West of Europe [60, 76], as a general trend in the Northern Hemisphere [76–80].

In the region under study, the five reference sites, Niembro and O Saviñao EMEP sites, show the typical seasonal cycle of boundary layer ozone for monthly daily mean (Figure 9) and monthly maximum mean (Figure 10) in low polluted regions of Northern Hemisphere. Both figures show a maximum on April, a relative minimum on July, and a minimum during winter. It is the same behavior of the ozone levels at Mace Head site and other Northern and Western European sites [5, 60, 65, 76, 81, 82]. In fact, there are evidences of the reduction of ozone levels at Mace Head after the year 2000 [5], which is also observed in the sites under study. At the same time, a significant contribution of transoceanic ozone was observed at Mace Head, as in the West Iberia [83], with some additional contribution of the stratospheric jet. The influence of ozone transported from the free troposphere toward the surface due to troposheric folding by high frequencies of troughs and cut-off lows in the Atlantic Iberian Coast must be taken into account, due to the high frecuency of these closed lows in the vicinity of the area under study, specially in summer [84, 85], and its clear impact over the surface ozone levels [55, 86–89]. This similar behavior of the sites under study can be explained because all these West Coast European sites are low sensitive to the reduction of NO_x emissions at Central Europe [76].

A different behavior is observed at Peñausende site (987 asl-m) showing a typical pattern of elevated inland sites, with maximum ozone in June, persistence during summer, and slow decrease to the winter minimum [60]. These elevated sites are affected by the ozone at the free troposphere or, at least, accumulated in the upper boundary layer.

Daily maximum monthly cycle (Figure 9) shows a similar behavior to the daily mean values cycle, especially at rural sites. Also B1 and C9 sites, with significant anthropogenic influence, show the highest values in winter, as the ozone destruction by the NO emissions does not affect the daily maximum.

4.3. Daily Cycle Variability of Tropospheric Ozone. Summer is the best season for studying the daily cycle of tropospheric ozone [90, 91], as the period with the most favorable meteorological conditions for ozone peaks: high solar radiation and thermal inversions. From the analysis of the summertime daily cycles in the selected sites (Figure 11), three different patterns were identified.

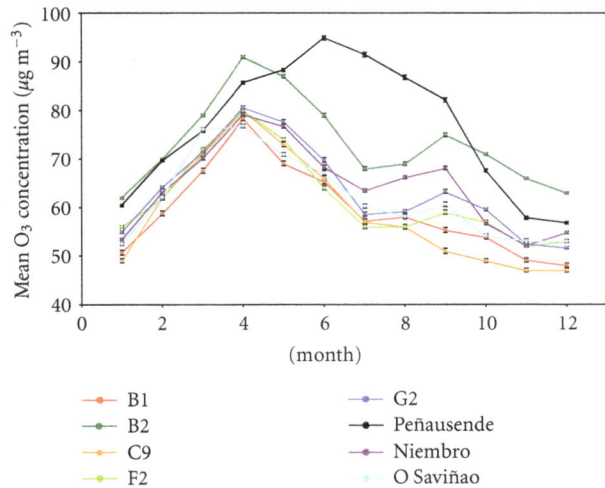

FIGURE 9: Series of monthly mean 24-h ozone concentrations throughout the 5-year-study period (2002–2006) for each site.

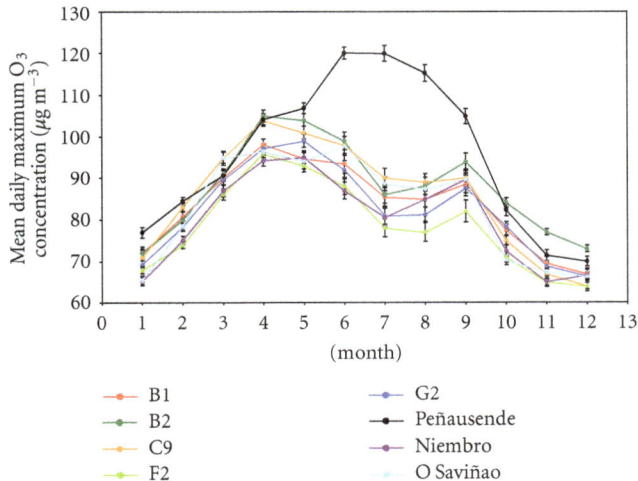

FIGURE 11: Series of hourly mean ozone concentrations in summertime (June, July, and August) throughout the 5-year-study period (2002–2006) for each site.

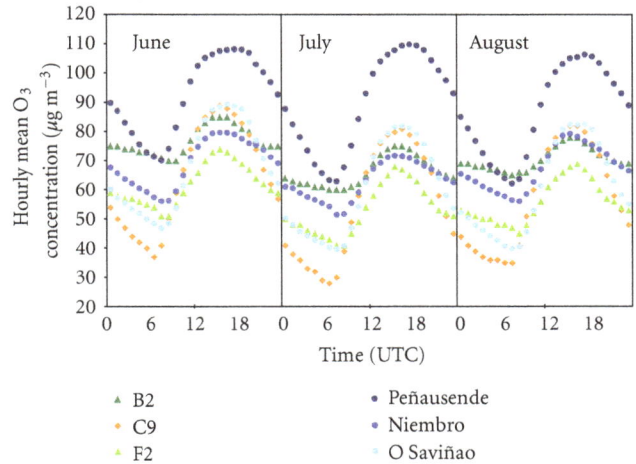

FIGURE 10: Series of monthly mean of daily maximum ozone concentrations throughout the 5-year-study period (2002–2006) for each site.

(a) *Oscillated Cycle.* A daily cycle with strong oscillations, starting with a minimum at sunrise, with a quick increment along the morning due to the surface mixing (breaking the nocturnal thermal inversion) and the photochemical production. This trend is slower after noon, as the mixing layer growth, reaching the ozone peak around 17 UTC; after then, ozone level quickly falls down. This cycle is typical at Peñausende and O Saviñao sites, two inland rural sites with significant altitudes.

(b) *Suburban Cycle.* A similar but less oscillating cycle, starting again with a minimum at sunrise, with a quick increment until 15 UTC, and a softer decrease along the afternoon. Although this cycle could be similar to the previous one, it is typical at suburban sites (B1, C9), and related to the local NO emissions

that destruct the ozone at night, and to a faster growth of the nocturnal stable layer, which is more favorable to the ozone elimination.

(c) *Nonoscillated Cycle.* A soft daily cycle, with a minimum around sunrise, but similar to the previous nocturnal levels. The ozone peak can be achieved between 14 and 16 UTC, falling down quickly until 22 UTC, when the nocturnal level keeps quite constant. B2 and Niembro sites clearly follow this pattern; F2 and G2 sites show a bit strong oscillations, as an intermediate behavior between soft and suburban cycle.

In this region, coastal sites show a soft daily cycle, due to the winds around them which promote instability [61, 92], mixing the surface ozone. In addition, if the NO_x nocturnal levels at these sites are low, the ozone destruction by NO is slow [59], increasing the ozone levels. This unstability nocturnal effect is especially significant at B2 site, because of its altitude favorable to strong nocturnal winds to the valley. In fact, in the hourly frequency of ozone peaks (Figure 12), small nocturnal ozone peaks can be observed, due to the mixing between surface and aloft layers with ozone [93].

About the absolute daytime ozone peaks, B2, G2, and Niembro sites show peaks around 14-15 UTC, which is earlier to the inland peaks at Peñausende (17-18 UTC). As daytime peaks mainly depend on photochemical production, and solar irradiance is similar in these sites, only the higher maximum summertime temperatures at Peñausende [94, 95] can explain this difference in ozone production.

At night, relative differences between coastal and inland sites are even more significant: B2, F2 and Niembro sites (Figure 12) show significant frequencies of nocturnal peaks (20–25%), whereas C9 and O Saviñao inland sites show less than 8%. These nocturnal peaks appear in the interval 21-00 UTC, when mixing layer depth is decreasing and nocturnal stable layer is growing [96]. At B2 and F2 sites,

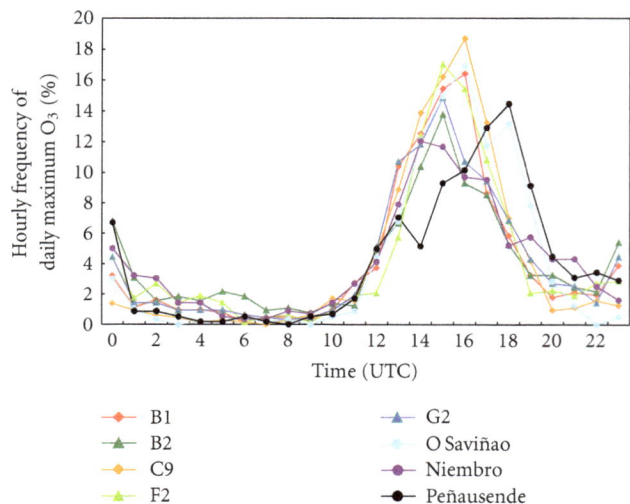

FIGURE 12: Frequency distribution of daily maximum hourly ozone in summertime (June–August) during the period 2002–2006.

this behavior can be explained by the downhill nocturnal winds, which carry ozone from upper layers [58, 93]. The pure coastal Niembro site is affected by the inland nocturnal breezes, which carry ozone from the top of hills close to this site.

Some nocturnal peaks can be also observed at Peñausende inland site, due to its high altitude and surrounding mountains, which develop the same phenomena as at B2 and F2 sites.

All these nocturnal differences are enhanced by the low destruction of ozone at rural sites, as B2, F2, Niembro, and Peñausende [97].

5. Conclusions

Long-term trends of tropospheric ozone studied in the past over diverse regions of Western Europe show a different behavior of Atlantic sites. Particularly, different trends were identified at several coastal regions of the Iberian Peninsula (Mediterranean, Cantabric, Southwestern). In this work, trends at northwestern part of this Peninsula from 2002 to 2007 were analyzed and compared to previous studies over boundary regions.

About annual trends, both mean and maximum values are quite stable, except in 2003, because of the influence of forest fire emissions over the Iberian Peninsula, and the summertime heat wave over Western Europe. 98th percentile shows a small ozone reduction, in agreement to the stronger global reduction over Europe. Smaller reduction over the region under study can be explained by the increment of NO_x emissions from Spain and Portugal during that period. Seasonal trend in this region is in agreement to trends in low polluted regions of Northern Hemisphere, because of the significant contribution of transoceanic pollution already identified in other reference sites (i.e., Mace Head). In fact, an inland site close to this region shows a different behavior.

About daily ozone cycle, from the analysis of summertime series, three different patterns were identified: oscillated, nonoscillated, and suburban, as intermediate between both. In this region, oscillated cycle was typical in inland elevated sites; nonoscillated cycle was observed in coastal sites (up to 30 km far from the sea), with more oscillation as the site was far from the coast; and suburban cycle was observed in coastal sites with NO local emissions; maximum daily ozone during summertime confirms this trend. Apart from it, the observation of relative peaks of nocturnal ozone at elevated sites is a specific phenomena that is usually associated to the injection of ozone from upper tropospheric layers; so it is not related to the photochemical activity in this region.

Considering just the available rural stations, the ozone trends observed in this study could be extrapolated to other rural areas of NW Iberian Peninsula, highlighting the Galician Coast and the northern coast of Portugal, because this geographical area has remarkably homogeneous topographical and climatological characteristics. In addition, the Galician-Portuguese Atlantic Coast is affected by the same phenomena of long-range pollutant transport, due to synoptic circulation patterns characteristic of this area. However, these ozone trends are linked to the interannual meteorological variability, which suffer significant changes during the period under study; in addition, changes in the photochemical precursor emissions over this region and its surroundings also affect those trends. The analysis of future ozone trends in this region should be very useful to confirm the influence of these factors.

Because of the small number of representative sites and the short time series available for this analysis, the systematic application of regional air quality models [98] over this region would help to improve and extend the understanding of regional surface ozone trends.

Acknowledgments

This work has been financially supported by the Spanish Research & Development Programme, Ministry of Science and Technology, under Project CTQ15481-PPQ and Endesa Generación S. A. Research grant of the "María Barbeito" Programme (Xunta de Galicia) to Á. Rodríguez is acknowledged.

References

[1] G. Brasseur, R. G. Prinn, and A. A. P. Pszenny, *Atmospheric chemistry in a changing world: an integration and synthesis of a decade of tropospheric chemistry research: the International Global Atmospheric Chemistry Project of the International Geosphere-Biosphere Programme*, Global Change-The IGBP Series, Springer, Berlin, Germany, 2003.

[2] J. H. Seinfeld and S. N. Pandis, *Atmospheric Chemistry and Physics: From Air Pollution to Climate Change*, John Wiley & Sons, New York, NY, USA, 1998.

[3] H. W. Y. Wu and L. Y. Chan, "Surface ozone trends in Hong Kong in 1985–1995," *Environment International*, vol. 26, no. 4, pp. 213–222, 2001.

[4] J. Lelieved, J. Van Aardenne, H. Fischer, M. De Reus, J. Williams, and P. Winkler, "Increasing ozone over the Atlantic Ocean," *Science*, vol. 304, no. 5676, pp. 1483–1487, 2004.

[5] R. G. Derwent, P. G. Simmonds, A. J. Manning, and T. G. Spain, "Trends over a 20-year period from 1987 to 2007 in surface ozone at the atmospheric research station, Mace Head, Ireland," *Atmospheric Environment*, vol. 41, no. 39, pp. 9091–9098, 2007.

[6] D. Jaffe and J. Ray, "Increase in surface ozone at rural sites in the western US," *Atmospheric Environment*, vol. 41, no. 26, pp. 5452–5463, 2007.

[7] National Expert Group on Transboundary Air Pollution (NETGAP), "Transboundary air pollution: acidification, eutrophication and ground-level ozone in the UK," 2001, http://www.freshwaters.org.uk/resources/documents/negtap_2001_final_report.php.

[8] R. G. Derwent, M. E. Jenkin, S. M. Saunders et al., "Photochemical ozone formation in north west Europe and its control," *Atmospheric Environment*, vol. 37, no. 14, pp. 1983–1991, 2003.

[9] S. Solberg, R. Bergström, J. Langner, T. Laurila, and A. Lindskog, "Changes in Nordic surface ozone episodes due to European emission reductions in the 1990s," *Atmospheric Environment*, vol. 39, no. 1, pp. 179–192, 2005.

[10] M. Amann, D. Derwent, B. Forsberg et al., *Health risks of ozone from long-range transboundary air pollution*, World Meteorological Organization (WMO), Regional Office for Europe, 2008.

[11] M. M. Millán, B. Artinano, L. Alonso, and M. Navazo, "The effect of meso-scale flows on regional and long-range atmospheric transport in the western Mediterranean area," *Atmospheric Environment - Part A General Topics*, vol. 25, no. 5-6, pp. 949–963, 1991.

[12] M. M. Millán, R. Salvador, E. Mantilla, and B. Artíñano, "Meteorology and photochemical air pollution in Southern Europe: experimental results from EC research projects," *Atmospheric Environment*, vol. 30, no. 12, pp. 1909–1924, 1996.

[13] M. M. Millán, R. Salvador, E. Mantilla, and G. Kallos, "Photooxidant dynamics in the Mediterranean basin in summer: results from European research projects," *Journal of Geophysical Research D*, vol. 102, no. 7, pp. 8811–8823, 1997.

[14] M. M. Millán, E. Mantilla, R. Salvador et al., "Ozone cycles in the western Mediterranean basin: interpretation of monitoring data in complex coastal terrain," *Journal of Applied Meteorology*, vol. 39, no. 4, pp. 487–508, 2000.

[15] M. J. Sanz and M. M. Millán, "The dynamics of aged airmasses and ozone in the western Mediterranean: relevance to forest ecosystems," *Chemosphere*, vol. 36, no. 4-5, pp. 1089–1094, 1998.

[16] L. Alonso, G. Gangoiti, M. Navazo, M. M. Millán, and E. Mantilla, "Transport of tropospheric ozone over the Bay of Biscay and the Eastern Cantabrian Coast of Spain," *Journal of Applied Meteorology*, vol. 39, no. 4, pp. 475–486, 2000.

[17] G. Gangoiti, M. M. Millán, R. Salvador, and E. Mantilla, "Long-range transport and re-circulation of pollutants in the western Mediterranean during the project Regional Cycles of Air Pollution in the West-Central Mediterranean Area," *Atmospheric Environment*, vol. 35, no. 36, pp. 6267–6276, 2001.

[18] G. Gangoiti, L. Alonso, M. Navazo et al., "Regional transport of pollutants over the Bay of Biscay: analysis of an ozone episode under a blocking anticyclone in west-central Europe," *Atmospheric Environment*, vol. 36, no. 8, pp. 1349–1361, 2002.

[19] M. J. Sanz, A. Carratalá, C. Gimeno, and M. M. Millán, "Atmospheric nitrogen deposition on the east coast of Spain: relevance of dry deposition in semi-arid Mediterranean regions," *Environmental Pollution*, vol. 118, no. 2, pp. 259–272, 2002.

[20] I. Toll and J. M. Baldasano, "Modeling of photochemical air pollution in the Barcelona area with highly disaggregated anthropogenic and biogenic emissions," *Atmospheric Environment*, vol. 34, no. 19, pp. 3069–3084, 2000.

[21] N. Barros, C. Borrego, I. Toll, C. Soriano, P. Jiménez, and J. M. Baldasano, "Urban photochemical pollution in the Iberian Peninsula: Lisbon and Barcelona airsheds," *Journal of the Air and Waste Management Association*, vol. 53, no. 3, pp. 347–359, 2003.

[22] S. Ortega, M. R. Soler, J. Beneito, and D. Pino, "Evaluation of two ozone air quality modelling systems," *Atmospheric Chemistry and Physics*, vol. 4, no. 5, pp. 1389–1398, 2004.

[23] P. Jiménez, R. Parra, S. Gassó, and J. M. Baldasano, "Modeling the ozone weekend effect in very complex terrains: a case study in the Northeastern Iberian Peninsula," *Atmospheric Environment*, vol. 39, no. 3, pp. 429–444, 2005.

[24] P. Jiménez, J. Lelieveld, and J. M. Baldasano, "Multiscale modeling of air pollutants dynamics in the northwestern Mediterranean basin during a typical summertime episode," *Journal of Geophysical Research D*, vol. 111, no. 18, Article ID D18306, 2006.

[25] M. Gonçalves, P. Jiménez-Guerrero, and J. M. Baldasano, "Contribution of atmospheric processes affecting the dynamics of air pollution in South-Western Europe during a typical summertime photochemical episode," *Atmospheric Chemistry and Physics*, vol. 9, no. 3, pp. 849–864, 2009.

[26] À. Ribas and J. Peñuelas, "Surface ozone mixing ratio increase with altitude in a transect in the Catalan Pyrenees," *Atmospheric Environment*, vol. 40, no. 38, pp. 7308–7315, 2006.

[27] M. R. Soler, J. Hinojosa, M. Bravo, D. Pino, and J. V. G. de Arellano, "Analyzing the basic features of different complex terrain flows by means of a Doppler SODAR and a numerical model: Some implications for air pollution problems," *Meteorology and Atmospheric Physics*, vol. 85, no. 1–3, pp. 141–154, 2004.

[28] M. R. Soler, S. Ortega, C. Soriano, D. Pino, M. Alarcón, and J. Aymamí, "Study of pollutant transport in complex terrain using different meteorological and photochemical modelling systems," in *9th International Conference on Harmonisation within Atmospheric Dispersion Modelling for Regulatory Purposes*, vol. 2, pp. 310–314, Garmisch-Partenkirchen, Germany, June 2004.

[29] S. Caballero, N. Galindo, C. Pastor, M. Varea, and J. Crespo, "Estimated tropospheric ozone levels on the southeast Spanish Mediterranean coast," *Atmospheric Environment*, vol. 41, no. 13, pp. 2881–2886, 2007.

[30] N. Castell, E. Mantilla, and M. M. Millan, "Analysis of tropospheric ozone concentration on a Western Mediterranean site: Castellon (Spain)," *Environmental Monitoring and Assessment*, vol. 136, no. 1–3, pp. 3–11, 2008.

[31] M. G. Evtyugina, T. Nunes, C. Pio, and C. S. Costa, "Photochemical pollution under sea breeze conditions, during summer, at the Portuguese West Coast," *Atmospheric Environment*, vol. 40, no. 33, pp. 6277–6293, 2006.

[32] M. G. Evtyugina, C. Pio, T. Nunes, P. G. Pinho, and C. S. Costa, "Photochemical ozone formation at Portugal West Coast under sea breeze conditions as assessed by master chemical mechanism model," *Atmospheric Environment*, vol. 41, no. 10, pp. 2171–2182, 2007.

[33] A. C. Carvalho, A. Carvalho, I. Gelpi et al., "Influence of topography and land use on pollutants dispersion in the Atlantic coast of Iberian Peninsula," *Atmospheric Environment*, vol. 40, no. 21, pp. 3969–3982, 2006.

[34] C. Borrego, A. I. Miranda, A. C. Carvalho, and C. Fernández, "Climate change impact on the air quality: the portuguese case," *Global Nest*, vol. 2, no. 2, pp. 199–208, 2000.

[35] G. Gangoiti, A. Albizuri, L. Alonso et al., "Sub-continental transport mechanisms and pathways during two ozone episodes in northern Spain," *Atmospheric Chemistry and Physics*, vol. 6, no. 6, pp. 1469–1484, 2006.

[36] V. Valdenebro, G. Gangoiti, A. Albizuri et al., "Evolution of the ozone episodes in Northern Iberia (Cantabric and Pyrenaic regions) under West European Atlantic blocking anticyclones," in *Proceedings of the 29th NATO/CCMS International Technical Meeting on Air Pollution Modeling and Its Application*, pp. 671–672, Aveiro, Portugal, 2008.

[37] A. Albizuri, "Clasificación de patrones meteorológicos y su relación con los episodios de ozono en la Comunidad Autónoma del País Vasco," in *Proceedings of the 4th Congress of Environmental Engineering*, pp. 441–451, Bilbao Exhibition Centre y Universidad del País Vasco, Bilbao, España, 2004.

[38] A. Albizuri, "Classification of meteorological patterns and its relation with the ozone episodes in the Basque Country," in *Proceedings of the 5th International Conference on Urban Air Quality (UAQ '05)*, R. S. Sokhi, M. Millán, and Y. N. Moussiopoulos, Eds., pp. 100–103, 2005.

[39] European Union, "Directiva 2008/50/CE del Parlamento Europeo y del Consejo, de 21 de Mayo de 2008, relativa al ozono en el aire ambiente y a una atmósfera más limpia en Europa," *Diario Oficial de la Comunidad Europea*, vol. L152, pp. 1–44, 2008.

[40] Consellería de Medio Ambiente e Desenvolvemento Sostible (Xunta de Galicia), "Calidade do Aire en Galicia - Estadísticos 2003–2008," Laboratorio de Medio Ambiente de Galicia.

[41] R. M. Peña, S. García, C. Herrero, and T. Lucas, "Spatial and temporal ozone pattern concentrations in a NW region of Spain," *Water, Air, and Soil Pollution*, vol. 117, no. 1–4, pp. 289–303, 2000.

[42] M. R. Méndez, J. A. Souto, J. Vilá-Guerau de Arellano, T. Lucas, and J. Casares, "Dispersion and transformation of nitrogen oxides emitted from a point source," in *Proceedings of the 1st Measurements and Modelling in Environmental Pollution Congress*, pp. 179–188, Madrid, Spain, 1997.

[43] M. Castellano, A. Franco, D. Cartelle, M. Febrero, and E. Roca, "Identification of NO_x and ozone episodes and estimation of ozone by statistical analysis," *Water, Air, and Soil Pollution*, vol. 198, no. 1–4, pp. 95–110, 2009.

[44] M. Gómez-Gesteira, L. Gimeno, M. deCastro et al., "The state of climate in NW Iberia," *Climate Research*, vol. 48, pp. 109–144, 2011.

[45] A. Martinez-Cortizas and A. Pérez-Alberti, *Atlas climático de Galicia*, Consellería de Medio Ambiente, Xunta de Galicia, Santiago de Compostela, Spain, 1999.

[46] European Union - European Commission, *Ozone position paper. Final version*, Ad hoc Working Group on Ozone Directive and Reduction Strategy Development, Office for Oficial Publication of the European Commission, Luxemburg, 1999.

[47] European Union, "Decisión 97/101/CE del Consejo Europeo, de 27 de enero de 1997, por la que se establece un intercambio recíproco de información y datos de las redes y estaciones aisladas de medición de la contaminación atmosférica en los Estados miembros," *Diario Oficial de la Comunidad Europea*, vol. L35, p. 14, 1997.

[48] European Union-European Commission, "Criteria for EUROAIRNET. The EEA air quality monitoring and information network," EEA technical report 12, Copenhagen, Denmark, 1999.

[49] European Union, "Directiva 92/72/CE del Consejo Europeo, de 21 de Septiembre de 1992, sobre la contaminación atmosférica por ozono," *Diario Oficial de la Comunidad Europea*, vol. L297, pp. 1–7, 1992.

[50] A. Fromage, "Prévision des pointes de pollution atmosphérique état de l'art dans le monde et perspectives pour la région Ile-de-France," Ecole des Mines de Paris - Institut Supérieur d'Ingénierie et de Gestion de l'Environnement, 1996.

[51] European Union, "Directiva 2002/3/CE del Parlamento Europeo y del Consejo, de 12 de Febrero de 2002, relativa al ozono en el aire ambiente," *Diario Oficial de la Comunidad Europea*, vol. L67, pp. 14–30, 2002.

[52] S Saavedra, A. Rodríguez, J. J. Taboada, J. A. Souto, and J. J. Casares, "Synoptic patterns and air masses transport during ozone episodes at northwestern Iberia," submitted to *Science of the Total Environment*.

[53] S. Solberg, P. Coddeville, C. Forster, O. Hov, Y. Orsolini, and K. Uhse, "European surface ozone in the extreme summer 2003," *Atmospheric Chemistry and Physics Discussions*, vol. 5, pp. 9003–9038, 2005.

[54] J. Fiala, L. Cernikowsky, F. de Leeuw, and P. Kurfuerst, "Air pollution by ozone in Europe in summer," Topic Report, European Environment Agency (EEA), 2003.

[55] L. Gimeno, E. Hernández, A. Rúa, R. García, and I. Martín, "Surface ozone in Spain," *Chemosphere*, vol. 38, no. 13, pp. 3061–3074, 1999.

[56] M. Coyle, R. I. Smith, J. R. Stedman, K. J. Weston, and D. Fowler, "Quantifying the spatial distribution of surface ozone concentration in the UK," *Atmospheric Environment*, vol. 36, no. 6, pp. 1013–1024, 2002.

[57] A. Chevalier, F. Gheusi, R. Delmas et al., "Influence of altitude on ozone levels and variability in the lower troposphere: a ground-based study for western Europe over the period 2001-2002," *Atmospheric Chemistry and Physics*, vol. 7, no. 16, pp. 4311–4326, 2007.

[58] M. L. Sánchez, B. De Torre, M. A. García, and I. Pérez, "Ozone concentrations at a high altitude station in the Central Massif (Spain)," *Chemosphere*, vol. 60, no. 4, pp. 576–584, 2005.

[59] J. Entwistle, K. Weston, R. Singles, and R. Burgess, "The magnitude and extent of elevated ozone concentrations around the coasts of the British Isles," *Atmospheric Environment*, vol. 31, no. 13, pp. 1925–1932, 1997.

[60] H. E. Scheel, H. Areskoug, H. Geiß et al., "On the spatial distribution and seasonal variation of lower-troposphere ozone over Europe," *Journal of Atmospheric Chemistry*, vol. 28, no. 1–3, pp. 11–28, 1997.

[61] A. Ribas and J. Peñuelas, "Temporal patterns of surface ozone levels in different habitats of the North Western Mediterranean basin," *Atmospheric Environment*, vol. 38, no. 7, pp. 985–992, 2004.

[62] R. O. Gilbert, *Statistical Methods for Environmental Pollution Monitoring*, Van Nostrand Reinhold, New York, NY, USA, 1987.

[63] F. A. A. M. De Leeuw, "Trends in ground level ozone concentrations in the European Union," *Environmental Science and Policy*, vol. 3, no. 4, pp. 189–199, 2000.

[64] Norwegian Institute for Air Research, "The development of European surface ozone. Implications for a revised abatement policy," EMEP/CCC - Report 1/2005.

[65] P. G. Simmonds, R. G. Derwent, A. L. Manning, and G. Spain, "Significant growth in surface ozone at Mace Head, Ireland, 1987–2003," *Atmospheric Environment*, vol. 38, no. 28, pp. 4769–4778, 2004.

[66] S. Solberg, "Monitoring of boundary layer ozone in Norway from 1977 to 2002," Norwegian Institute for Air Research, NILU OR 85/2003.

[67] J. E. Jonson, D. Simpson, H. Fagerli, and S. Solberg, "Can we explain the trends in European ozone levels?" *Atmospheric Chemistry and Physics*, vol. 6, no. 1, pp. 51–66, 2006.

[68] Ministerio de Medio Ambiente y Medio Rural y Marino, "Inventario Nacional de Emisiones de Contaminantes a la Atmósfera, Calidad y Evaluación Ambiental," 2009, http://www.marm.es/es/calidad-y-evaluacion-ambiental/temas/sistema-espanol-de-inventario-sei-/.

[69] European Environment Agency (EEA), "Annual European Community LRTAP Convention emission inventory report 1990-2006," EEA Technical Report, European Environment Agency (EEA), 2008.

[70] A. M. Ramos, M. N. Lorenzo, and L. Gimeno, "Compatibility between modes of flow frequency variability and Circulation Types: a case study of the NW Iberian Peninsula," *Journal of Geophysical Research*, vol. 115, 2010.

[71] M. N. Lorenzo, A. M. Ramos, J. J. Taboada, and L. Gimeno, "Changes in present and future circulation types frequency in northwest Iberian Peninsula," *PLoS One*, vol. 6, no. 1, 2011.

[72] M. Demuzere, R. M. Trigo, J. V. G. De Arellano, and N. P. M. Van Lipzig, "The impact of weather and atmospheric circulation on O_3 and PM10 levels at a rural mid-latitude site," *Atmospheric Chemistry and Physics*, vol. 9, no. 8, pp. 2695–2714, 2009.

[73] P. S. Monks, "A review of the observations and origins of the spring ozone maximum," *Atmospheric Environment*, vol. 34, no. 21, pp. 3545–3561, 2000.

[74] R. Vingarzan, "A review of surface ozone background levels and trends," *Atmospheric Environment*, vol. 38, no. 21, pp. 3431–3442, 2004.

[75] R. G. Derwent and T. J. Davies, "Modelling the impact of NO(x) or hydrocarbon control on photochemical ozone in Europe," *Atmospheric Environment*, vol. 28, no. 12, pp. 2039–2052, 1994.

[76] R. G. Derwent, D. S. Stevenson, W. J. Collins, and C. E. Johnson, "Intercontinental transport and the origins of the ozone observed at surface sites in Europe," *Atmospheric Environment*, vol. 38, no. 13, pp. 1891–1901, 2004.

[77] P. G. Simmonds, S. Seuring, G. Nickless, and R. G. Derwent, "Segregation and interpretation of ozone and carbon monoxide measurements by air mass origin at the TOR station Mace Head, Ireland from 1987 to 1995," *Journal of Atmospheric Chemistry*, vol. 28, no. 1–3, pp. 45–59, 1997.

[78] S. Liu, M. Trainer, F. Fehsenfeld et al., "Ozone production in the rural troposphere and the implications for regional and global ozone distributions," *Journal of Geophysical Research*, vol. 92, pp. 4191–4207, 1987.

[79] Y. Wang, D. J. Jacob, and J. A. Logan, "Global simulation of tropospheric O_3-NO_x-hydrocarbon chemistry - 3. Origin of tropospheric ozone and effects of nonmethane hydrocarbons," *Journal of Geophysical Research D: Atmospheres*, vol. 103, no. 3339, pp. 10757–10767, 1998.

[80] S. A. Penkett, M. J. Evans, C. E. Reeves et al., "Long-range transport of ozone and related pollutants over the North Atlantic in spring and summer," *Atmospheric Chemistry and Physics Discussion*, vol. 4, pp. 4407–4454, 2004.

[81] R. G. Derwent, P. G. Simmonds, S. Seuring, and C. Dimmer, "Observation and interpretation of the seasonal cycles in the surface concentrations of ozone and carbon monoxide at Mace Head, Ireland from 1990 to 1994," *Atmospheric Environment*, vol. 32, no. 2, pp. 145–157, 1998.

[82] D. C. Carslaw, "On the changing seasonal cycles and trends of ozone at Mace Head, Ireland," *Atmospheric Chemistry and Physics*, vol. 5, no. 12, pp. 3441–3450, 2005.

[83] M. A. García, M. L. Sánchez, I. A. Pérez, and B. De Torre, "Ground level ozone concentrations at a rural location in northern Spain," *Science of the Total Environment*, vol. 348, no. 1–3, pp. 135–150, 2005.

[84] R. Nieto, L. Gimeno, L. de la Torre et al., "Climatological features of cutoff low systems in the Northern Hemisphere," *Journal of Climate*, vol. 18, no. 16, pp. 3085–3103, 2005.

[85] R. Nieto, M. Sprenger, H. Wernli, R. M. Trigo, and L. Gimeno, "Identification and climatology of cut-off lows near the tropopause," *Annals of the New York Academy of Sciences*, vol. 1146, pp. 256–290, 2008.

[86] A. W. Delcloo and H. De Backer, "Five day 3D back trajectory clusters and trends analysis of the Uccle ozone sounding time series in the lower troposphere (1969–2001)," *Atmospheric Environment*, vol. 42, no. 19, pp. 4419–4432, 2008.

[87] A. Colette and G. Ancellet, "Impact of vertical transport processes on the tropospheric ozone layering above Europe.: Part II: climatological analysis of the past 30 years," *Atmospheric Environment*, vol. 39, no. 29, pp. 5423–5435, 2005.

[88] M. Beekmann, G. Ancellet, S. Blonsky et al., "Regional and global tropopause fold occurrence and related ozone flux across the tropopause," *Journal of Atmospheric Chemistry*, vol. 28, no. 1–3, pp. 29–44, 1997.

[89] J. R. Holton, P. H. Haynes, M. E. McIntyre, A. R. Douglass, R. B. Rood, and L. Pfister, "Stratosphere-troposphere exchange," *Reviews of Geophysics*, vol. 33, no. 4, pp. 403–439, 1995.

[90] D. Xu, D. Yap, and P. A. Taylor, "Meteorologically adjusted ground level ozone trends in Ontario," *Atmospheric Environment*, vol. 30, no. 7, pp. 1117–1124, 1996.

[91] D. Danalatos and S. Glavas, "Diurnal and seasonal variations of surface ozone in a Mediterranean coastal site, Patras, Greece," *Science of the Total Environment*, vol. 177, pp. 291–301, 1996.

[92] PORG (Photochemical Oxidants Review Group), "Ozone in the United Kingdom, Fourth Report," Prepared at the Request of the Department of the Environment, Transport and the Regions, London, UK, 1997.

[93] J. A. Salmond and I. G. McKendry, "Secondary ozone maxima in a very stable nocturnal boundary layer: observations from the Lower Fraser Valley, BC," *Atmospheric Environment*, vol. 36, no. 38, pp. 5771–5782, 2002.

[94] C. Dueas, M. C. Fernández, S. Caete, J. Carretero, and E. Liger, "Assessment of ozone variations and meteorological effects in an urban area in the Mediterranean Coast," *Science of the Total Environment*, vol. 299, no. 1–3, pp. 97–113, 2002.

[95] F. J. Acero, V. L. Mateos, J. A. García, and M. Núñez, "Concentraciones de ozono troposférico en Extremadura durante la ola de calor del 2003," XXVIII Jornadas Científicas de la Asociación Meteorológica Española, 2004.

[96] M. L. Sanchez, B. de Torre, M. A. García, and I. Pérez, "Ground-level ozone and ozone vertical profile measurements close to the foothills of the Guadarrama mountain range (Spain)," *Atmospheric Environment*, vol. 41, no. 6, pp. 1302–1314, 2007.

[97] G. P. O'Hare and R. Wilby, "A review of ozone pollution in the United Kingdom and Ireland with an analysis using Lamb

weather types," *The Geographical Journal*, vol. 161, no. 1, pp. 1–20, 1995.

[98] A. Monteiro, A. Strunk, A. Carvalho et al., "Investigating a high ozone episode in a rural mountain site," submitted to *Environmental Pollution*.

15

Impact of Parameterization of Physical Processes on Simulation of Track and Intensity of Tropical Cyclone Nargis (2008) with WRF-NMM Model

Sujata Pattanayak, U. C. Mohanty, and Krishna K. Osuri

Centre for Atmospheric Sciences, Indian Institute of Technology, Delhi, Hauz Khas, New Delhi 110016, India

Correspondence should be addressed to Sujata Pattanayak, sujata05@gmail.com

Academic Editors: L. Leslie and C. Ottle

The present study is carried out to investigate the performance of different cumulus convection, planetary boundary layer, land surface processes, and microphysics parameterization schemes in the simulation of a very severe cyclonic storm (VSCS) Nargis (2008), developed in the central Bay of Bengal on 27 April 2008. For this purpose, the nonhydrostatic mesoscale model (NMM) dynamic core of weather research and forecasting (WRF) system is used. Model-simulated track positions and intensity in terms of minimum central mean sea level pressure (MSLP), maximum surface wind (10 m), and precipitation are verified with observations as provided by the India Meteorological Department (IMD) and Tropical Rainfall Measurement Mission (TRMM). The estimated optimum combination is reinvestigated with six different initial conditions of the same case to have better conclusion on the performance of WRF-NMM. A few more diagnostic fields like vertical velocity, vorticity, and heat fluxes are also evaluated. The results indicate that cumulus convection play an important role in the movement of the cyclone, and PBL has a crucial role in the intensification of the storm. The combination of Simplified Arakawa Schubert (SAS) convection, Yonsei University (YSU) PBL, NMM land surface, and Ferrier microphysics parameterization schemes in WRF-NMM give better track and intensity forecast with minimum vector displacement error.

1. Introduction

Tropical cyclones are serious threats to human life and property. Even with the recent rapid improvements in numerical weather prediction, tropical cyclone forecasting remains a challenging problem to atmospheric modeling groups. The Bay of Bengal is a potentially active region for development of cyclonic storms and an average of five tropical cyclones annually forms over the region, representing 5.85% of the global frequency [1]. Moreover, the Bay of Bengal storms are exceptionally devastating, especially when they make landfall [2]. This is mainly due to a densely populated and low lying coastline. With the Bay of Bengal tropical cyclones being the deadliest natural disasters in the Indian subcontinent, it significantly impacts the socioeconomic conditions of the countries bordering the east coast of India.

Cumulus convection, surface fluxes of heat, moisture, and momentum and vertical mixing in the PBL play important roles in the development of tropical cyclones [3]. Convection has long been recognized as a process of central importance in the development of cyclonic storms. The scales of convective clouds are too small to be resolved by the numerical models and hence need to be parameterized in terms of variables defined at each grid point. A number of parameterization schemes have been developed over the years but each have their respective limitations [4, 5]. Performance of a numerical model in tropical cyclone forecasts depends on how well the convection is parameterized in the model [6, 7]. These studies have led to an increased understanding of the importance of the boundary layer and convective processes in the tropical cyclone development. An extensive study is carried out on the impact of the parameterization of physical processes in the simulation of two severe cyclonic storms developed over the Bay of Bengal using the MM5 model [7]. This study with MM5 model suggested that the combination of MRF and Grell (or Betts-Miller) for PBL and convection schemes, respectively, consistently give better results than the other combinations. The sensitivity experiments of convection, boundary layer, and moisture processes using the MM5 model for the prediction of the Orissa Super Cyclone 1999 is carried out by Bhaskar Rao and Hari Prasad [8] and the study suggested that convective processes plays an important role in the cyclone track prediction while the PBL controls intensification. A comparison study of four PBL parameterization schemes in simulation of Hurricane Bob (1991) is presented [9] using the MM5 model. This study suggested that significant sensitivity is seen in the central pressure and maximum surface wind (10 m). The precipitation forecast in hurricanes can be just as sensitive to the formulation of the different PBL schemes. The customization of Advanced Research WRF (WRF-ARW) model is also carried out [10] for the tropical cyclones over North Indian Ocean which suggested that the combination of KF cumulus convection scheme along with the YSU PBL is providing better track and intensity forecast. All the above-mentioned studies clearly demonstrate the impact of the parameterization of physical processes in different fields of studies using the MM5 and WRF-ARW models. So, there is a need to investigate the impact of the parameterization of physical processes in simulation of tropical cyclones over the Bay of Bengal using NMM dynamic core of the WRF (WRF-NMM) model.

In the present study, NCEP mesoscale model WRF-NMM is used to simulate a very severe cyclone Nargis with sensitivity experiments carried out to explore the impact of physical parameterizations on track and intensity prediction. The sensitivity of the model simulations to initial conditions is also explored using the optimum combination of physical parameterizations.

A brief description of the model as well as the parameterization schemes used in the study is presented in Section 2. The synoptic situation for the above-mentioned cyclone used in the present study is described in Section 3. Various numerical experiments and data used are described in Section 4. Model simulated results along with the evaluation of performance of the model with different initial conditions are presented in Section 5 with the conclusions in Section 6.

2. Model Description

The WRF-NMM version 3.0.1 developed by National Center for Environmental Prediction (NCEP)/National Oceanic and Atmospheric Administration (NOAA) is designed to be a flexible, state-of-the-art mesoscale modeling system. It is a fully compressible, nonhydrostatic model with a hydrostatic option [11–13]. Its vertical coordinate is a hybrid sigma-pressure coordinate. The grid staggering is the Arakawa E-grid. The dynamics conserve a number of first- and second-order quantities including energy and enstrophy [14]. Forward-backward time integration scheme is used for the horizontally propagating fast waves and implicit scheme

is used for the vertically propagating sound waves. Adams-Bashforth scheme for horizontal advection and Crank-Nicholson scheme for vertical advection are used by the model. The same time step is used for all terms. The Geophysical Fluid Dynamic Laboratory (GFDL) long-wave and short-wave radiation schemes are incorporated in the model. Additionally, to represent deep, moist convection in the model, various parameterizations schemes are included. All the schemes considered for this study are reasonably independent and hence useful for sensitivity experiments.

2.1. Planetary Boundary Layer Schemes. The Yonsei University (YSU) PBL scheme is a revised vertical diffusion package with a nonlocal turbulent mixing coefficient in the boundary layer. The major ingredient of the revision is the inclusion of an explicit treatment of entrainment processes at the top of the PBL. The YSU PBL increases boundary layer mixing in the thermally induced free convection regime and decreases it in the mechanically induced forced convection regime, which alleviates the well-known problems in the Medium-Range Forecast (MRF) PBL. Excessive mixing in the mixed layer in the presence of strong winds is resolved. Overly rapid growth of the PBL in the case of the Hong and Pan (1996) [15] is also rectified. Consequently, the YSU scheme does a better job in reproducing the convective inhibition.

The Mellor Yamada Janjic (MYJ) PBL scheme is a one-dimensional prognostic turbulent kinetic energy scheme with local vertical mixing [12, 13]. An advantage of the MYJ scheme is that it allows advection of turbulent regions during the forecast. The top of the layer depends on the TKE as well as the buoyancy and shear of the driving flow.

The NCEP Global Forecast System (NCEP GFS) PBL scheme is a nonlocal vertical diffusion scheme [16] and further described in Hong and Pan (1996). The PBL height is determined using an iterative bulk-Richardson approach working from the ground upward whereupon the profile of the diffusivity coefficient is specified as a cubic function of the PBL height. Coefficient values are obtained by matching the surface layer fluxes. A counter-gradient flux parameterization is included.

2.2. Cumulus Parameterization Schemes. The Simplified Arakawa Schubert (SAS) scheme is based on Arakawa and Schubert (1974) [17] as simplified by Grell (1993) [18] and with a saturated downdraft. The major modification is done in entrainment relation to avoid the costly calculation that is necessary to find the entrainment parameter of cloud detraining at the model levels. It is very simplistic and computationally highly efficient convective parameterization scheme leads to a very realistic simulation of the mesoscale convective systems. The scheme uses a stability closure, assumes a large cloud size, parameterizes moist downdrafts, and does not assume unrealistically large lateral mixing to simulate penetrative convection [19].

The Kain-Fristch (KF) is a deep and shallow subgrid scheme using a mass-flux approach with downdrafts [20, 21]. Mixing is allowed at all vertical levels through entrainment and detrainment. This scheme removes convective available

Impact of Parameterization of Physical Processes on Simulation of Track and Intensity of Tropical Cyclone
Nargis (2008) with WRF-NMM Model

151

potential energy (CAPE) through vertical reorganization of mass at each grid point. The scheme consists of a convective trigger function (based on grid-resolved vertical velocity), a mass flux formulation, and closure assumptions.

The Betts Miller and Janjic (BMJ) scheme is an adjustment-type scheme for deep and shallow convection relaxing towards reference profile of temperature and specific humidity determined from thermodynamic considerations [22, 23]. The scheme's structure favors activation in cases with substantial amounts of moisture in low and mid-levels and positive CAPE. The representation is accomplished by constraining the temperature and moisture fields by the convective cloud field.

Grell-Devenyi (GD) scheme is a multiclosure, multiparameter ensemble method. It is an ensemble average of typically more than 100 types of clouds, which includes different closures such as CAPE removal, quasiequilibrium and moisture convergence, and variants of cumulus parameterization such as changes in the parameters for entrainment, cloud radius, maximum cap, and precipitation efficiency.

The detailes of the model specifications used for the present study are presented in Table 1.

3. Synoptic Situation of Tropical Cyclone Nargis

Nargis was a category 4 tropical cyclone that caused worst natural disaster in the recorded history of Mynmar. In the last week of April 2008, an area of deep convection persisted near a low level circulation in the Bay of Bengal about 1150 km east-southeast of Chennai. With good outflow and low vertical wind shear, the system slowly organized into a depression at 0300 UTC 27 April 2008. The system intensified into deep depression stage at 1200 UTC 27April 2008 with a minimum central MSLP of 998 hPa and the maximum sustained surface winds of 30 kts. After 12 hours at 0000 UTC 28 April 2008, the system intensified into cyclonic storm with a minimum central MSLP of 994 hPa and the maximum sustained surface winds of 35 kts. The system further intensified into severe cyclonic storm with a minimum central MSLP of 986 hPa and the maximum sustained surface winds of 55 kts at 0900 UTC 28 April 2008 and moved in northward direction. Then around 0300 UTC 29 April 2008, the system became a very severe cyclonic storm (VSCS) with a minimum central MSLP of 980 hPa and the maximum sustained surface winds of 65 kts. The storm remained in VSCS for a period of 93 hours, that is, up to 0000 UTC 03 May 2008. The observed minimum central pressure was 962 hPa with the pressure drop of 40 hPa and the maximum sustained surface winds of 90 kts. The storm crossed southwest coast of Myanmar around 1200 UTC 02 May 2008. The system remained on the land for further 24 hours and caused extensive devastation to coastal areas.

4. Numerical Experiments and Data Used

The mesoscale model WRF-NMM described in section 2 is integrated up to 123 hours in a single domain with the

TABLE 1: Details of the WRF-NMM model specifications.

Model	NCEP mesoscale model WRF-NMM V3.0.1
Dynamics	Nonhydrostatic with terrain following hybrid pressure sigma vertical coordinate.
Map projection	Rotated lat-lon
Resolution	9 km
No. of vertical levels	51
Horizontal grid scheme	Arakawa E-grid
Time integration scheme	Horizontal: forward-backward scheme
	Vertical: Implicit scheme
Lateral boundary condition	NCEP/NCAR GFS forecast
Radiation scheme	Long wave: GFDL
	Short wave: GFDL
Planetary boundary layer parameterization schemes	(1) NCEP GFS
	(2) Yonsei University (YSU)
Cumulus parameterization schemes	(1) Kain-Fritsch
	(2) Betts-Miller-janjic
	(3) Grell-Devenyi
	(4) Simplified Arakawa Schubert
Land surface physics	(1) NMM
	(2) Thermal diffusion
	(3) Noah
	(4) RUC
Microphysics	(1) Ferrier
	(2) WSM 5-class
	(3) WSM 6-class graupel
	(4) Thompson

horizontal resolution of 9 km. The model has 51 levels up to a height of 30 km in the vertical. A number of numerical experiments producing 123 hours forecasts (for each experiment) are carried out with the possible combination of four cumulus convection schemes, two PBL schemes, four land surface physics schemes and four microphysics schemes. The four convection schemes are Kain-Fritsch [20, 21, 24], Betts-Miller-Janjic [25], Grell-Devenyi [26] and Simplified Arakawa-Schubert [17–19], which thereafter referred as K, B, G and S respectively. The two PBL schemes are NCEP Global Forecast System [15, 16] and Yonsei University [27], which thereafter referred as NC and Y respectively. The four land surface physics schemes are NMM [28], Thermal Diffusion [27], Noah [29] and RUC [30, 31], which thereafter referred as N, T, NO and R respectively. The four microphysics schemes are Ferrier (New ETA) [32], WRF Single-Movement (WSM) 5-class [33, 34], WSM 6-class graupel [33, 35, 36] and Thompson et al. [37], which thereafter referred as F, W5, W6 and T respectively. The experiments are categorized into two main groups, choosing different parameterization schemes of convection, PBL, land surface and microphysics for the best possible combination.

Then in the second group, the best two combinations are re-investigated with 5 additional sets of initial conditions of the same cyclone case yielding 6 groups of simulation results. Results obtained from all possible experiments are examined by comparing with the verification analysis and observations to find the best combinations towards forecasting the track and intensity of the above mentioned cyclone.

The initial and lateral boundary conditions to a limited area model are usually provided from the large scale analysis of different NWP centers in the world. The NCEP/GFS analysis and forecasts ($1° \times 1°$ horizontal resolution) have been used to provide the initial and lateral boundary conditions to the model. The TC Nargis was intensified into cyclonic storm at about 0000 UTC 28 April 2008 and hence chosen as the initial time for the model simulations. Furthermore, the evaluation of performance of the model is carried out with the results obtained from model integration at different initial conditions. For this purpose, six simulations have been carried out from the initial conditions of 0000 UTC 28 April 2008, 1200 UTC 28 April 2008, 0000 UTC 29 April 2008, 1200 UTC 29 April 2008, 0000 UTC 30 April 2008 and 1200 UTC 30 April 2008 with the optimum model configuration. Also, each simulation is done up to 0300 UTC 03 May 2008.

5. Results and Discussions

The results as obtained with different combinations of parameterization schemes producing 123 hours forecasts for Nargis (as described above) are presented in this section to examine the performance of the parameterization of physical processes in the prediction of track and intensity of the tropical cyclone.

5.1. Sensitivity Experiments with Convection Schemes. In this subsection, four experiments are carried out with the variation of the parameterization scheme for convection as Kain-Fritsch (K), Betts-Miller-Janjic (B), Grell-Devenyi (G), and Simplified Arakawa-Schubert (S) in combination with YSU (Y) scheme for PBL, NMM (N) scheme for land surface and Ferrier (F) microphysics scheme. Model-simulated track positions are presented in Figure 1(a) to facilitate the evaluation. The results indicate that B and S schemes are producing similar type of results in terms of track prediction. The movement of the cyclone with K scheme is much faster than any other schemes. The G scheme is providing the reasonable prediction of the track position. The results indicate that the movement of the tropical cyclone is sensitive to the convective process.

Sensitivity in model simulation is seen, with MSLP varying by up to 30 hPa and maximum winds by 31 kts among the above four experiments. The observed minimum central MSLP was 962 hPa and the maximum wind was 90 kts. The minimum central MSLP and the maximum surface winds for each convection schemes are calculated and presented in Figures 1(b) and 1(c), respectively. The experiment utilizing the K scheme yields an intensity that is quantitatively much closer to observations than the other forecasts with a minimum MSLP of 967.5 and maximum surface wind of 62 kts.

The caveat to these values being that the maximum intensity occurs 30 hours prior to observations. The experiment with B convection scheme predicted a minimum central MSLP of 981 hPa and maximum wind of 52 kts. The experiment with the G convective scheme resulted in a minimum central MSLP of 997.5 hPa and maximum surface winds of 31 kts. Similarly, the experiment with S scheme produced the MSLP of 973 hPa and maximum wind of 52 kts. The time of maximum intensity of the B and S convective schemes nearly matches observations. Although the G convective scheme produced a track most similar to the observed value, its intensity forecast was much worse than the other schemes.

Since cumulus convection schemes play an important role in the development of tropical cyclones, hence to further examine the implication of utilizing different convective schemes, the structure of tropical cyclone is examined. Figure 2 represents the temperature anomaly and horizontal wind structure at the most intense time of the cyclone. The results clearly suggested that S and B convective schemes are giving similar type of result with a clear representation of the intense structure of the storm; however, G and K convective schemes fail to represent the same. But, at the same time, the K scheme is producing nearly same intensity as observed. Next, the results of PBL sensitivity experiments are presented with K, B, and S convection scheme.

5.2. Sensitivity Experiments with PBL Schemes. As per the results noted in Section 5.1, the sensitivity of the forecasts to two different PBL schemes YSU (Y) and NCEP GFS (NC) is considered. Hence, another three more experiments are carried out with NC scheme producing a total of 6 experiments for the PBL schemes. The model-simulated track, MSLP, and maximum wind are presented in Figures 3(a), 3(b), and 3(c), respectively. The storm movement is well predicted by the Y scheme than the NC scheme for both B and S convection schemes. However, the K scheme behaves in a different manner with a different PBL scheme. Significant sensitivity is seen in intensity prediction with different PBL schemes. The NC scheme produces a much higher intensity storm than the Y scheme, but it always occur before the observed maximum intensity time. The combination of S + NC + N + F and K + NC + N + F produces the MSLP of 967 hPa and 964 hPa, respectively, whereas S + Y + N + F and K + Y + N + F produces the MSLP of 973 hPa and 968 hPa, respectively. The results indicate that with the NC scheme the MSLPs decreases by nearly 4–6 hPa. This clearly suggested that NC scheme leads to more intense storm, but at an earlier time (nearly 42 hrs for K scheme and 24 hrs for S scheme) than observed. However, the Y scheme simulates intensity reasonably well. Hence, the two best combinations "B + Y + N + F" and "S + Y + N + F" from six combinations are chosen for further investigation.

5.3. Sensitivity Experiments with Land Surface Schemes. As discussed in the previous subsections, the combination of S and B schemes for convection and Y scheme for PBL produce the better simulation of cyclone Nargis. Hence, another six more experiments are carried out with the available land

Impact of Parameterization of Physical Processes on Simulation of Track and Intensity of Tropical Cyclone
Nargis (2008) with WRF-NMM Model

153

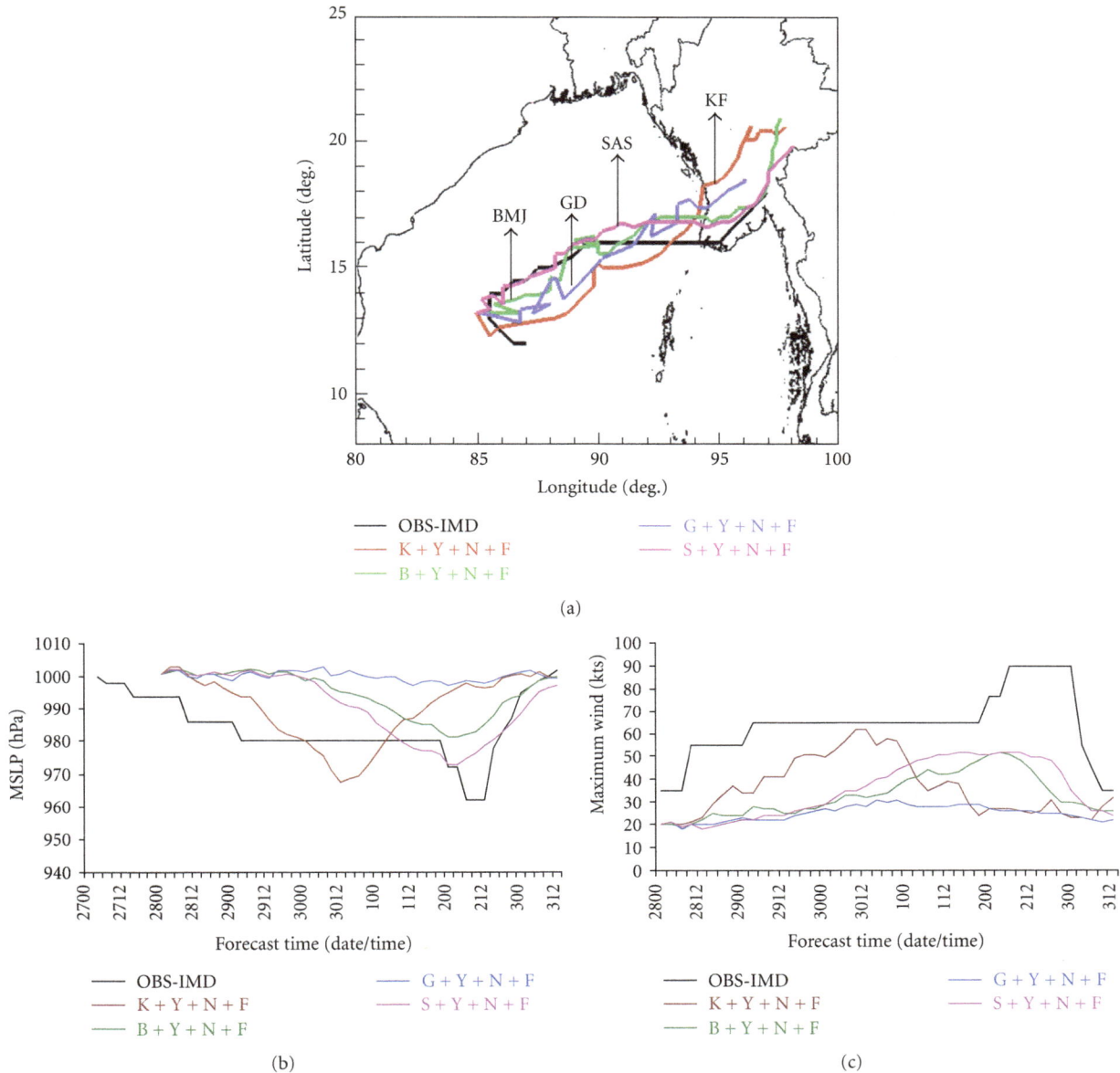

FIGURE 1: (a) Track of the cyclone Nargis, (b) MSLP (hPa), and (c) maximum wind (kts) with 4 different cumulus convective schemes, YSU PBL, NMM land surface, and Ferrier microphysics option.

surface physics schemes to determine the role of surface fluxes of heat and moisture in forecasts of the tropical cyclone. The model-simulated track, MSLP and maximum wind are presented in Figures 4(a), 4(b) and 4(c), respectively along with the IMD observation as part of further analysis of the two best combinations of parameterizations. Track of the cyclone is very well simulated with all the land surface processes. However, significant sensitivity is seen in the intensity prediction and the MSLP varies from 993 hPa to 973 hPa for the various combinations of land surface schemes. It is also noticed that NO and N land surface schemes produce similar results in terms of MSLP and maximum wind in both S and B convection schemes. But the S scheme produces more accurate value towards observation

than that of B scheme. Noah (NO) land surface is also producing similar type of result as NMM (N) with both S and B convection schemes in terms of MSLP and maximum wind. But, the S convection scheme is producing more realistic value and comparable to observation value than the B scheme. Hence, the combinations of S + Y + N + F and S + Y + NO + F are selected for further study with microphysics schemes based on their performances.

5.4. Sensitivity Experiments with Microphysics Schemes. As per the results noted in previous subsections, the combination of S scheme for convection, Y scheme for PBL, and N and NO schemes for land surface processes produces the

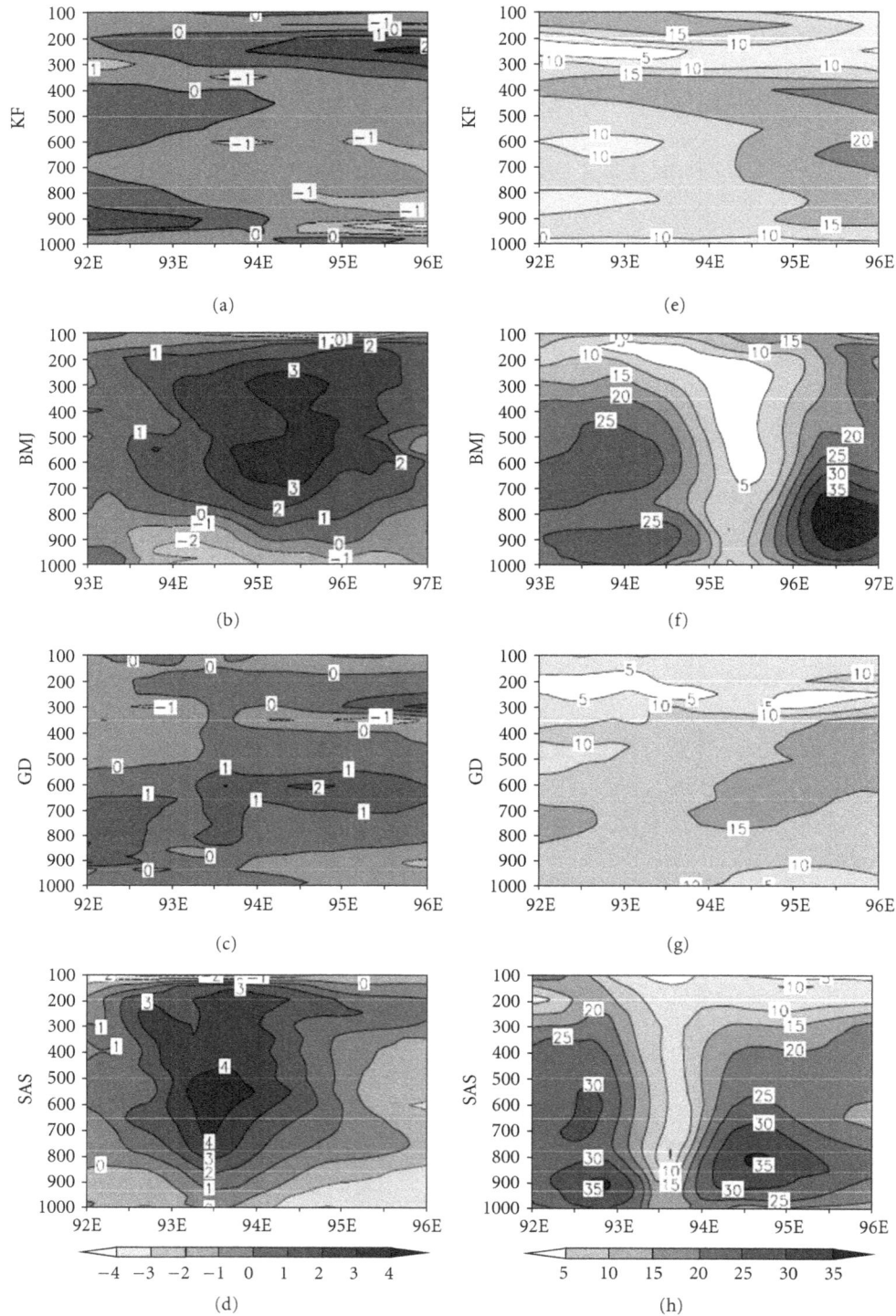

FIGURE 2: Temperature anomaly (°C) (left panel) and horizontal wind (ms^{-1}) (right panel) at the most intense period of the cyclone Nargis with different cumulus convective schemes.

better simulation result. Hence, in order to investigate the performance of microphysics schemes, another six experiments are carried out with the different options for microphysics as W5, W6, and T with two experiments each (as we are taking 2 land surface options) make the total eight

experiments (two for F) along with S cumulus convection scheme and Y PBL scheme. The model-simulated track, MSLP, and maximum wind are shown in Figures 5(a), 5(b), and 5(c), respectively. The track simulations W5, W6, and T schemes are well matched with observational data as

Impact of Parameterization of Physical Processes on Simulation of Track and Intensity of Tropical Cyclone
Nargis (2008) with WRF-NMM Model

155

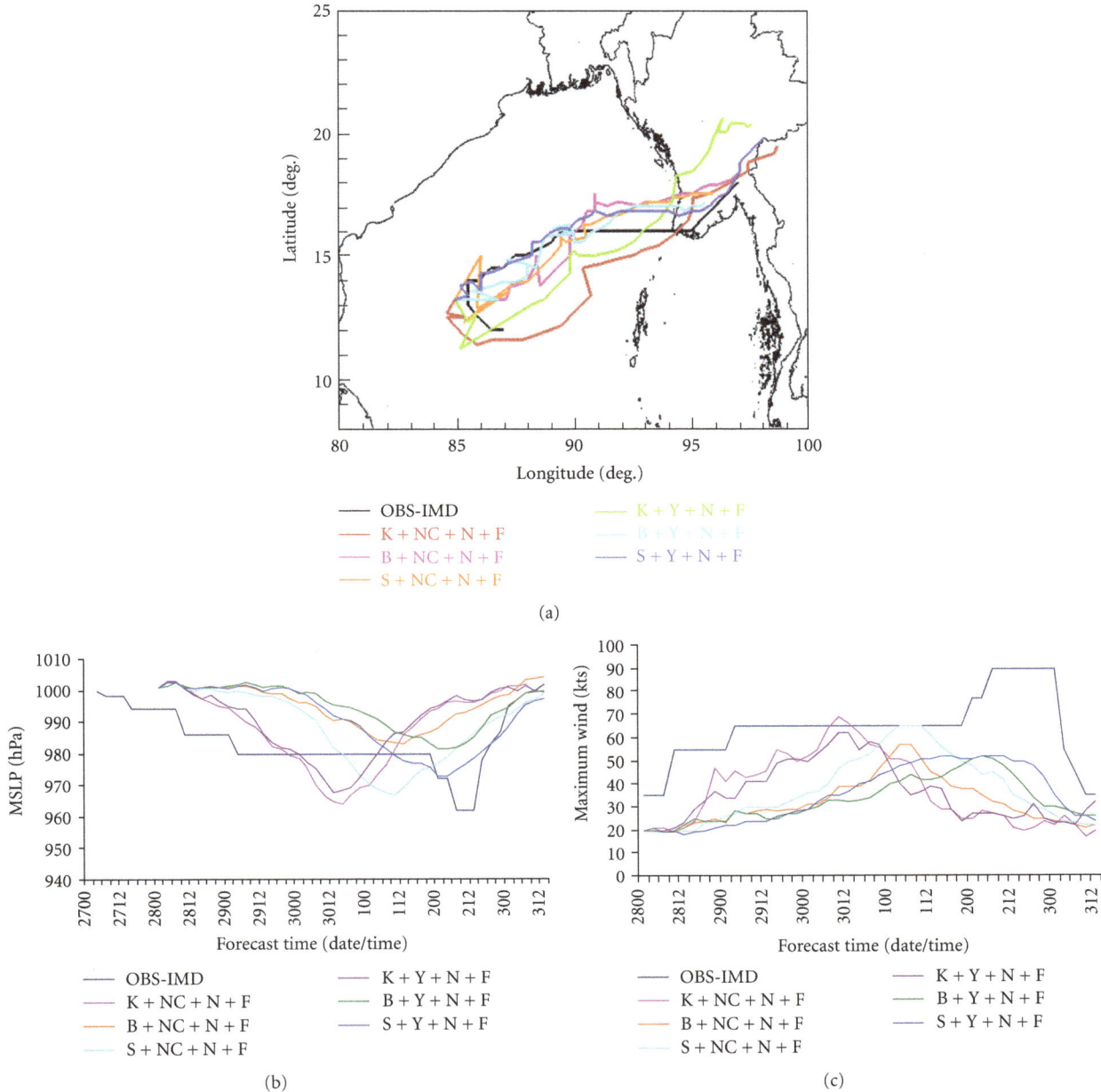

FIGURE 3: (a) Track, (b) MSLP (hPa), and (c) maximum wind (kts) of the cyclone NARGIS with three best cumulus convections, 2 different PBL, NMM land surface, and Ferrier microphysics option.

provided by IMD than the F scheme. However, the intensity predictions with W5, W6, and T schemes are very poorly represented. The minimum MSLP as predicted by S + Y + N + F is 973 hPa, whereas it is 985 hPa, 984 hPa, and 990 hPa with S + Y + N + W5, S + Y + N + W6, and S + Y + N + T schemes, respectively. Again, the minimum MSLP as predicted by S + Y + NO + F is 976 hPa, where as it is 985 hPa, 984 hPa, and 989 hPa with S + Y + NO + W5, S + Y + NO + W6, and S + Y + NO + T schemes, respectively. Hence, it may be concluded that the combinations of S + Y + N + F and S + Y + NO + F are producing better result than any other combination. However, it may be noted that the N land

surface scheme is giving slightly better result than the NO scheme in terms of movement and intensity of the storm.

5.5. *Precipitation.* The results as obtained from the previous subsections clearly show that, the S convection scheme, Y PBL scheme, N and NO land surface schemes, and F microphysics scheme are producing better result than any other combinations. Figure 6 shows 24 hrs accumulated precipitation as obtained from Tropical Rainfall Measuring Mission (TRMM 3B42) datasets, which is a merger of TMI, other microwave radiometers (SSMI, AQUA), and IR radiometers calibrated using rain gauges and TRMM's

(a)

(b)

(c)

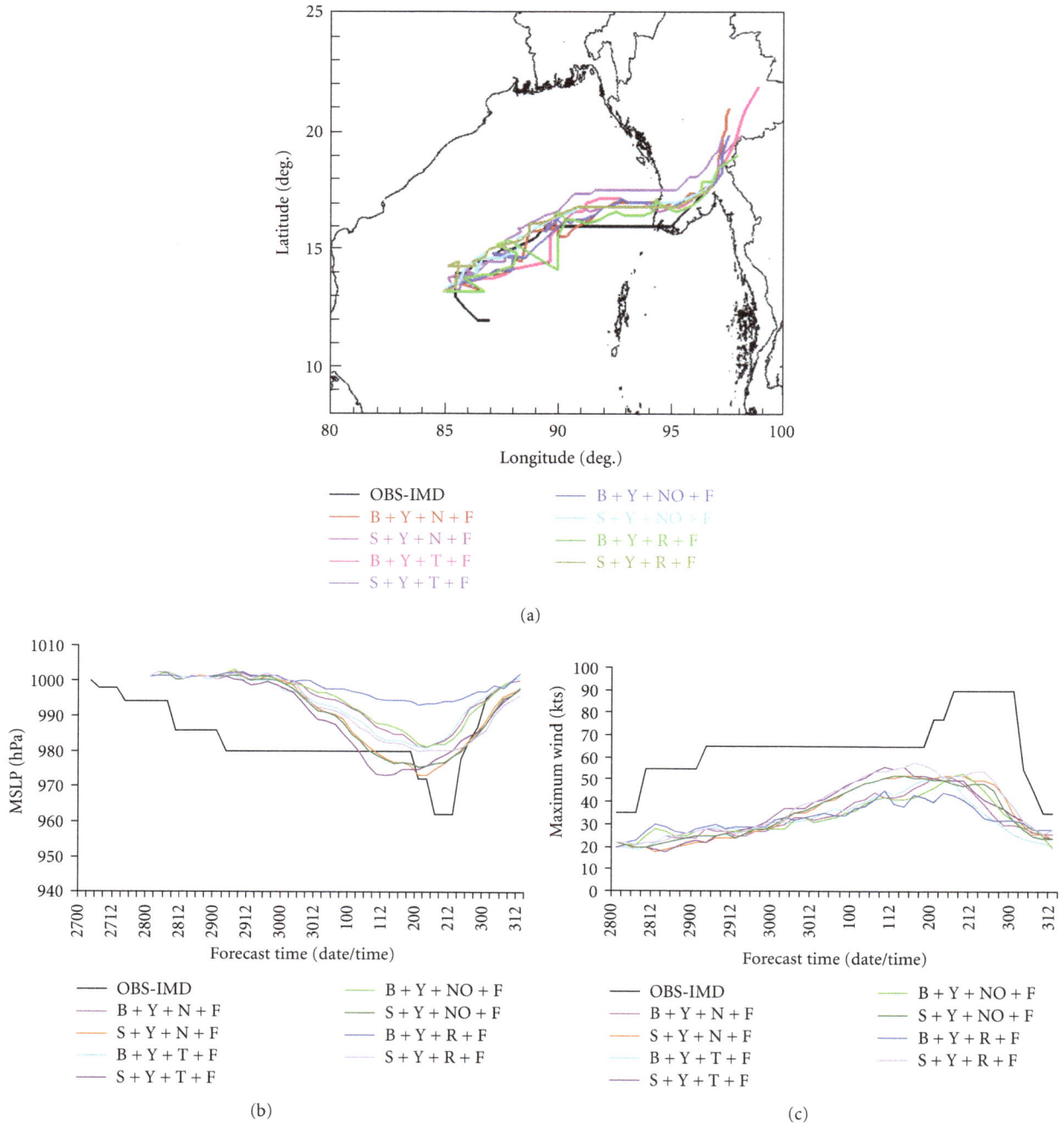

FIGURE 4: (a) Track, (b) MSLP (hPa), and (c) maximum wind (kts) of the cyclone NARGIS with two best cumulus convections, best PBL, 4 different land surface schemes, and Ferrier microphysics option.

precipitation radar and carried out by National Aeronautics and Space Administration (NASA) and model simulations. The precipitation data are obtained from the NASA web site (http://disc2.nascom.nasa.gov/Giovanni/tovas/). The left panel is from TRMM observed precipitation, middle panel is for model simulation with S + Y + N + F combination, and right panel is for S + Y + NO + F combination. The spatial distribution of precipitation is found to be nearly same with both N and NO land surface schemes. However, N scheme is able to produce peak precipitation in terms of

both amount and time of occurrence and comparable with observed precipitation than the NO scheme, which has been clearly demonstrated in subsequent section.

5.6. *Evaluation of Performance of the Model with Different Initial Conditions.* As discussed above (Section 4), the model performance is evaluated with the best two combinations after a detailed investigation of different combinations of convection, PBL, land surface, and microphysics schemes.

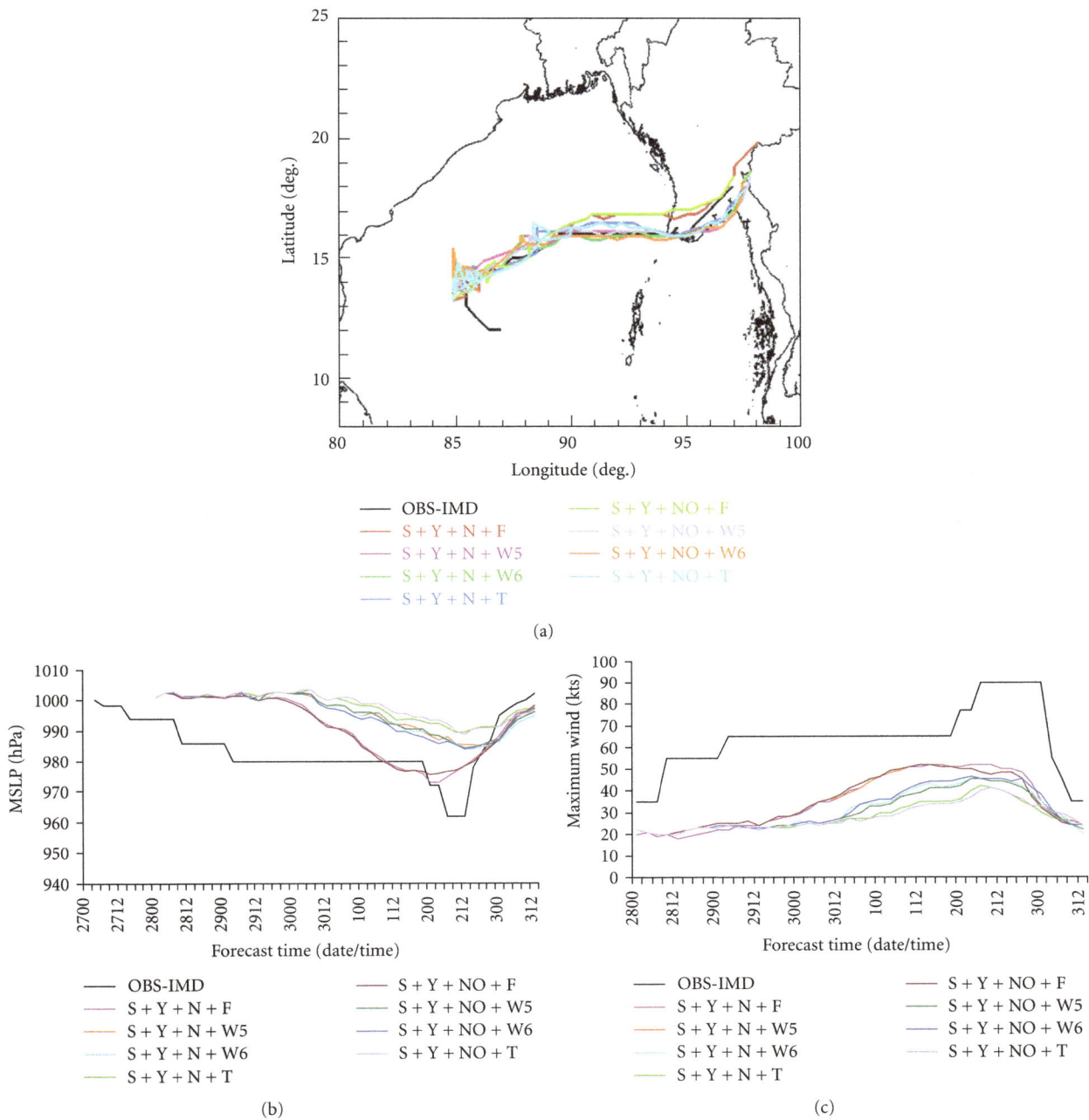

FIGURE 5: (a) Track, (b) MSLP (hPa), and (c) maximum wind (kts) of the cyclone NARGIS with best cumulus convections, best PBL, 2 best land surface schemes, and 4 different microphysics options.

For this purpose, starting from 0000 UTC 28 April 2008 and in every 12 hour interval, the model is integrated up to 0300 UTC 03 May 2008 for each simulation. Thus, another ten experiments (five experiments for each combination) are carried out from the initial condition of 1200 UTC 28 April 2008, 0000 UTC 29 April 2008, 1200 UTC 29 April 2008, 0000 UTC 30 April 2008, and 1200 UTC 30 April 2008.

5.6.1. Simulation of Track and Intensity. The model simulated track positions, MSLP, and maximum wind along with the

IMD observations are presented in Figure 7. Figures 7(a), 7(b), and 7(c) are the results as obtained with S convection, Y PBL, F microphysics, and N land surface which shows that the track and intensity is well simulated by the model. Figures 7(d), 7(e), and 7(f) are the results as obtained with S convection, Y PBL, F microphysics, and NO land surface processes. The track simulations with S + Y + N + F and S + Y + NO + F combinations are providing similar types of results. But, a lot of difference is found in intensity prediction. The S + Y + NO + F combination

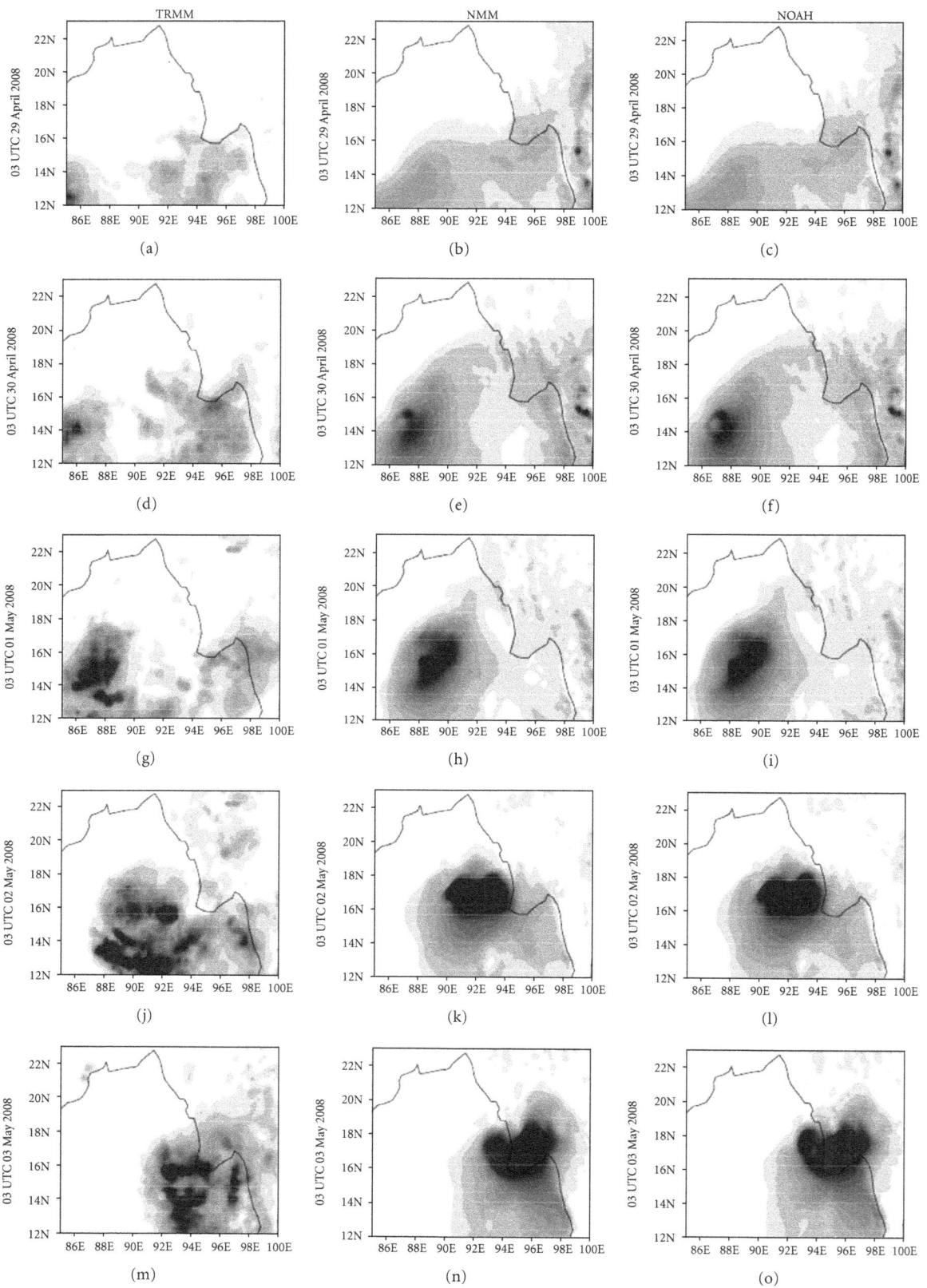

FIGURE 6: 24 hrs accumulated rainfall from TRMM (left panel), NMM land surface (middle panel), and NOAH land surface (right panel) valid at corresponding time.

Impact of Parameterization of Physical Processes on Simulation of Track and Intensity of Tropical Cyclone
Nargis (2008) with WRF-NMM Model

159

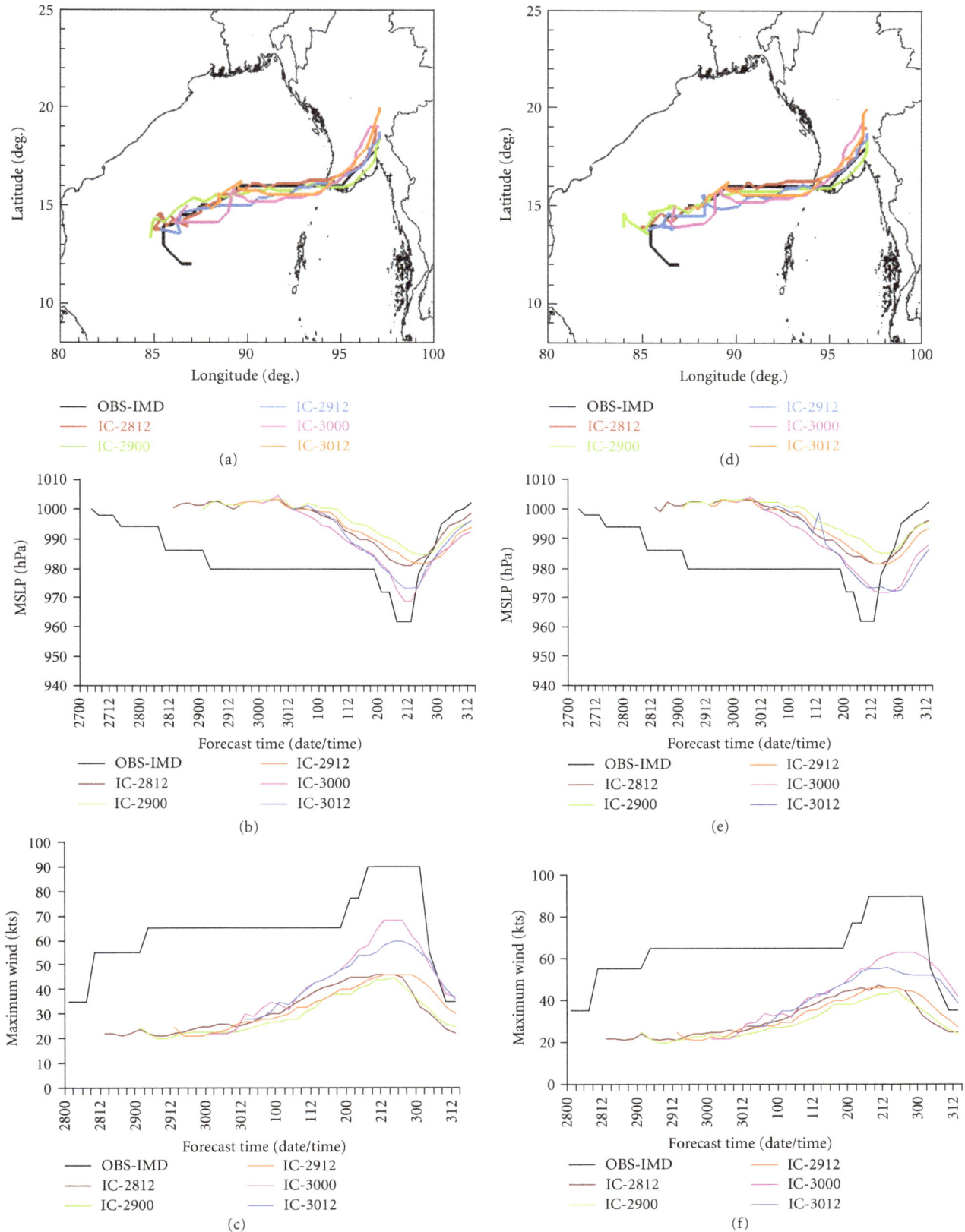

FIGURE 7: (a) Track, (b) MSLP (hPa), and (c) maximum wind (kts) of the cyclone NARGIS with S cumulus convections, Y PBL, NMM land surface scheme, and Ferrier microphysics option at different initial conditions; (d), (e), and (f) are same as (a), (b), and (c) but with NOAH land surface scheme.

(a)   (b)

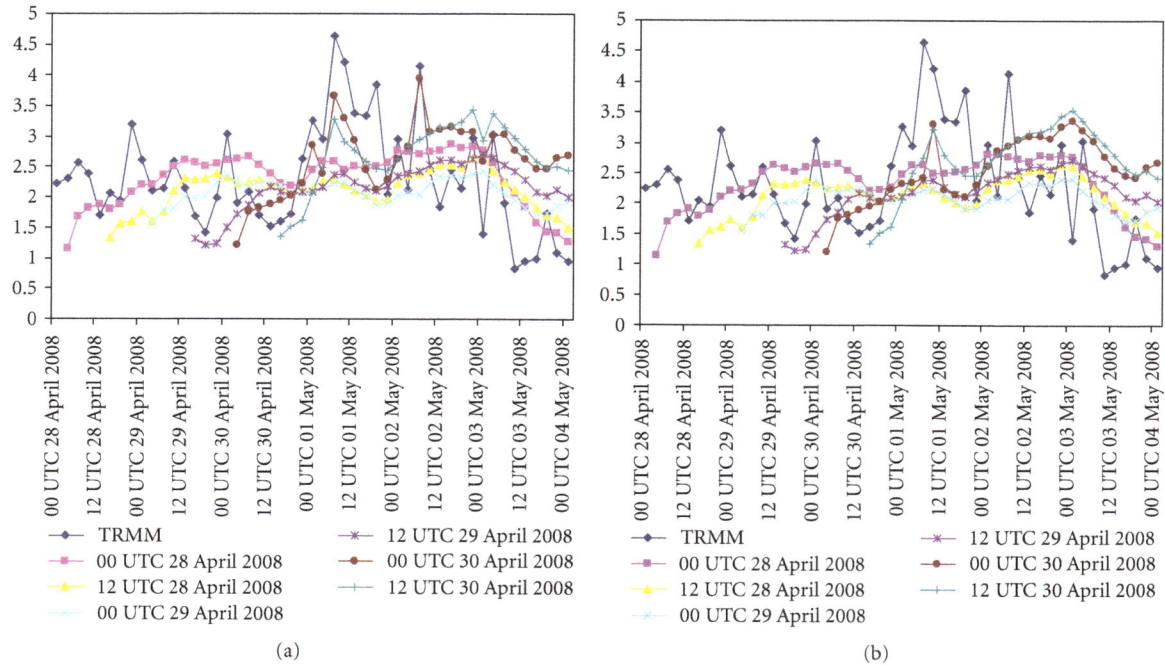

FIGURE 8: Time-series of area-averaged rainfall (cm hr^{-1}) from TRMM and model simulation at different initial conditions. (a) Comparison among TRMM and model simulation with NMM land surface scheme and (b) same as (a) but with NOAH land surface scheme.

TABLE 2: Mean absolute track errors (km) with two optimum physics combinations with different initial conditions.

Initial conditions	Land surface	00 hr	12 hrs	24 hrs	36 hrs	48 hrs	60 hr	72 hrs	84 hrs	96 hrs
1200 UTC 28 Apr 08		118.68	145.58	22.28	33.97	68.96	57.90	54.00	77.90	144.4
0000 UTC 29 Apr 08		185.75	75.73	55.06	101	10.71	53.44	42.75	53.44	63.8
1200 UTC 29 Apr 08	NMM	31.78	19.76	170.6	61.33	106	106.8	15.4	76.8	
0000 UTC 30 Apr 08		59.61	204.7	54.85	88.95	55.5	0.0	0.0		
1200 UTC 30 Apr 08		38.8	77.25	39	69.57	22.23	43.35			
Mean error		*86.924*	*104.604*	*68.358*	*70.964*	*52.68*	*52.298*	*28.0375*	*69.38*	*104.1*
1200 UTC 28 Apr 08		118.68	192.93	95	73.67	109.65	172.84	64.75	87.76	150.24
0000 UTC 29 Apr 08		185.75	45.37	96.91	56.67	29.62	88.29	47.87	71.24	120.11
1200 UTC 29 Apr 08	NOAH	31.78	9.29	160.48	81.62	78	88.4	54.46	46.1	
0000 UTC 30 Apr 08		59.61	204.7	54.85	91.5	59.5	0.0	0.0		
1200 UTC 30 Apr 08		38.8	93.43	39	69.57	22.23	43.35			
Mean error		*86.924*	*109.144*	*89.248*	*74.606*	*59.8*	*78.576*	*41.77*	*68.367*	*135.175*
% of improvement		*0*	*4.34*	*30.6*	*5.1*	*13.5*	*50.3*	*49*	*−1.4*	*30*

produces the less intensity prediction and also results in 03 hrs delay in time.

The mean absolute track error (MATE) (km) with the two optimum physics combinations with different initial conditions are evaluated up to 96 hrs of simulation. The mean MATEs are also calculated for the same period. The 24 hrs result shows that there is an improvement of 30.6% with S + Y + N + F combination than S + Y + NO + F options. Similarly, 48, 72, and 96 hrs results clearly show an improvement of 13.5%, 49%, and 30%, respectively, with the S + Y + N + F combination than S + Y + NO + F options. The detailes of the MATEs are presented in Table 2.

The landfall point errors (LEs) and landfall time errors are also calculated with the two optimum physics combinations with different initial conditions. Results show that S + Y + N + F combination is giving less landfall point error than the S + Y + NO + F combination, though the landfall time error is same in both the schemes. The detailes of the LEs are presented in Table 3.

5.6.2. Simulation of Precipitation Pattern. Figures 8(a) and 8(b) demonstrated the time series of area averaged precipitation simulation with S + Y + N + F and S + Y + NO + F

Impact of Parameterization of Physical Processes on Simulation of Track and Intensity of Tropical Cyclone
Nargis (2008) with WRF-NMM Model

161

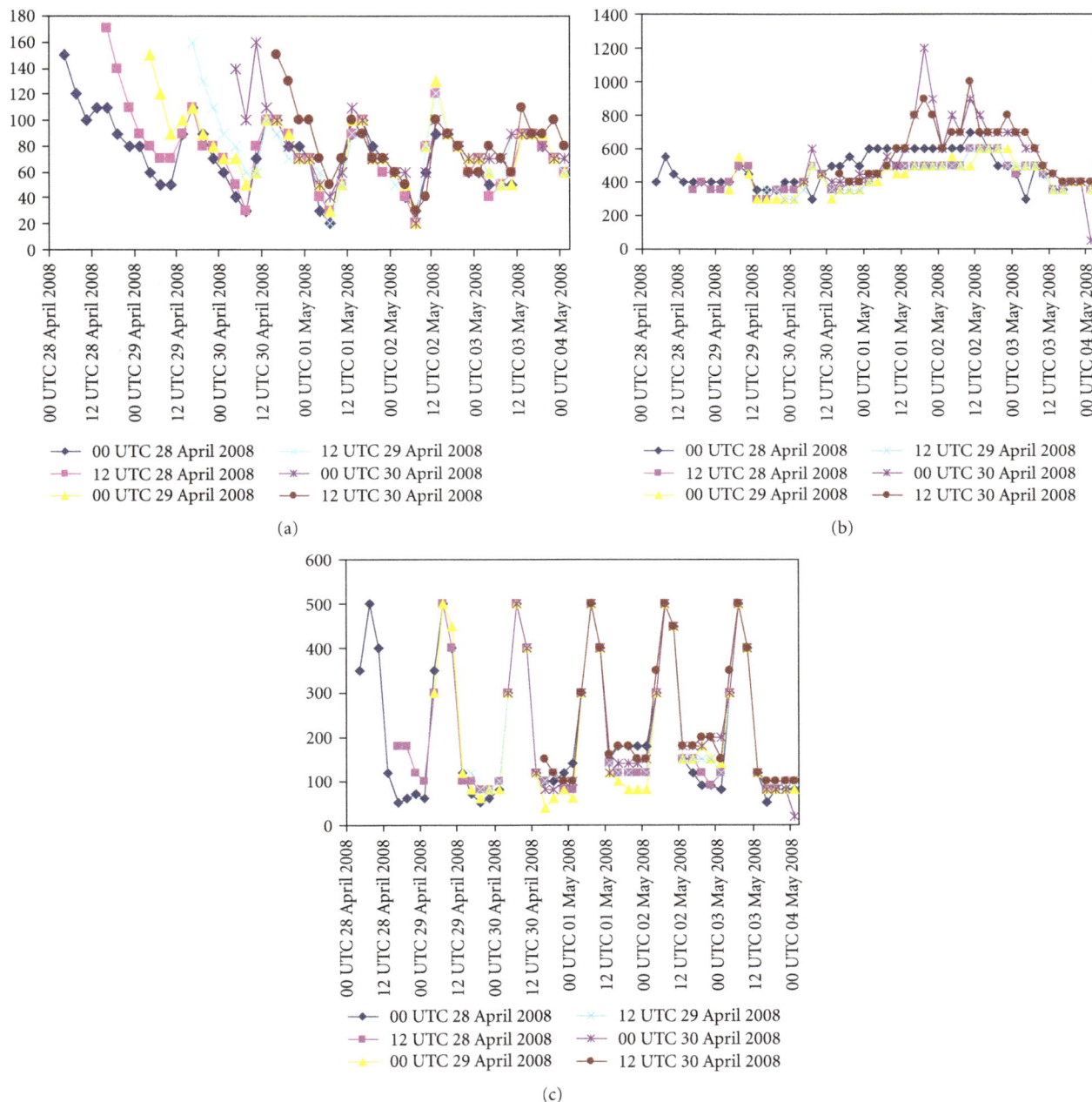

(a)

(b)

(c)

FIGURE 9: Time series of model-simulated (a) ground heat flux (Wm^{-2}), (b) latent heat flux (Wm^{-2}), and (c) sensible heat flux (Wm^{-2}) at different initial conditions with NMM land surface scheme.

TABLE 3: Landfall point errors (km) and landfall time errors (hrs) with two optimum physics combination at different initial conditions.

Initial conditions	Landfall point error (km) with different land surface		Landfall time error (hrs) with different land surface	
	NMM	NOAH	NMM	NOAH
0000 UTC 28 Apr 08	92.4	123.3	−9	−9
1200 UTC 28 Apr 08	54.8	64.2	−6	−6
0000 UTC 29 Apr 08	7.8	7.8	−6	−6
1200 UTC 29 Apr 08	32	34.3	−6	−6
0000 UTC 30 Apr 08	53.4	54.8	0	0
1200 UTC 30 Apr 08	53.4	75.8	0	0

(a)

(b)

(c)

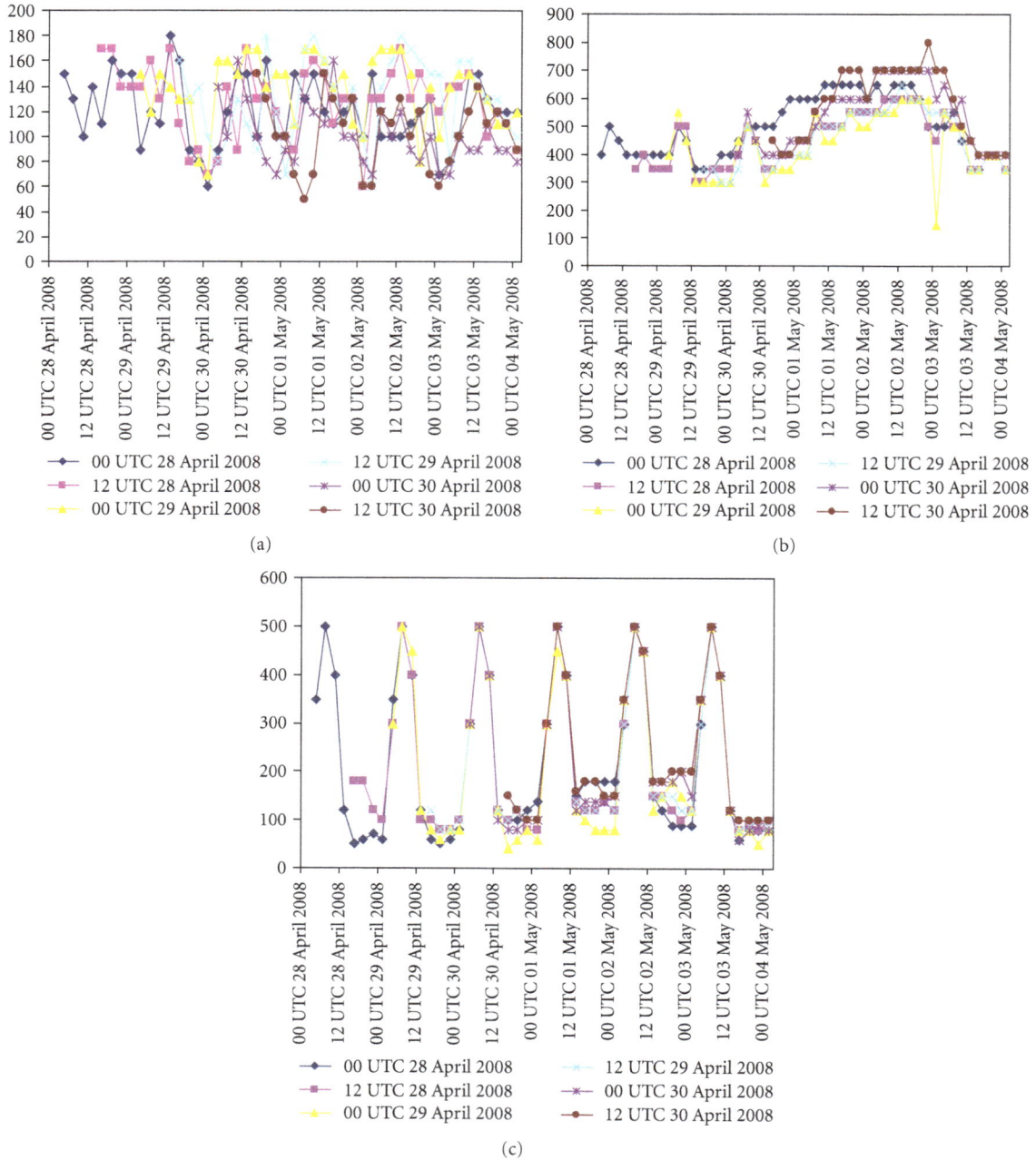

FIGURE 10: Time series of model simulated (a) ground heat flux (Wm^{-2}), (b) latent heat flux (Wm^{-2}) and (c) sensible heat flux (Wm^{-2}) at different initial conditions with NOAH land surface scheme.

combinations, respectively. For the both combinations, the model is integrated at different initial conditions as described above. Also, the time series of area-averaged TRMM precipitation is considered for better comparison. A lot of improvement is seen with N scheme than the NO scheme. Two peak intensities are found from TRMM precipitation at 0600 UTC 01 May 2008 and 0600 UTC 02 May 2008. The model with N scheme is able to simulate the peak precipitation than the NO scheme. At 0600 UTC 01 May 2008, TRMM produced the averaged precipitation of 4.6 mm and model could simulate the precipitation of

3.7 mm and 2.4 mm with N and NO schemes, respectively. Similarly, at 0600 UTC 02 May 2008, TRMM produced averaged precipitation of 4.1 mm and model could simulate the precipitation of 3.9 mm and 2.9 mm with N and NO schemes, respectively. Hence, it may be concluded that the N scheme well-simulates the precipitation than the NO scheme.

5.6.3. *Some Characteristic Features of Nargis.* It has been attempted to study the structure of Nargis in terms of simulation of heat fluxes, vertical velocity, and absolute vorticity

Impact of Parameterization of Physical Processes on Simulation of Track and Intensity of Tropical Cyclone
Nargis (2008) with WRF-NMM Model

163

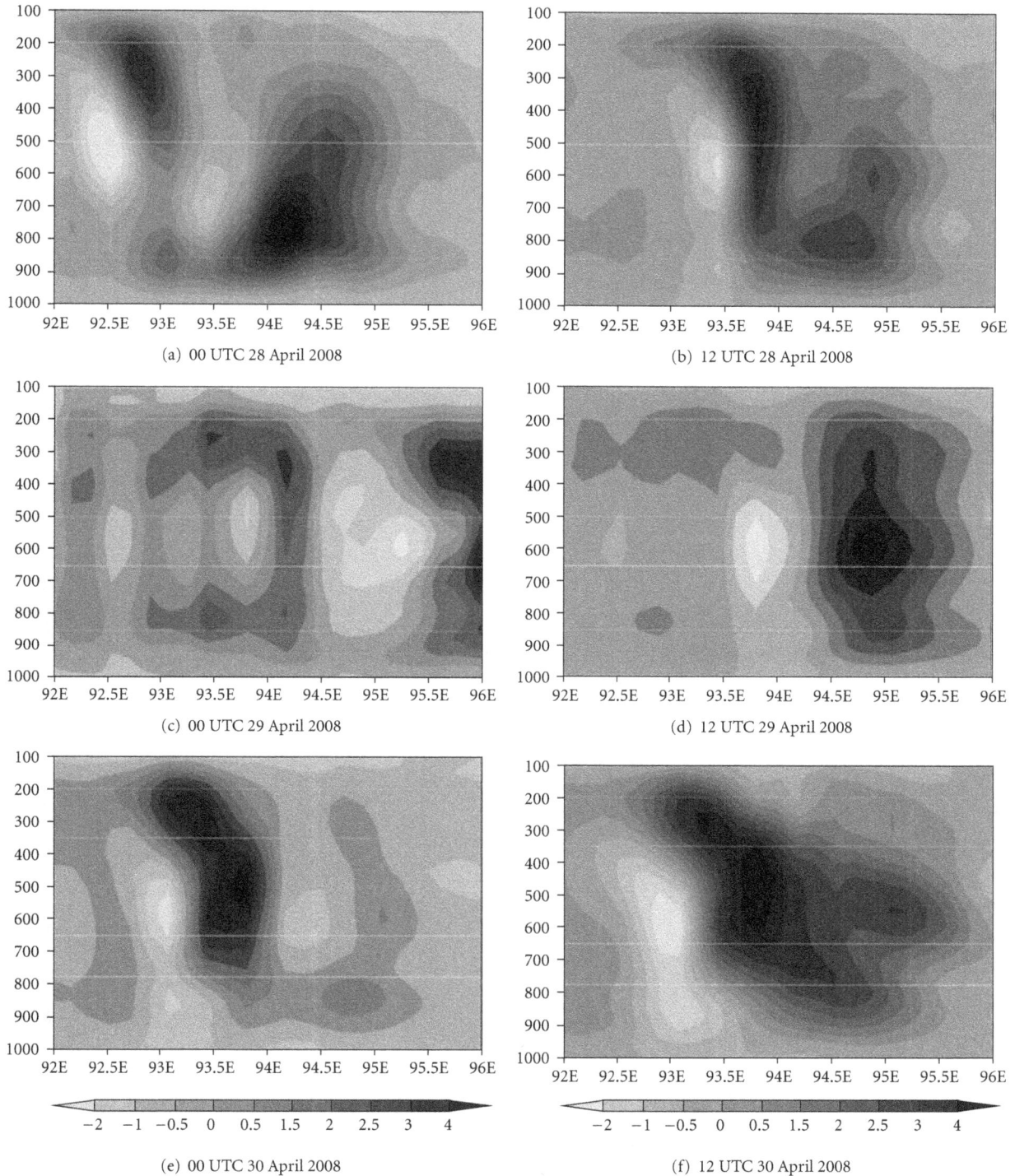

(a) 00 UTC 28 April 2008

(b) 12 UTC 28 April 2008

(c) 00 UTC 29 April 2008

(d) 12 UTC 29 April 2008

(e) 00 UTC 30 April 2008

(f) 12 UTC 30 April 2008

FIGURE 11: Model-simulated vertical velocity (ms^{-1}) at the peak intense time with different initial conditions with NMM land surface scheme.

at different initial conditions. Figures 9(a), 9(b), and 9(c) represent the model-simulated ground heat fluxes (GHF-), latent heat fluxes (LHF-), and sensible heat fluxes (SHF-) with N scheme and with different initial conditions. Figures 10(a), 10(b), and 10(c) represent the model-simulated

ground heat fluxes, latent heat fluxes, and sensible heat fluxes with NO scheme and with different initial conditions. The latent heat flux is one of the dominant components of the air-sea energy exchange processes associated with tropical cyclones. The model simulation with N scheme produced the

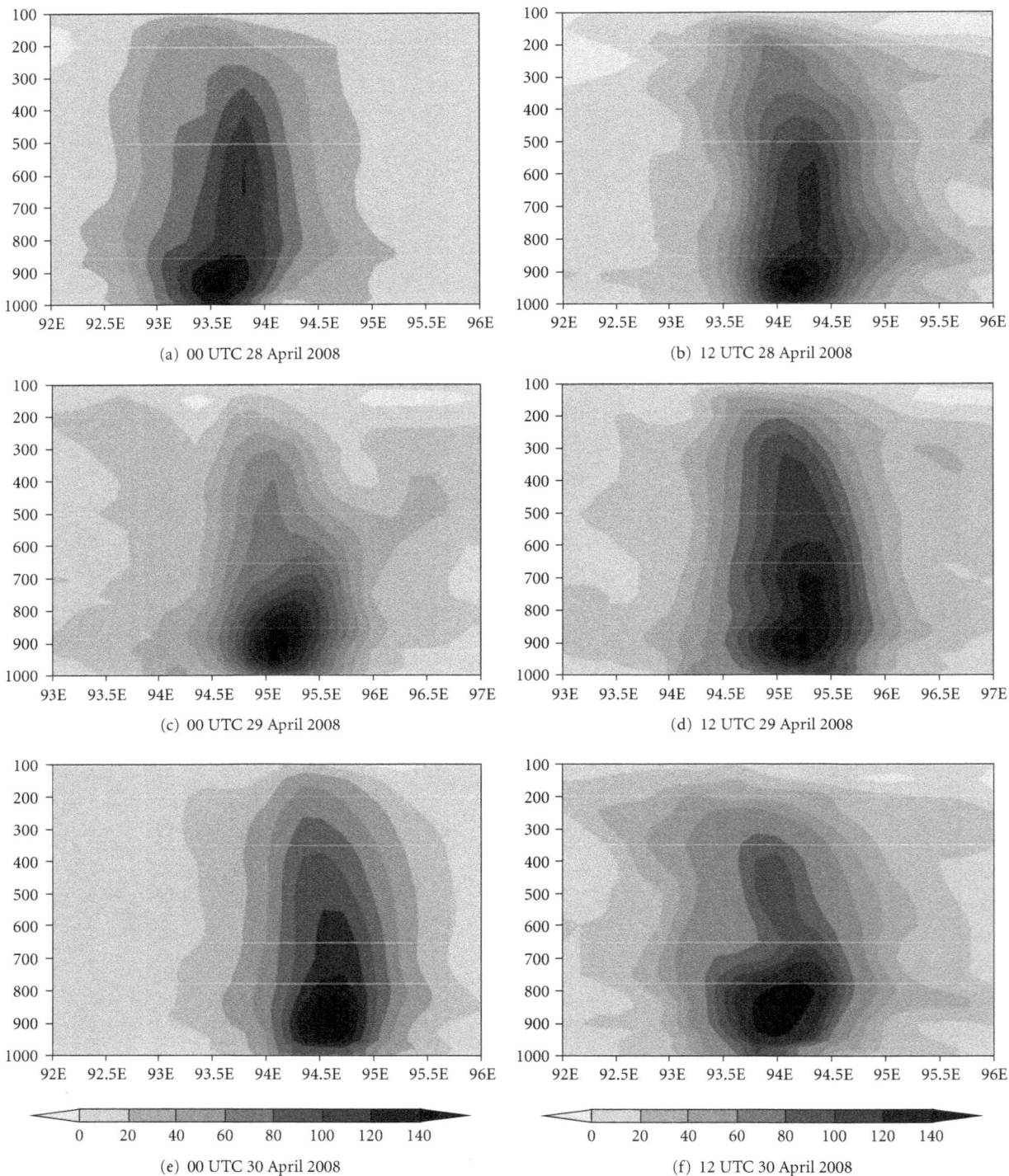

(a) 00 UTC 28 April 2008

(b) 12 UTC 28 April 2008

(c) 00 UTC 29 April 2008

(d) 12 UTC 29 April 2008

(e) 00 UTC 30 April 2008

(f) 12 UTC 30 April 2008

FIGURE 12: Model-simulated absolute vorticity ($\times 10^{-5}$ s^{-1}) at the peak intense time with different initial conditions with NMM land surface scheme.

LHF of 1200 Wm^{-2}, whereas the simulation with NO scheme produced the LHF of 800 Wm^{-2}.

Furthermore, the vertical structure of the storm has been demonstrated with the optimum combination, that is, with S convection, Y planetary boundary layer, N land surface, and F microphysics scheme. Figure 11 represents the model-simulated vertical velocity at the peak intense time of the system with different initial conditions. The strong updraft and downdraft are noticed from model simulation. The maximum value of 5 ms^{-1} is seen in the middle level and

Impact of Parameterization of Physical Processes on Simulation of Track and Intensity of Tropical Cyclone
Nargis (2008) with WRF-NMM Model

165

updraft is extended up to 150 hPa. Figure 12 represents the model-simulated absolute vorticity at the peak intense time with different initial conditions. The positive vorticity of order of $20–140 \times 10^{-5}$ S^{-1} is extended up to 100 hPa. Also, strong positive vorticity is found up to 400 hPa.

All the above results and discussions clearly demonstrate that the S + Y + N + F combination is the optimum combination among all the other combinations in terms of predicting track, intensity, precipitation, and structure of the storm.

6. Conclusions

From the present study on the impact of parameterization schemes for simulation of tropical cyclone, the following broad conclusions are drawn.

The model is sensitive to cumulus convection, planetary boundary layer, and microphysics parameterization schemes. The results from sensitivity experiments with different schemes for cumulus convection indicate that the movement of the cyclone is quite sensitive to the convection processes. The Simplified Arakawa Schubert convection scheme gives better track positions with minimum vector displacement and landfall errors. The result has been clearly demonstrated from the simulation of inner core structure of the storm through temperature anomaly and horizontal wind pattern.

The results from sensitivity experiments with different PBL schemes indicate that the PBL plays an important role in the intensification of the storm. The NCEP GFS scheme gives early intensification of the storm. However, YSU scheme well-simulated the intensification of the storm which is more comparable with the observed value and intense period of the storm. Also, track is well simulated with YSU scheme.

The results from different experiments with land surface physics schemes show that the NMM and NOAH land surface schemes are producing similar type of results and performing well than any other schemes. However, the NMM scheme is giving better result in terms of track and intensity prediction of the storm than the NOAH scheme. Similarly, the results from sensitivity experiments with different microphysics schemes show that the Ferrier scheme is providing better result in terms of track and intensity prediction than other schemes considered in this study.

Further, the results on optimum suitable combination of physical processes in WRF-NMM system are confirmed with additional five different initial values as illustrated in this study. The mean vector displacement error at 24, 48, 72, and 96 hrs are improved by 30%, 13%, 49%, and 30%, respectively, with the optimum combination. The time of occurrence of maximum rainfalls is well captured. Also, the structure of the storm is well predicted with the optimum combination. The results indicate that the combination of Simplified Arakawa Schubert for cumulus convection, Yonsei University planetary boundary layer, NMM land surface, and Ferrier microphysics schemes are providing better result in terms of simulation of track, intensity, and structure of the cyclone than other combinations considered in this study.

Acknowledgments

The Indian National Center for Ocean Information Services (INCOIS) is acknowledged for affording financially under INDOMOD project for this research. The authors gratefully acknowledge the NCEP for providing the NMM model and real-time large-scale analysis as well as forecasts of Global Forecast System (GFS). The authors acknowledge the India Meteorological Department (IMD) for providing observational datasets for intensity and best-fit track data of the storm.

References

[1] D. V. Bhaskar Rao, C. V. Naidu, and B. R. Srinivasa Rao, "Trends and fluctuations of the cyclonic systems over North Indian Ocean," Mausam, vol. 52, pp. 37–46, 2001.

[2] D. De Angelis, "World of tropical cyclones—North Indian Ocean," Marines Weather Log, vol. 20, pp. 191–194, 1976.

[3] R. A. Anthes, Tropical Cyclones: Their Evolution, Structure and Effects, Meteorological Monographs, American Meteorological Society, Boston, Mass, USA, 41 edition, 1982.

[4] W. M. Frank, "The cumulus parameterization problem," Monthly Weather Review, vol. 111, pp. 1859–1871, 1983.

[5] J. Molinari and M. Dudek, "Parameterization of convective precipitation in mesoscale numerical models: a critical review," Monthly Weather Review, vol. 120, no. 2, pp. 326–344, 1992.

[6] R. A. Anthes, "A cumulus parameterization scheme utilizing a one-dimensional cloud model," Monthly Weather Review, vol. 105, no. 3, pp. 270–286, 1977.

[7] M. Mandal, U. C. Mohanty, and S. Raman, "A study on the parameterization of physical processes on prediction of tropical cyclones over the Bay of Bengal with NCAR/PSU mesoscale model," Natural Hazards, vol. 31, no. 2, pp. 391–414, 2004.

[8] D. V. Bhaskar Rao and D. Hari Prasad, "Sensitivity of tropical cyclone intensification to boundary layer and convective processes," Natural Hazards, vol. 41, no. 3, pp. 429–445, 2007.

[9] S. A. Braun and W.-K. Tao, "Sensitivity of high resolution simulation of hurricane Bob (1991) to planetary boundary layer parameterizations," Monthly Weather Review, vol. 128, no. 12, pp. 3941–3961, 2000.

[10] Krishna K. Osuri, U. C. Mohanty, A. Routray, M. A. Kulkarni, and M. Mohapatra, "Customization of WRF-ARW model with physical parameterization schemes for the simulation of tropical cyclones over North Indian Ocean," Natural Hazard. In press.

[11] Z. I. Janjic, "Nonsingular Implementation of the Mellor-Yamada level 2.5 scheme in the NCEP mesoscale model," National Centers for Environmental Prediction Office Note, vol. 437, p. 61, 2001.

[12] Z. I. Janjic, "A Non-hydrostatic model based on a new approach," Meteorology and Atmospheric Physics, vol. 82, no. 1–4, pp. 271–285, 2003.

[13] Z. I. Janjic, "The NCEP WRF core and further development of its physical package," in Proceedings of the 5th International SRNWP Workshop on Non-Hydrostatic Modeling, Bad Orb, Germany, October, 2003.

[14] Z. I. Janjic, "Non-linear advection schemes and energy cascade on semi-staggered grids," Monthly Weather Review, vol. 112, no. 6, pp. 1234–1245, 1984.

[15] S.-Y. Hong and H.-L. Pan, "Non-local boundary layer vertical diffusion in a medium range forecast model," *Monthly Weather Review*, vol. 124, no. 10, pp. 2322–2339, 1996.

[16] I. Troen and L. Mahrt, "A simple model of the atmospheric boundary layer: sensitivity to surface evaporation," *Boundary Layer Meteorology*, vol. 37, no. 1-2, pp. 129–148, 1986.

[17] A. Arakawa and W. H. Schubert, "Interaction of a cumulus cloud ensemble with the large scale environment," *Journal of Atmospheric Science*, vol. 31, no. 3, pp. 674–701, 1974.

[18] G. A. Grell, "Prognostic evaluation of assumption used by cumulus parameterization," *Monthly Weather Review*, vol. 121, no. 3, pp. 764–787, 1993.

[19] H.-L. Pan and W.-S. Wu, "Implementing a mass flux convection parameterization package for the NMC medium-range forecast model," NMC Office Note 409, National Centers For Environmental Prediction, Environmental Modeling Center, Washington, DC, USA, 1995.

[20] J. S. Kain, "The Kain-Fritsch convective parameterization: an update," *Journal of Applied Meteorology*, vol. 43, no. 1, pp. 170–181, 2004.

[21] J. S. Kain and J. M. Fritsch, "Convective parameterization for mesoscale models: the Kain-Fritsch scheme," in *The Representation of Cumulus Convection in Numerical Models*, K. A. Emanuel and D. J. Raymond, Eds., p. 246, American Meteorological Society, 1993.

[22] A. K. Betts and M. J. Miller, "A new convective adjustment scheme. Part II: single column tests using GATE wave, BOMEX, ATEX, and Arctic air-mass data sets," *Quarterly Journal of Royal Meteorological Society*, vol. 112, no. 473, pp. 693–709, 1986.

[23] Z. I. Janjic, "The step-mountain eta coordinate model: further developments of the convection, viscous sub-layer and turbulence closure schemes," *Monthly Weather Review*, vol. 122, no. 5, pp. 927–945, 1994.

[24] J. S. Kain and J. M. Fritsch, "A one-dimensional Entraining/Detraining plume model and its application in convective parameterization," *Journal of Atmospheric Science*, vol. 47, no. 23, pp. 2784–2802, 1990.

[25] Z. I. Janjic, "Comments on development and evaluation of a convection scheme for use in climate models," *Journal of Atmospheric Science*, vol. 57, no. 21, pp. 3686–3686, 2000.

[26] G. A. Grell and D. Dévényi, "A generalized approach to parameterizing convection combining ensemble and data assimilation techniques," *Geophysical Research Letter*, vol. 29, no. 14, pp. 1693–1697, 2002.

[27] W. C. Skamaraock, J. B. Klemp, J. Dudhia et al., "A description of the Advanced Research WRF Version 2," NCAR Technical Note, 2005.

[28] M. B. Ek, K. E. Mitchell, Y. Lin et al., "Implementation of NOAH land surface model advances in the NCEP operational mesoscale Eta model," *Journal of Geophysical Research*, vol. 108, no. D22, p. 8851, 2003.

[29] F. Chen and J. Dudhia, "Coupling an advanced land-surface/hydrology model with the Penn State/NCAR MM5 modeling system. Part I: model description and implementation," *Monthly Weather Review*, vol. 129, no. 4, pp. 569–585, 2001.

[30] T. G. Smirnova, J. M. Brown, S. G. Benjamin et al., "Performance of different soil model configurations in simulating ground surface temperature and surface fluxes," *Monthly Weather Review*, vol. 125, no. 8, pp. 1870–1884, 1997.

[31] T. G. Smirnova, J. M. Brown, S. G. Benjamin, and D. Kim, "Parameterization of cold season processes in the MAPS land-surface scheme," *Journal of Geophysical Research*, vol. 105, no. D3, pp. 4077–4086, 2000.

[32] B. S. Ferrier, Y. Lin, T. Black, E. Rogers, and G. DiMego, "Implementation of a new grid-scale cloud and precipitation scheme in the NCEP Eta model," in *Proceedings of the 15th Conference on Numerical Weather Prediction*, pp. 280–283, American Meteorological Society, San Antonio, Tex, USA, 2002.

[33] S.-Y. Hong, H.-M. H. Juang, and Q. Zhao, "Implementation of prognostic cloud scheme for a regional spectral model," *Monthly Weather Review*, vol. 126, no. 10, pp. 2621–2639, 1998.

[34] S.-Y. Hong, J. Dudhia, and S.-H. Chen, "A Revised approach to ice microphysical processes for the bulk parameterization of clouds and precipitation," *Monthly Weather Review*, vol. 132, no. 1, pp. 103–120, 2004.

[35] Y.-L. Lin, R. D. Farley, and H. D. Orville, "Bulk parameterization of the snow field in a cloud model," *Journal of Climate and Applied Meteorology*, vol. 22, no. 6, pp. 1065–1092, 1983.

[36] J. Dudhia, "Numerical study of convection observed during the winter monsoon experiment using a mesoscale two-dimensional model," *Journal of Atmospheric Science*, vol. 46, no. 20, pp. 3077–3107, 1989.

[37] G. Thompson, R. M. Rasmussen, and K. Manning, "Explicit forecasts of winter precipitation using an improved bulk microphysics scheme, Part I: description and sensitivity analysis," *Monthly Weather Review*, vol. 132, no. 2, pp. 519–542, 2004.

16

Characterisation of Particulate Matter Emitted from Cofiring of Lignite and Agricultural Residues in a Fixed-Bed Combustor

Nattasut Mantananont,[1] Savitri Garivait,[1] and Suthum Patumsawad[2]

[1] *The Joint Graduate School of Energy and Environment, King Mongkut's University of Technology Thonburi, Bangkok 10140, Thailand*
[2] *Department of Mechanical Engineering, King Mongkut's University of Technology North Bangkok, Bangkok 10800, Thailand*

Correspondence should be addressed to Suthum Patumsawad, stt@kmutnb.ac.th

Academic Editors: A. S. Hursthouse and T. Tanisaka

This study is focused on the emission of fixed bed combustor batch operated. Real-time analyser ELPI (electrical low-pressure impactor) system was used to size-segregated particulate matter emission ranging from 40 nm to 10 μm. The results show that total number concentration were 3.4×10^3, 1.6×10^4, and 1.5×10^5 particles/cm$^3 \cdot$ kg$_{fuel}$, while total mass of particles were 12.2, 8.0, and 6.5 mg/Nm$^3 \cdot$ kg$_{fuel}$ for combustion of lignite, rice husk and bagasse, respectively. But it can be noticed that cofiring released more particulate matter. Meanwhile it was found that the effect of ratio of over-fired air to total air supply is more pronounced, since decrease in this ratio, the amount of particles are decreased significantly. For particle size distribution, it can be observed that submicron-sized particles dominate and the most prevailing size is in the range: 50 nm < D_p < 100 nm, for lignite and agricultural residues. However, during cofiring of fuel mixture at 70% rice husk mass concentration, it is found that there are two major fractions of particle size; 40 nm < D_p < 70 nm and 0.2 μm < D_p < 0.5 μm. The analysis of particle morphology showed that the isolate shape of submicron particle produced during lignite combustion is characterised by different geometries such as round, capsule, rod, flake-like, whereas the spherical shape is obtained with combustion of rice husk.

1. Introduction

Airborne particulate matter (PM) is one of the major pollutants affecting negatively the atmospheric environment, combustion system, and human health. For its impact on atmospheric environment, it is known that sub-micron-sized particle (e.g., 0.1–1 μm) whether in form of solid or droplet plays a role to decrease visibility [1]. In problematic of combustion system [2], there were reports that serious corrosion problems were found in the cooler part of the flue gas path. From SEM-EDS analysis, it indicated that the corroded tube was covered with oxide layer having rich of Fe, K, Cl, Si, and S, which these elements mostly contain in submicron particle.

The major fuel for energy production in Thailand is lignite; however, its amount is limited in a long term. As agricultural countries, Thailand produces large amount of agricultural residues such as rice husk, and bagasse [3]. In the light of these, the energy production by co-firing lignite/residues

becomes a promising option. PM emission during cocombustion of coal/biomass/wastes has broadly been investigated [4–10], but they are mostly processed by densification and burnt as pellet or briquette. Meanwhile, in Thailand, study of emission from cocombustion of domestic lignite, biomass, and waste has been investigated [11] but they focused only on gaseous emission and combustion efficiency associated with combustion condition regardless to the measurement of particulate matter. In fact, the characteristics of particulate emitted either from combustion of Thai lignite, rice husk, and bagasse or from cofiring of Thai lignite/rice husk have not been investigated in Thailand up to now.

This study is, therefore, focused on the PM emission from lab-scale fixed bed combustor batch operated. The point of study includes total number/mass concentration of PM, and determination of particle morphology. The effect of the fuel mixture and the ratio of overfired air to total air supply (OFA/TA) on PM characteristics has also been addressed.

FIGURE 1: fuel physical (a) lignite, (b) rice husk, and (c) bagasse.

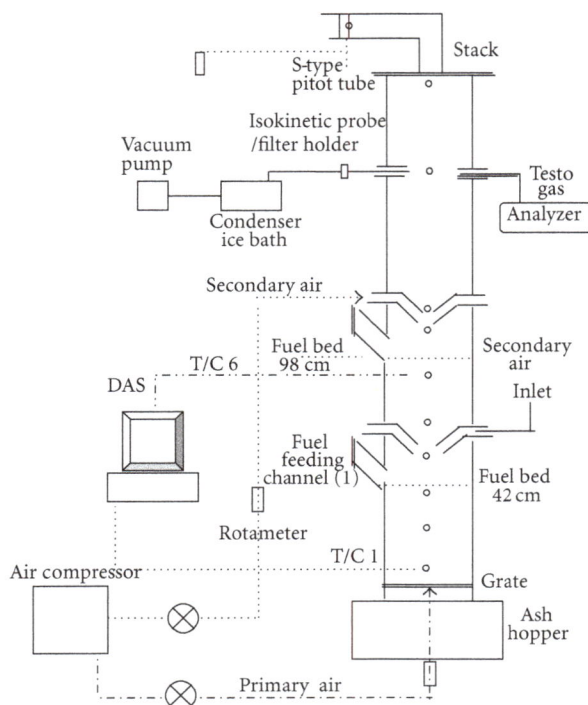

FIGURE 2: Fixed-bed combustor and particulate sampling system.

2. Experimental Setup

2.1. Fuel Preparation and Properties. In this study, domestic lignite and two agricultural residues; rice husk and bagasse have been selected. Their physical are depicted by Figure 1. Lignite was supplied by the Electricity Utility in Thailand and was crushed and sieved to 3–5 mm in diameter range. Rice husk and bagasse were received from rice mills and sugar cane factory, respectively, and used in as-received characters, as shown in Figure 1. Since bagasse physical is inhomogeneous in size/shape (e.g., short-long line, thinner-thicker shape, or powder portion) including low density ($60 \, \text{kg/m}^3$) in comparison with lignite, making difficult to well mixing with lignite, so combustion of bagasse alone is present in this paper. However, rice husk seemed better to mix with lignite; hence cofiring of lignite and rice husk can be tested.

For cofiring of lignite and rice husk, they were mixed together before loading to reactor. The fuel mixtures of lignite and rice husk are 30/70 and 60/40 by mass concentration. Burning 100% of lignite, rice husk and bagasse are performed as baseline data. Fuel properties are shown in Table 1.

2.2. Experimental Rig. The experiments were conducted in a lab-scale fixed bed combustion system which is a vertical cylindrical chamber of 120 mm internal diameter, 2680 mm height, insulating with 45 mm of refractory material, and 20 mm of rock wool. The grate is located at the bottom. Eleven thermocouples (types K chromel-alumel) were used to measure the temperature along the reactor including combustion zone and freeboard. Air supply was divided into two parts, named as overfired and underfired air. Underfired air was fed beneath the grate while overfired air was put at 840 mm height above the grate. The ratio of overfired air to total air (OFA/TA) was increased from 0 to 0.3. Figure 2 shows the schematics of experimental setup.

2.3. Particulate Sampling and Analysis. Real-time analyser: ELPI (electrical low-pressure impactor) was used to size-segregated PM emission ranging from 40 nm to 10 μm. Sampling probe was inserted in fixed-bed reactor located at 2680 mm above the grate. Twenty-five millimetre of diameter of Thirteen-Teflon filters without greasing was used to sampling particulate per one case. Particle sized larger than 10 μm was trapped at 13th stage of ELPI and only particle sized below 10 μm was allowed through size-segregation 1st–12th stages of ELPI. The sampling rate was fixed constantly at 10 L/min entire the sampling time. Number of particle is a function of measured current and mass concentration of particle was calculated based on 1 g/cm^3 of particle density. Total number and total mass of particle were obtained by integration of value in each stage (i.e., 1st–12th).

SEM (scanning electron microscopy) model: JSM-6301F was used to study the particle morphology. The analysis of particle morphology covered the significant mode of particulate, especially submicron particle, nucleation mode (D_p; 70–100 nm), or 2nd stage of filter, accumulation mode (D_p; 0.2–0.31 μm) or 4th stage including supermicron particle (D_p; 3–5 μm) or 10th stage.

TABLE 1: Fuel properties.

Fuel	Proximate analysis (% wt, as received)				Ultimate analysis (% wt, dry ash free)					HHV (MJ/kg)
	Moisture	Volatile matter	Fixed carbon	Ash	C	H	O	N	S	
Lignite	9.0	41.4	33.1	11.2	62.7	4.1	27.0	1.1	5.2	18.9
Rice husk	8.5	57.5	17.2	16.8	49.1	6.3	44.4	0.2	—	14.1
Bagasse	14.2	68.0	20.3	2.5	43.1	5.1	51.4	0.3	—	17.8

	Oxide composition of fuel ash (% wt)										
	SiO_2	Al_2O_3	Fe_2O_3	CaO	TiO_2	MgO	SO_3	P_2O_5	Na_2O	MnO_2	K_2O
Lignite	34.38	16.50	12.80	13.70	0.36	1.87	17.05	0.16	0.90	0.11	2.72
Rice husk	92.70	0.14	2.00	0.54	0.02	0.35	0.37	0.43	0.07	0.19	2.50
Bagasse	42.90	23.80	16.90	2.20	2.50	2.10	0.60	1.30	0.60	ND	3.20

TABLE 2: Total number/mass concentration of particles.

Fuel	Fuel density (kg/m^3)	Total number (particles/cm^3 · kg$_{fuel}$)	Total mass concentration (mg/Nm^3 · kg$_{fuel}$)
Lignite	736	3.4×10^3	12.2
Rice husk	90	1.6×10^4	8.0
Bagasse	60	1.5×10^5	6.5

Total air flow rate: 300 LPM and OFA/TA: 0.3.

3. Results and Discussion

3.1. Total Number/Mass Concentration of Particle Emitted from Combustion. The results are shown in Table 2. The total number of particle emitted from combustion of lignite, rice husk, and bagasse are 3.4×10^3, 1.6×10^4, and 1.51×10^5 particles/cm^3 · kg$_{fuel}$, respectively, while total mass of particles are 12.2, 8.0, and 6.5 mg/Nm^3 · kg$_{fuel}$. These results indicate roughly that combustion of low bulk density fuel may be one of the causes to generate higher the emitted particle (compared to lignite).

However, comparison with 1.8×10^{13}, 1×10^{13}, 1.7×10^{13} particle/kg released from combustion in self-built burning stove of wheat straw, corn straw, and rice straw, respectively [13]. It seems that the low of density fuel (i.e., rice husk or bagasse) may not be a priority concerned with high emission of PM but the combustion technology or operating condition seems more importance.

Other interesting is that bagasse and rice husk have higher volatile yield than coal, therefore, the main combustion process is marked by devolatilisation rate of fuel and homogeneous (gas-phase) reaction dominated, which later favours particle formation via gas-to-solid pathway (e.g., condensation). According to this phenomenon, this could be observed from the reverse relationships between particle number and particle mass concentration. For instance, most prevailing size of particle of bagasse combustion is at d_p 70 nm of 80% cumulative of total number of particle. Because this high content of submicron particles is less significant to contribute the overall mass loading, thus low mass concentration does. In addition, low mass of emitted particle also indicates the lower particle density of residues.

3.2. Total Number/Mass Concentration of Particle Emitted from Cofiring of Lignite and Rice Husk. Cofiring of lignite and rice husk was performed under various mass fraction and

ratio of overfired air to total air and results are shown in Table 3.

It can be seen that cofired lignite and rice husk result in increase of both particle number and mass concentration compared to burning of either lignite or rice husk. This synergy effect could be from the difference in fuel properties and physical which needed further investigation and analysis. While mass fraction concentration has affected to PM emission, an increase in lignite mass concentration leads to decrease in PM emission. However, total number/mass concentration of particle is decreased dramatically at overfired air to total air ratio of 0.1. The result of particle number of fuel mixture (8.7×10^3) is in between those of lignite and rice husk (3.4×10^3 and 1.6×10^4) but mass concentration is much lower. This could be said that PM emission at this condition probably is very fine particle.

3.3. Particle Size Distribution (PSD) of Combustion. As from the results, it was found that the most prevailing particle size was in the range of 50 nm to 100 nm for combustion of lignite and bagasse. It was accounted to 60 and 80% of total particle for lignite and bagasse, respectively. Meanwhile, for rice husk combustion, it was obviously seen that there were two groups of particle size range; 50–100 nm and 0.5–1.0 μm. This could be inferred that ultrafine or fresh particle was collided and agglomerated to form fine particle.

3.4. Particle Size Distribution (PSD) of Cofiring of Lignite and Rice Husk. The results from cofiring lignite with rice husk show the same effect as burning of either lignite or agricultural residues. The major fraction of particle size is 40–70 nm but the number of particle is higher. However, the increase in rice husk mass fraction, 40 to 70%, leads to release larger particle size. At 70% rice husk mass fraction, there are

Stage no. (D_p)	S2 (70–100 nm)	S4 (0.2–0.31 μm)	S10 (3–5 μm)

Fuel mass fraction 100% lignite (by mass)

(a)

60% lignite 40% rice husk

(b)

30% lignite 70% rice husk

(c)

100% rice husk

Nucleation mode   Accumulation mode   Coarse mode

(d)

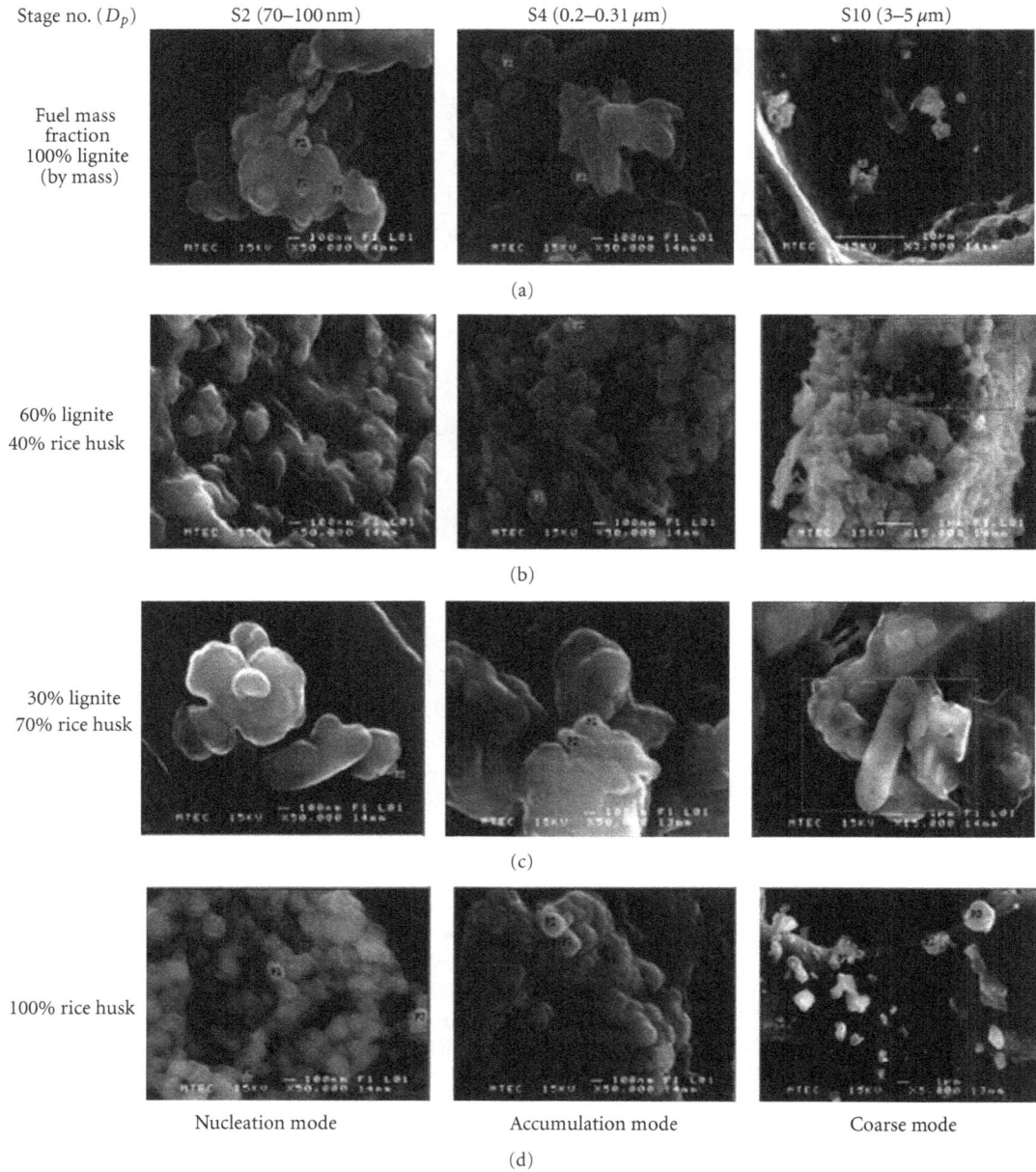

FIGURE 3: Particle morphologies.

two modes of particle size range; 40–70 nm and 0.2–0.5 μm, which are agreed with the results of rice husk combustion.

3.5. Particle Morphology. SEM was used to investigate particle morphology. Particle shape derived from combustion of lignite, rice husk, and cofiring of lignite/rice husk are illustrated by Figures 3(a)–3(d). It can be seen from Figure 3(a) that the isolate shape of submicron particle produced during lignite combustion is characterised by different geometries such as round, capsule, rod, flake-like, whereas the spherical shape is obtained from rice husk combustion (see Figure 3(d)).

For cofiring mode, Figure 3(c) (left and middle) shows that co-firing of high mass fraction of rice husk (70%) enables to modify structure of submicron particle from "small-roundly shaped" to "large-amorphously shaped," in comparison to rice husk burning case, which finally results in increasing of the average diameter of particle.

4. Conclusion

Characterisation of particulate matter emitted from firing and cofiring of lignite and agricultural residues, rice husk and bagasse, has been investigated in fixed-bed combustor batch operated. Parameters concerned in this study are comprised

TABLE 3: Total number/mass concentration of particle.

Lignite/Rice husk mass fraction	OFA/TA	Fuel Density (kg/m^3)	Burning Rate (g/sec) [12]	Total number (particles/cm^3 · kg$_{fuel}$)	Total mass concentration (mg/Nm^3 · kg$_{fuel}$)
30/70	0.3	120	0.44	2.2×10^5	20.5
60/40	0.3	220	0.42	6.3×10^4	15.3
60/40	0.1	220	0.57	8.7×10^3	2.1

Total air flow rate: 300 LPM.

of total number/total mass concentration and particle morphology. It can be summarised the results as follows.

(1) Total number concentration was 3.4×10^3, 1.6×10^4, and 1.5×10^5 particles/cm^3 · kg$_{fuel}$, while total mass of particles was 12.2, 8.0 and 6.5 mg/Nm^3 · kg$_{fuel}$ for combustion of lignite, rice husk, and bagasse, respectively.

(2) In cofiring of lignite and rice husk, the results show synergy effect which released particulate matter is higher than burning either lignite and rice husk. The increase in rice husk mass fraction tends to increase the amount of particle. Nevertheless, it was found that the effect of ratio of overfired air to total air supply is more pronounced, since decrease in this ratio, from 0.3 to 0.1, the amount of particles are decreased significantly.

(3) During cofiring fuel mixture at 70% of rice husk mass fraction, it is found that there are two major fraction of particle size; $40 < D_p < 70$ nm to $0.2 < D_p < 0.5$ μm. This indicates the possibility of agglomeration of ultrafine particle increasing the average diameter of particle.

(4) The analysis of particle morphology shows that the isolate shape of submicron particle produced during lignite combustion is characterised by different geometries such as round, capsule, rod, flake-like, whereas the spherical shape is obtained with rice husk combustion.

Acknowledgments

The authors gratefully acknowledge the financial support provided by The Joint of Graduate School of Energy and Environment (JGSEE) and the technical support of the Department of Mechanical Engineering at King Mongkut's University of Technology North Bangkok (KMUTNB).

References

[1] W. John, "Multimodal size distributions of inorganic aerosol during SCAQS," in *Proceedings of the International Specialty Conference*, Los Angeles, Calif, USA, 1993.

[2] J. F. Frandsen, L. Moiraghi, S. van Lith, P. A. Jensen, and P. Glarborg, Release of metals, sulphur and chlorine and residual ash formation during biomass combustion on a grate.

[3] *Biomass*, Biomass Clearing House, Energy for Environment Foundation, 2006.

[4] B. K. Gullett, K. Raghunathan, and J. E. Dunn, "The effect of cofiring high-sulfur coal with municipal waste on formation of polychlorinated dibenzodioxin and polychlorinated dibenzofuran," *Environmental Engineering Science*, vol. 15, no. 1, pp. 59–69, 1998.

[5] T. Nussbaumer, "Combustion and co-combustion of biomass," in *Proceedings of the 12th European Conference and Technology Exhibition on Biomass for Energy, Industry and Climate Protection*, Amsterdam, The Netherlands, June 2002.

[6] W. Maenhaut, M. T. Fernandez-Jimenez, T. Lind et al., "In-stack particle size and composition transformations during circulating fluidized bed combustion of willow and forest residue," *Nuclear Instruments and Methods in Physics Research B*, vol. 150, no. 1-4, pp. 417–421, 1999.

[7] I. Obernberger and T. Brunner, *Fly Ash and Aerosol Formation in Bio-Combustion Processes-an Introduction*, vol. 6 of *Thermal Biomass Utilization*, 2005.

[8] R. Korbee, M. K. Cieplik, and J. H. A. Kiel, Release of aerosol forming species during combustion in pulverized fuel systems.

[9] L. S. Johansson, C. Tullin, B. Leckner, and P. Sjovall, "Particle emissions from biomass combustion in small combustors," *Biomass and Bioenergy*, vol. 25, no. 4, pp. 435–446, 2003.

[10] C. Y. H. Chao, P. C. W. Kwong, J. H. Wang, C. W. Cheung, and G. Kendall, "Co-firing coal with rice husk and bamboo and the impact on particulate matters and associated polycyclic aromatic hydrocarbon emissions," *Bioresource Technology*, vol. 99, no. 1, pp. 83–93, 2008.

[11] K. Suksankraisorn, S. Patumsawad, P. Vallikul, B. Fungtammasan, and A. Accary, "Co-combustion of municipal solid waste and Thai lignite in a fluidized bed," *Energy Conversion &Management*, vol. 45, no. 6, pp. 947–962, 2004.

[12] S. Brunello, I. Flour, P. Maissa, and B. Bruyet, "Kinetic study of char combustion in a fluidized bed," *Fuel*, vol. 75, no. 5, pp. 536–544, 1996.

[13] H. Zhang, X. Ye, T. Cheng et al., "A laboratory study of agricultural crop residue combustion in China: emission factors and emission inventory," *Atmospheric Environment*, vol. 42, no. 36, pp. 8432–8441, 2008.

17

The Influence of Wildfires on Aerosol Size Distributions in Rural Areas

E. Alonso-Blanco,[1] A. I. Calvo,[2] R. Fraile,[1] and A. Castro[1]

[1] Department of Physics (IMARENAB), University of León, León 24071, Spain
[2] Centre for Environmental and Marine Studies (CESAM), University of Aveiro, Aveiro 3810-193, Portugal

Correspondence should be addressed to R. Fraile, roberto.fraile@unileon.es

Academic Editor: Beat Schmid

The number of particles and their size distributions were measured in a rural area, during the summer, using a PCASP-X. The aim was to study the influence of wildfires on particle size distributions. The comparative studies carried out reveal an average increase of around ten times in the number of particles in the fine mode, especially in sizes between 0.10 and 0.14 μm, where the increase is of nearly 20 times. An analysis carried out at three different points in time—before, during, and after the passing of the smoke plume from the wildfires—shows that the mean geometric diameter of the fine mode in the measurements affected by the fire is smaller than the one obtained in the measurements carried out immediately before and after (0.14 μm) and presents average values of 0.11 μm.

1. Introduction

The size of atmospheric aerosols depends greatly on the sources and sinks, as well as on the meteorological processes dominating during their lifetime [1–4]. Biomass burning is one of the most important contributors of aerosols to the atmosphere, releasing large amounts of particles and gases, and causing alterations in atmospheric composition at a local and at a global scale. The effects aerosols may trigger in the atmosphere depend on the type of material burnt, the combustion phase, the relative humidity, and the wind conditions [5]. A growing number of studies focus on these changes, because aerosols undergo complex interactions in the atmosphere and condition global energy balance [6].

Many studies have found an increased number of aerosols in the fine or accumulation mode during wildfires [7–9]. An in-depth analysis of the emissions released by biomass burning can be found in Reid et al. [5]. This study showed that out of all the particles released during these phenomena, between 80% and 90% correspond to this mode, with a count median diameter (CMD) of 0.13 μm. These particles are mainly composed of organic material; inorganic elements only account for approximately 10% of the mass [7].

The water-absorption capacity of these particles varies greatly. The hygroscopicity of aerosols originated by biomass burning depends on the internal composition of the organic and inorganic material of the submicrometer particles [10–12]. In consequence, studies on aerosol size distributions must take into account hygroscopicity. The ability of aerosols to absorb water vapor from the atmosphere is an important feature, but very difficult to study [13]. The water content in the particles affects factors such as size, total mass, acidity, the amount of water-soluble substances they contain, light-dispersion properties, chemical reactivity, the ability to function as condensation nuclei in a cloud, and their permanence in the atmosphere.

The ability of aerosols generated by biomass burning to form condensation nuclei has been the focus of several studies. Authors such as Warner and Twomey [14], Eagan et al. [15], and Roberts et al. [16] have found that, at supersaturations higher than 0.5%, the particles generated by wildfires function as condensation nuclei (CCN). More recently, Petters et al. [17] studied the hygroscopic properties of aerosols freshly emitted from laboratory biomass burning experiments. They conclude that at the point of emission, most particles are CCN active and do not depend upon

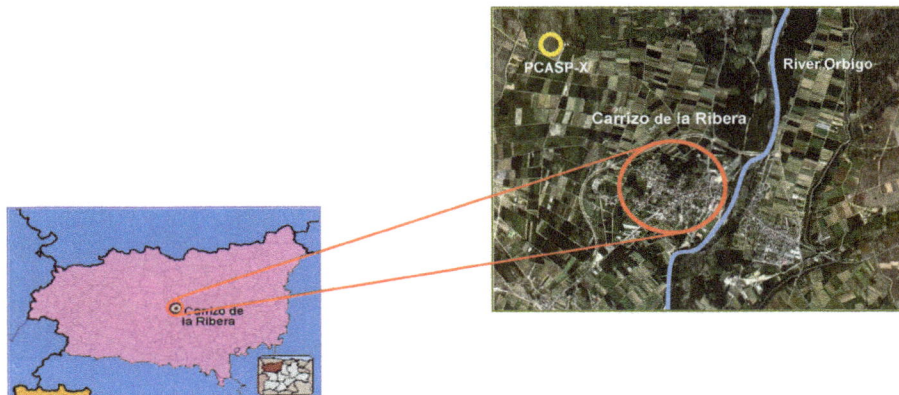

FIGURE 1: Location of Carrizo de la Ribera in the province of León. Location of the PCASP-X probe and the river Órbigo, with respect to the town.

conversion in the atmosphere to more hygroscopic compositions before they can participate in cloud formation and undergo wet deposition.

In this paper, we will analyze the characteristics of particle size distributions in a rural area in the summer months, and the changes in these distributions caused by the arrival of aerosols from biomass burning, mainly wildfires affecting no-tree woodland, but sometimes also trees, and crops or crop stubble. These wildfires are particularly frequent in the rural study zone. The wildfires considered are all the fires registered in the province of León, Spain. The distance with the probe installed at the sampling site varies between a few kilometres and a maximum of 70 km.

In the case of wildfires, the source of the particles is well defined and located; but in the study of aerosol size distributions, we must also take into account the origin and the type of air mass that carries the particulate matter when it reaches the sampling site [18].

2. Study Zone

The measurements have been carried out in the district of Carrizo de la Ribera, in the middle of the province of León, Spain, north-west of the capital (Figure 1). This rural area lies at 873 m above sea level and has 2,554 inhabitants. Farming is the main economic activity, and hop the most widely grown crop. The weather is monitored by a station from the Spanish National Weather Agency installed in that town (42°35′N, 5°39′W).

The study zone lies on the left bank of the River Órbigo, over a fluvial terrace. The climate is of the continentalized Mediterranean type, with a marked seasonality. Precipitation is scattered irregularly along the year, with minimum values in the summer and the highest values in spring and autumn.

Temperatures are fresh, with an annual average of 10.5°C. The winters are cold, with frequent frosts. The summers are warm, with maximum temperatures that may be over 35°C, although nocturnal temperatures are moderate, as it is an area of irrigated farming systems where land is mainly watered by flooding.

TABLE 1: Refractive Index of Atmospheric Aerosol Particle at a He-Ne Laser Wavelength ($0.6328\,\mu$m) for Rural Aerosol Model (according to Kim and Boatman, [20]).

Relative humidity (%)	Rural aerosol
0	1.530–$6.60 \times 10^{-3}\,i$
50	1.520–$6.26 \times 10^{-3}\,i$
70	1.501–$5.60 \times 10^{-3}\,i$
80	1.443–$3.70 \times 10^{-3}\,i$
90	1.399–$2.22 \times 10^{-3}\,i$
99	1.359–$9.16 \times 10^{-4}\,i$

3. Materials and Methods

3.1. Measurement Equipment. A laser spectrometer (Passive Cavity Aerosol Spectrometer Probe, PMS Model PCASP-X) was installed in a field 2 km from Carrizo de la Ribera (42°35′59″ N, 5°50′50″ W) to determine aerosol size spectra in the rural areas. This device measures particles ranging between 0.1 and 10 μm considering their light-dispersion properties in a wave length of 633 nm between the angles of 35° and 135°. The spectrometer measures 31 channels, that is, 31 discrete particle size intervals. The probe was calibrated by the manufacturer using polystyrene latex particles of a known size. The refractive index of latex beads (1.59–0 i) is different from that of atmospheric particles, resulting in an aerosol size distribution that is "latex equivalent".

Here, we are presenting PCASP-X size distributions corrected using Mie theory and implemented with a computer code developed by Bohren and Huffman [19], according to the refractive index typical of rural aerosols, whose value varies depending on the relative humidity [20]. The refractive indices were calculated interpolating the real part and the imaginary part in accordance with the relative humidity at the time of measurement (values between 17% and 100%). Table 1 shows the refraction indices used for the interpolation.

It was also necessary to carry out a number of corrections on the number of counts sampled by the spectrometer in each channel. First, the flow measurement value was set in relation to the altitude of the sampling point; the probe was installed at 834 m introducing a correction factor of 0.905. In addition, each measurement was corrected according to the activity registered. Finally, the last correction was determined by the duration of the samplings, which was 600 seconds. These corrections are described in more detail in Calvo et al. [6].

The study period covers the 122 days of the months of June, July, August, and September of the year 2000. Eight measurements were carried out daily, measuring the ambient particle size spectrum in the rural study area automatically, during 15-minute intervals every 3 hours.

The probe was installed next to a weather station registering automatically data on precipitation, pressure, temperature, relative humidity, and wind speed and direction. A wind profiler was also used (Sodar SR1000), with a pulse frequency of 5 tones, around 2.150 Hz, pulse power of 300 W, a pulse-repetition period of 8 s, and a maximum range of 1.250 m. This device registered automatically data from the three wind components. The data registered by the weather station and the data registered by the Sodar were all stored on a computer every 30 minutes.

3.2. Thermal Inversions and Circulation Weather Type Classification.

The thermal inversions have been calculated using data from soundings in La Coruña (43.36°N, 8.41°W, altitude 67 m), Madrid (40.50°N, 3.58°W, altitude 633 m), and Santander (43.48°N, 3.80°W, altitude 59 m) provided by the University of Wyoming (http://weather.uwyo.edu/upperair/sounding.html).

In order to identify the type of weather associated to a particular synoptic situation, a Circulation Weather Type classification (CWTs) was designed based on Jenkinson and Collison [21] and Jones et al. [22]. These procedures were initially developed to define objectively Lamb Weather Types [23] for the British Isles. The daily circulation affecting the Iberian Peninsula is described using a set of Indices associated to the direction and vorticity of the geostrophic flow. The Indices used were the following: southerly flow (SF), westerly flow (WF), total flow (F), southerly shear vorticity (ZS), westerly shear vorticity (ZW), and total shear vorticity (Z). These Indices were computed using sea level pressure (SLP) values obtained for the 16 grid points, distributed around the Iberian Peninsula. This method allows for a maximum of 26 different CWTs. Following Trigo and DaCamara [24] in their study for Portugal, this study does not have an unclassified class, but rather opts for disseminating the fairly few cases (<2%) with possible unclassified situations among the retained classes. This classification has already been used on the Iberian Peninsula, for example, in the study of lightning [25], splash erosion [26], or aerosol size distribution in precipitation events [27].

3.3. Data Analysis.

The Department for the Environment of the regional government of the Junta of Castile and León provided the database with the number of wildfires, the district where each fire occurred, the date of detection and extinction (with exact date and time), the land area affected

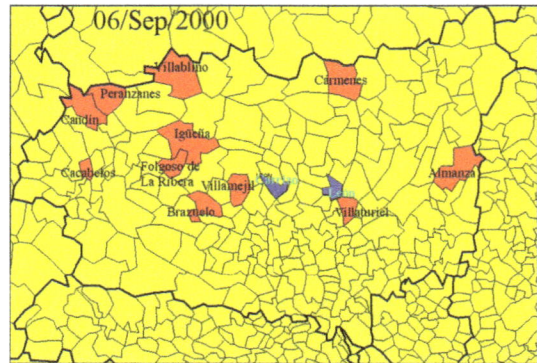

FIGURE 2: Example of a map representing the districts in the province of León, Spain, with active wildfires on 6th September 2000. The probe was installed in Carrizo.

in hectares (ha), and the type of vegetation burnt during the summer months (June, July, August, and September). This information was used to draw for each day a map of the province of León highlighting all the districts affected by wildfires. One of these maps is shown in Figure 2.

This material and the data on wind direction at surface level and at a certain altitude, registered by the weather station and the Sodar, were used to identify the measurements carried out by the aerosol probe that could have been affected by the transport of the smoke plumes from any nearby wildfires. The changes in the number of particles revealed the arrival of a plume at the sampling site. In general, a plume affected only 1 out of the 8 measurements carried out daily, as these measurements were carried out every 3 hours. However, in the case of large fires several measurements may be affected in one day or even over several subsequent days, depending on the duration of the fire and the intensity of the wind.

The particle measurements of the 8 daily registers were compared with the time interval of the wildfire, from detection to extinction. The measurements with very large numbers of particles were thus identified. In some cases, there were 20-fold increases with respect to the other measurements registered the same day. After identifying the measurements that were possibly affected by the arrival of a smoke plume to the probe, the data were confirmed with an analysis of the wind direction and speed and the distance between the wildfire and the probe. The increases in the measurements that could not be explained by these variables were excluded from the study.

It must be taken into account that on some days high levels of particles may be explained by the arrival of smoke plumes from wildfires in other provinces close to the province of León, especially Zamora and Orense. However, the data on the wildfires in those provinces during the study period were not made available.

Two different databases were built using the information obtained: one includes the measurements affected by the smoke plumes carried by the wind towards the probe (41 measurements in 28 different days out of the 122 study days); the other one comprises the measurements that were not

Table 2: Meteorological study of the months of June, July, August, and September 2000, with data on maximum, minimum and average temperatures, relative humidity, wind intensity, and total precipitation registered.

Months	T_{max} (°C)	T_{min} (°C)	$T_{av.}$ (°C)	HR (%)	Wind (m/s)	P_{total} (mm)
June	31.2	2.2	16.9	62	1.6	8
July	32.4	2.0	17.1	65	1.6	12
August	31.7	2.3	16.7	63	1.4	3
September	30.9	3.2	14.5	71	1.3	26.8

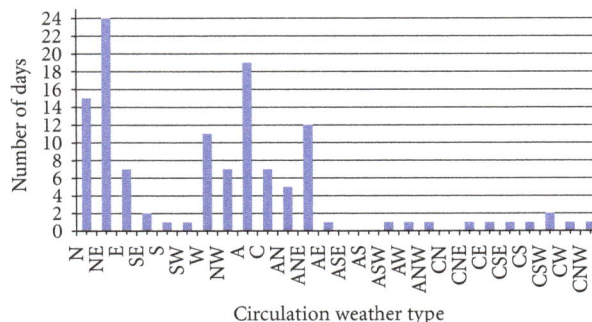

Figure 3: Circulation Weather Type classification in the months of June, July, August, and September 2000.

affected by the smoke plumes from surrounding fires (935 measurements in 121 different days out of the 122 study days). Eight measurements were carried out every single day, so there are both affected and non-affected measurements on most days.

The aerosol size distribution and the daily CWT were analyzed, considering the affected measurements and the non-affected measurements separately.

In the measurements that were not affected by the wildfires the study focuses on the influence of the weather types on the number of particles and the count median diameter of the fine mode. Soundings provided data on thermal inversions, mainly radiative inversions, on the days where the number of particles was higher than average.

The distributions of the measurements affected by wildfires were studied in detail. Measurements contaminated by the fires were compared with those not contaminated, and a detailed analysis was carried out of the geometric diameter of the fine mode and the number of particles in this size range during the 8 daily registers. The evolution of the accumulation mode and of the coarse mode has been analyzed too, as well as the relative humidity at three different stages: in the measurements before the ones affected by the plumes from the fires, in the ones affected by the fires, and in the ones after the fires.

4. Results and Discussion

4.1. Meteorological Study and Circulation Weather Types.
Table 2 shows the meteorological features of the months of June, July, August, and September 2000. The average monthly temperatures are typical of the summer months. The highest value is reached in July with 17.1°C. The maximum temperatures are high in all four months, over 30°C, and the minimum temperatures are around 2°C.

The monthly precipitation accumulated in the study period is low. The highest value was reached in September with 26.8 mm. The average relative humidity is around 65%, with 71% in September. The average wind speed was low, with a monthly average under 2 m/s. The high relative humidity registered, mainly during the night, are due to the irrigation system used in the fields close to the town and also to the fact that the River Órbigo flows close by at a distance of about 2.5 km.

The CWT classification shows that during the months from June to September (Figure 3), the dominant CWTs are the following three: the purely north-eastern type (NE), the anti-cyclonic type (A) controlled by the geostrophic vorticity, and the purely northern type (N), with 24, 19, and 15 days, respectively. The purely directional weather types are also relatively frequent: the Eastern type (E), the Western type (W), and the Northwestern type (NW), two hybrid types with northern (AN) and northeastern direction (ANE), and one nondirectional type, the cyclonic type (C). During the summer of the year 2000, the anticyclonic weather type and the flows with a northern component were dominant over the Iberian Peninsula, following the same trend as the one found in other studies on weather types and precipitation in the whole province of León [28].

4.2. Characterization of the Wildfires.
The province of León is the largest in the region of Castile and León, Spain, with an area of 15,581 km². The climate is of the Mediterranean type with continental influence and with some areas affected by the Atlantic Ocean too. The result is a wide range of different landscapes, making León one of the provinces in Spain that suffers the highest number of wildfires with the largest areas burnt.

Most of these wildfires are deliberate. According to statistics provided by the Junta of Castile and León, 90% of wildfires are caused by humans, either deliberately or as the result of negligence.

In the summer of the year 2000, a total of 465 wildfires were registered, an average of four per day. The fires are mostly detected during the central hours of the day, between 1000 UTC and 1800 UTC, and only very few fires start during the night.

Most wildfires were classified as of medium size, burning between 1 and 500 ha. In general, it was found that the wildfires that scorch more than 30 ha last longer than the day when they were detected, and some may even be active over several days, in the case of very large fires (those affecting more than 500 ha).

During the study period, it was observed that the number of wildfires increased gradually: August was the month with the highest number of fires, 201, followed by September, with 171, whereas June and July had less than 70 fires each

TABLE 3: Number of fires per month and monthly surface burnt according to the type of vegetation affected (trees, nontree woodland, or no-forest land). Total land area affected during the summer of 2000.

Months	No. of Fires	Area Burnt (ha)			
		Trees	Nontree woodland	No-forest land	Total area burnt
June	26	70	389	21	450
July	67	59	612	57	728
August	201	703	8871	275	9848
September	171	922	8075	582	9580

TABLE 4: District where the fire occurred, date of detection and extinction, land area burnt of each type of vegetation (trees, non-tree woodland, and no-forest land) and total land area affected by the large fires (over 500 ha) in the summer of 2000.

Municipality	First detected		Date of extinction		Area burnt (ha)			
	Day	Time (UTC)	Day	Time (UTC)	Trees	Nontree woodland	No-forest mass	Total area burnt
Encinedo (42°16'15"N, 6°35'39"W)	18/08/2000	0311	22/08/2000	1930	—	1034	154	1188
Truchas (42°15'40"N, 6°26'07"W)	18/08/2000	1650	23/08/2000	0800	179	1539	—	1718
Castrocalbón (42°11'45"N, 5°58'43"W)	21/08/2000	1110	22/08/2000	1800	62	773	—	835
Lucillo (42°24'38"N, 6°18'16"W)	31/08/2000	1500	03/09/2000	1900	—	1255	—	1255
Castrillo de Cabrera (42°20'25"N, 6°32'39"W)	02/09/2000	0600	02/09/2000	1930	—	920	—	920
Villablino (42°55'59"N, 6°19'00"W)	06/09/2000	1126	13/09/2000	0700	269	761	—	1,030
Barjas (42°36'40"N, 6°58'43"W)	11/09/2000	0830	15/09/2000	1805	33	542	—	575
Bembibre (42°36'54"N, 6°25'12"W)	14/09/2000	1605	16/09/2000	1915	104	426	—	530
Villablino (42°55'59"N, 6°19'00"W)	17/09/2000	1300	18/09/2000	1900	—	1031	—	1031

(Table 3). The lowest number was registered in June, with only 26 wildfires.

The total balance of hectares burnt was 20,636, affecting forest areas, no-tree woodland, and crops and crop stubble, with 87% of the land area corresponding to low-vegetation areas. The highest number of hectares burnt was registered in August and September, with over 9,500 ha per month.

The surface area affected (S) varies greatly during the summer. In June and July, it is mostly outbreaks of wildfires that are detected (S < 1 ha) or medium-size fires (500 ha > S > 1 ha), with no large fires at all (S > 500 ha). In contrast, in the months of August and September most of the fires are medium-sized with nearly 500 ha burnt, and there are 9 large fires, 5 in August and 4 in September, which together account for 45% of the total area burnt in the summer of the year 2000. The two largest fires took place on the 18th of August 2000 in the districts of Encinedo and Truchas (about 70 km from the study zone), which together burnt nearly 3,000 ha (Table 4).

The smoke plumes of these wildfires, medium and large, may be carried away by the wind to far off places, at least at a regional scale, and in this study they are responsible for the huge increases in the number of aerosols registered by the probe in the rural study zone.

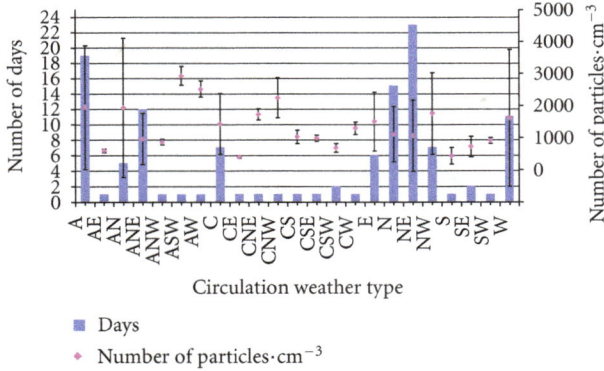

FIGURE 4: Number of particles registered and standard deviation for each Circulation Weather Type in 121 days in the months of June, July, August and September, in accordance with the days with each particular weather type.

4.3. Analysis of the Measurements "Not Contaminated" by Wildfires

4.3.1. Analysis of the Weather Types.
In order to study the influence of the weather types onto the number of aerosols in the study zone, we have analyzed the measurements that were not contaminated by aerosols from the wildfires. For each weather type, we have studied the average number of total particles and the standard deviation, as well as the number of days with those weather types (Figure 4). 23% of the study days (27 out of 121) present an average content of particles cm^{-3} between 2,000 and 3,000, and the weather types registered were the anticyclonic type (A), northern anticyclonic (AN), western anti-cyclonic (AW), southwestern anti-cyclonic (ASW), and the northwestern cyclonic type (CNW). In other words, high pressures seem to favor higher numbers of aerosols, with the arrival of maritime air masses. However, the standard deviation observed on the number of particles varies greatly in the anti-cyclonic type (A) and the northern hybrid (AN). This means that with these weather types the number of particles depends greatly on the air mass. On the other hand, the following weather types have little influence in the number of aerosols, as they register mean values of less than 1,000 particles cm^{-3}: eastern anti-cyclonic (AE); northeastern and northwestern anti-cyclonic (ANE & ANW), eastern, south-eastern and south-western cyclonic types (CE, CSE and CSW); the purely directional southern, south-eastern and south-western types (S, SE and SW). The standard deviations are also very low, except in the north-eastern anti-cyclonic weather type (ANE). These weather types account for only 18% of the days. However, they favor the arrival of continental air masses to the Iberian Peninsula, especially air masses from the north of Africa. In the remaining weather types, the average values lie between 1,000 and 2,000 particles cm^{-3}.

Next, a comparative analysis was carried out on the relationship between the count median diameter of the fine mode (CMD$_f$) and the weather types (Table 5). Four

TABLE 5: Count median diameter of the fine or accumulation mode (CMD$_f$) for each Circulation Weather Type.

CMD$_f$ (μm)	Circulation Weather Type
<0.13	ANW-ASW-AW-C-CE-CS-CSE-CSW-E-S-SW
0.13–0.14	A-AN-CW-N-NE-NW-SE-W
0.14–0.15	ANE-CNW
>0.15	AE-CNE

size ranges have been defined to enable us to interpret the results more easily. The particles with diameters smaller than 0.13 μm are mainly detected in cyclonic weather types with southern and western wind components. The air masses that reach the Iberian Peninsula in these situations come from the north of Africa (Saharan air masses) or from the Atlantic Ocean, carrying with them marine aerosols. The particles with CMD$_f$ sizes between 0.13 and 0.14 μm show dominant northern and western components, although anti-cyclonic situations may generate particles in this size range too. These weather types are clearly influenced by maritime and continental air masses from Western Europe. In the case of diameters between 0.14 and 0.15 μm, the weather types found are the north-eastern anti-cyclonic type and the northwestern cyclonic type, that is, the air masses that arrive at the Iberian Peninsula are both maritime and continental air masses from Eastern Europe. The largest diameters (>0.15 μm) correspond to continental air masses (CNE) and the Eastern anti-cyclonic weather type (AE) with maritime aerosols from the Mediterranean.

To sum up, the air masses from the north of Africa carry smaller particles (smaller than 0.13 μm) than the air masses brought to the Iberian Peninsula from continental Europe (larger than 0.13 μm). With the weather types typical of maritime air masses, the sizes registered show that, in general, the aerosols are smaller than 0.14 μm. In other words, even though continental air masses contribute fewer particles than maritime air masses, the aerosols they carry are larger. The arrival of European air masses has been studied by authors such as Alonso et al. [29], Gangoiti et al. [30], Viana et al. [31], or, more recently, by Escudero et al. [32].

4.3.2. Temporal Evolution of the Number of Particles.
Figure 5 shows the daily evolution of the number of particles and the standard deviation during the whole study period. Positive linear correlations were found, thus revealing that the number of particles in the study zone increases gradually over time as the summer goes by. Both Pearson correlations are significant for a significance level of 0.05. In general, the average number of particles registered daily is less than 2000 particles cm^{-3}, with big differences in the measurements. There were two periods at the end of the summer with oscillations between 2000 and 7000 particles cm^{-3} (second half of August and second half of September). This increase may be due to the high number of wildfires in August and September (some days have been recorded up to 50,000 particles cm^{-3}), which account for most of the land area burnt that summer in the province of León.

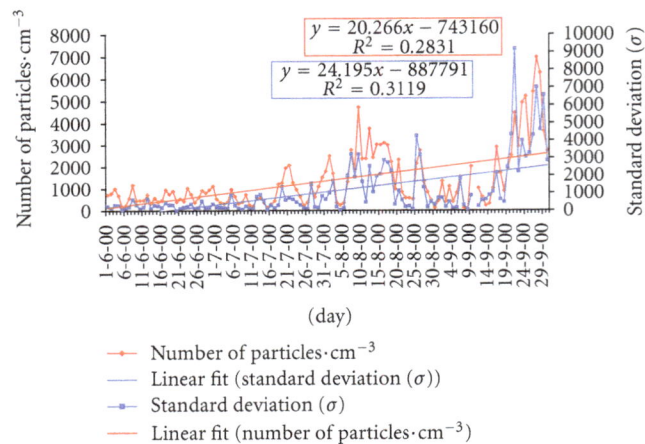

FIGURE 5: Average number of particles and standard deviation in the months of June, July, August, and September 2000.

It must not be forgotten that the province of León is surrounded by two other provinces which are often affected by wildfires too, Orense, in the region of Galicia, to the north-west of León (the distance between both capitals is 274 km), and Zamora, in the region of Castile and León, to the south of León (the distance between the two capitals is 133 km). Both provinces are important emitters of particulate matter to the atmosphere, contributing to the increase in the number of aerosols and to changes in their size distributions.

During the 27 days comprised by these two periods (second half of August and second half of September) there were frequent thermal inversions at altitudes of less than 1000 meters (AGL), both radiative and subsidence inversions. Both types of inversion often occurred the same day (Table 6). These inversions hinder the vertical dispersion of aerosols, so their number remains high in the lower layers of the atmosphere. The days with more than 2000 particles cm^{-3} correspond mainly to the anticyclonic weather type (A), registered in six study days, followed by the cyclonic type (C), the purely directional northern (N), and north-eastern types (NE), with three days each. Therefore, the increase in the daily numbers of particles in the months of August and September should not be attributed only to the large number of wildfires raging those days, but also to the clear influence of the anti-cyclonic weather type, which favours a situation of great atmospheric stability enhancing the formation of thermal inversions.

4.4. Influence of Wildfires on the Measurements of Aerosol Size Distributions. The multi-log-normal function has been used to characterize the size distributions of aerosol particles [33, 34]. This enables a simple comparison between several data sets of aerosol particles to be carried out. The multi-log-normal concept is thoroughly described in the literature and the overall outcome has proven to be useful whenever parameterizations are required [35, 36]. The particle size

distribution is assumed to consist of several lognormal modes:

$$\frac{dN(D)}{d\left(\log\left(D_p\right)\right)}$$
$$= \sum_{i=1}^{n} \frac{N_i}{\sqrt{2\Pi}\log\left(\sigma_{g,i}\right)} \exp\left[-\frac{\left(\log\left(D_p\right) - \log\left(\overline{D}_{pg,i}\right)\right)^2}{2\left(\log\sigma_i\right)^2}\right].$$
(1)

The three parameters that characterize an individual mode i are the mode number concentration N_i (cm^{-3}), the mode geometric variance $\sigma_{g,i}^2$ (dimensionless), and the mode geometric mean diameter $\overline{D}_{pg,i}$ (μm). D_p is the particle diameter and n is the number of individual modes. The least-square method was used to estimate the lognormal parameters N_i, $\sigma_{g;i}$, and $\overline{D}_{pg,i}$ [37].

For the count distribution, the geometric mean diameter, D_g, is customarily replaced by the count median diameter or CMD. The geometric mean is the arithmetic mean of the distribution of log D_p, which is a symmetrical normal distribution, and hence, its mean and median are equal. The median of the distribution of log D_p is also the median of the distribution of D_p, as the order of values does not change in converting to logarithms. Thus, for a lognormal count distribution, D_g = CMD.

The size distributions analyzed using the PCASP-X were found to be bimodal, that is, they have a fine and a coarse mode. The daily number of particles in the coarse mode was very low, around 5 particles cm^{-3}, so only data for the fine mode are considered in the results.

The influence of wildfires on the particle size distributions was studied taking two separate samples: on the one hand, the measurements not affected by the fires, taken before and after the plume crossed the study zone; and on the other hand, a sample of measurements affected by the smoke plumes from wildfires in a radius of 70 km around the probe. Comparative analyses were then carried out using these data to identify changes in atmospheric aerosols and their evolution during these events.

The comparative analysis between the monthly aerosol size distributions including all measurements and including only data not contaminated by wildfires (Figure 6) reveals a clear increase in the number of particles between 0.1 and 0.2 μm. This increase is not noticeable in the months of June and July because few wildfires are registered—26 fires in June and 67 in July, with 500 and 728 ha burnt, respectively. These fires contaminated only 6 measurements (4 in June and 2 in July) out of the 41 total measurements contaminated by particulate matter from smoke plumes between June and September. In contrast, in the months of August and September the increase in the number of particles is very sharp. We claim that this is due to the higher number of wildfires registered, 201 in August and 171 in September, burning an area of 19,428 ha distributed equally between the two months. The smoke plumes contaminated 35 measurements of the probe, 16 in August, and 19 in September, out of the 41 total.

TABLE 6: Weather types and radiative and subsidence thermal inversions during the days with an average number of particles exceeding 2000 particles cm^{-3} (—) means the data are not available.

Day	No. of particles cm^{-3}	Circulation weather type	Madrid 0000 UTC Inversions		A Coruña 0000 UTC Inversions		Santander 0000 UTC Inversions	
			Radiative (m AGL)	Subsidence (m AGL)	Radiative (m AGL)	Subsidence (m AGL)	Radiative (m AGL)	Subsidence (m AGL)
21/07/00	2084 ± 748	C	208			662	252	
01/08/00	2501 ± 1041	NE	94	494		837		767
07/08/00	2775 ± 3243	E		664		894		
09/08/00	4723 ± 3233	NE	235			462		277
10/08/00	2376 ± 1671	N	361			602	441	634
11/08/00	2380 ± 505	NW	113		—	—	77	798
12/08/00	3755 ± 2562	A				720		
13/08/00	2431 ± 1092	A					102	
14/08/00	3032 ± 1991	A	177		—	—	111	
15/08/00	3014 ± 2090	ANE	—	—				
16/08/00	3101 ± 2911	NE	190			794		
17/08/00	3000 ± 2766	N	132			960		
18/08/00	2240 ± 2304	CNW	161			733	85	
20/08/00	2330 ± 1155	N	65				69	
25/08/00	2157 ± 4291	C	46					324
26/08/00	2760 ± 3202	AN			—	—		
09/09/00	2028 ± 886	E		664	68	606	103	436
16/09/00	2896 ± 2179	C	186					945
22/09/00	2931 ± 2227	ASW	292					111
23/09/00	4912 ± 4005	W	192			295	267	897
24/09/00	5201 ± 3079	A	184	278				
25/09/00	2654 ± 3305	A	268			341	16	
26/09/00	5405 ± 4341	AN	92			968	372	
27/09/00	6972 ± 7035	A	109					
28/09/00	6259 ± 4654	W	—	—	—	—	—	
29/09/00	3625 ± 6597	NW	—	—	—	—	—	
30/09/00	2736 ± 2870	NW						

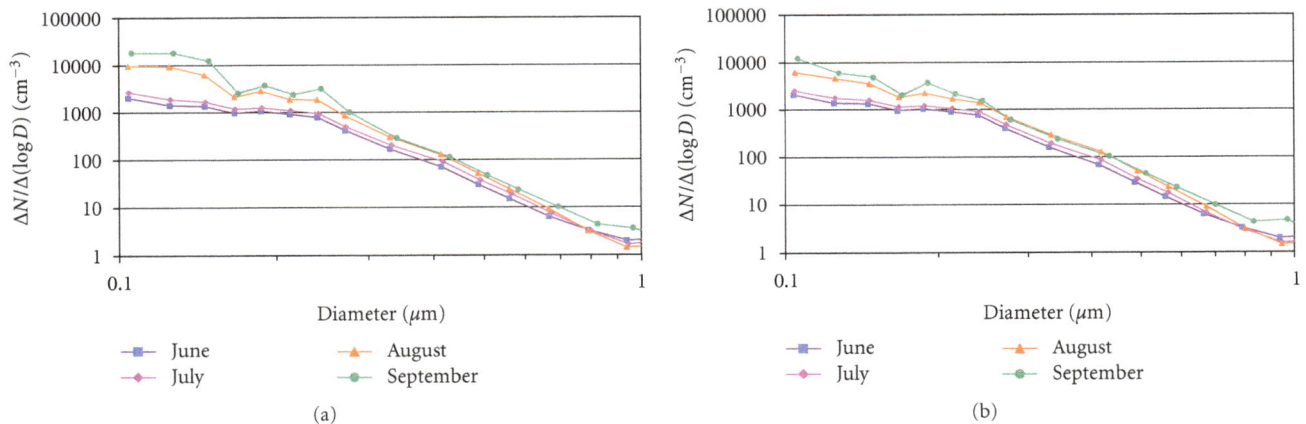

FIGURE 6: Mean size distributions in the months of June, July, August, and September 2000 in (a) the measurements registered with wildfires and in (b) measurements not contaminated by wildfires. Only the sizes between 0.1 and 1 μm are shown.

The increase in this size range (0.1 to 0.2 μm) is observed in all the measurements contaminated by smoke in the study zone. Figure 7 illustrates two examples: (a) between 4th August 2000 at 0700 UTC and 7th August 2000 at 2200 UTC and (b) between 22nd of August 2000 at 1300 UTC and 24th August 2000 at 1600 UTC, with 22 and 18 size distribution measurements, respectively. Both examples include measurements prior to the arrival of the smoke plume, the contaminated measurements, and the ones carried out after the smoke plume, showing a clear increase in the number of particles smaller than 0.2 μm in the contaminated measurements.

To determine the influence of the wildfires on the 8 daily particle measurements, we carried out a comparative analysis of the evolution of the geometric mean diameter of the fine mode (CMD_f) and the total number of particles in affected and nonaffected measurements (Figure 8). Contaminated and noncontaminated data follow the same trend along the 8 daily measurements: there is a clear difference between the measurements taken during the night (from 1900 UTC, beginning of dusk, to 0700 UTC, beginning of dawn) and the ones taken during daylight (from 0700 UTC to 1900 UTC). In both cases, the geometric mean diameter registered during the night tends to increase until 0.15 μm in contaminated measurements, and until 0.16 μm in noncontaminated measurements, around 0400 UTC. From that time on, there is a decrease in the geometric mean diameter of the fine mode in both data sets, until 0.11 μm in data contaminated by fires, and 0.12 μm in non-contaminated data, at around 1300 UTC. This situation remains stable until 1900 UTC, when the diameter increases again. The results show that on days with wildfires the particle sizes are smaller than on days with no wildfires, both during the day and during the night.

As for the number of particles in the fine mode, we find two completely different situations when comparing affected and non-affected measurements. The data from contaminated measurements show that during the night and in the early hours (from 2200 UTC to 1000 UTC) the number of particles registered is between 2000 and 3500 particles cm^{-3}. From that time on, the number of particles increases, reaching its maximum at 1300 UTC, with nearly 7,000 particles cm^{-3}, coinciding with the hours registering the lowest value in the geometric mean diameter of the fine mode and with the time when the fires are most active. Initial values are back by 1900 UTC. On the other hand, in the data not contaminated by aerosols from wildfires, the number of particles in the fine mode remains stable during the whole day, around 1000 particles cm^{-3}. In conclusion, the increase in the number of aerosols in the smallest fraction of the fine mode is clearly due to the contamination by particulate matter from wildfires, because there is no other anthropogenic source of aerosols nearby, as we are in a rural area.

The aerosol size distributions in the 41 measurements contaminated by particulate matter from the wildfires were compared with the distributions in the 82 measurements taken immediately before and after the ones affected (Figure 9). Contaminated measurements had an average number of 17,000 particles cm^{-3} in the fine mode, compared

to only 1,500 particles cm^{-3} on average in non-contaminated measurements. This represents an average increase in the number of particles in the fine mode of 1000%, with values of nearly 2000% in diameter ranges between 0.10 and 0.14 μm.

The concentration of PM10 particulate matter was estimated considering 1.35 g cm^{-3} as the density of particles from biomass burning [38] and 1 g cm^{-3} for ambient air in rural environments. The average concentrations in measurements not contaminated by wildfires are around 14 μg m^{-3}, compared to an average of 86 μg m^{-3} in those measurements contaminated by aerosols from wildfires. The pollution caused by wildfires exceeds the threshold value of 50 μm m^{-3} established by the Spanish and European Regulation on Air Quality (Royal Decree 1073/2002 of 18th October). These remarkable increases cannot be explained by high pressure systems with atmospheric stability lasting several days, nor by the simultaneous presence of radiative and subsidence inversions. The main cause is the arrival of smoke plumes from wildfires in a radius of 70 km from the probe.

The number of particles in the fine mode increases considerably, and to know exactly what happens to the size of these particles when the probe is measuring aerosols from the wildfires, we compared the mean CMD of the fine or accumulation mode before, during, and after the smoke plumes. The influence of humidity on the growth of these particles was also analyzed (Table 7). When compared with the measurement prior to the arrival of the smoke plume, in the affected measurement, there is a clear decrease in the size of particles of around 19%, dropping from average values of CMD_f of 0.14 μm to 0.11 μm. This occurs as the number of particles with these sizes increases with the arrival of the smoke plume.

It was observed that when the relative humidity in the contaminated measurement is higher than the one in the previous measurement, a humidity of around 80%, the geometric mean diameter of the fine mode increases from 0.10 μm to 0.14 μm. In these cases, we claim that the aerosols derived from biomass burning undergo hygroscopic growth absorbing atmospheric water vapor, thus increasing their size. In 1959 Orr et al. [39] already found that when the relative humidity (RH) rises above 40%, even weakly soluble aerosol particles can absorb water from the air. As a consequence, this additional water increases the particle size. Once the measurements are not affected by the smoke plume from the wildfires, the CMD tends to stabilize returning to the values prior to the arrival of the plume from the fire.

5. Conclusions

In the summer season, the analyzed air masses from the north of Africa contributed smaller particles (smaller than 0.13 μm) than the ones brought by air masses from continental Europe (larger than 0.13 μm). With weather types including maritime air masses, the size ranges registered reveal that in general the aerosols are smaller than 0.14 μm. That is continental air masses contribute fewer particles than maritime air masses, but these particles are larger.

TABLE 7: Count median diameter of the fine mode (CMD$_f$) and relative humidity (RH) for the measurements before the fire, the ones affected by the fire and the ones after the fire.

Measurements prior to those affected by the fire				Measurements affected by the fire				Measurements after those affected by fire			
Day	Time (UTC)	CMD$_f$ (μm)	RH (%)	Day	Time (UTC)	CMD$_f$ (μm)	RH (%)	Day	Time (UTC)	CMD$_f$ (μm)	RH (%)
19/06/2000	1600	0.10	40	20/06/2000	2200	0.18	79	20/06/2000	1600	0.14	43
19/06/2000	1600	0.10	40	20/06/2000	0100	0.20	88	20/06/2000	1600	0.14	43
19/06/2000	1600	0.10	40	20/06/2000	0400	0.16	87	20/06/2000	1600	0.14	43
19/06/2000	1600	0.10	40	20/06/2000	0700	0.16	66	20/06/2000	1600	0.14	43
08/07/2000	1000	0.17	48	08/07/2000	1600	0.11	40	09/07/2000	2200	0.16	89
21/07/2000	1600	0.12	30	22/07/2000	2200	0.19	69	22/07/2000	0400	0.13	83
04/08/2000	0700	0.20	53	04/08/2000	1300	0.11	45	05/08/2000	2200	0.12	74
05/08/2000	1000	0.16	45	05/08/2000	1600	0.10	44	06/08/2000	2200	0.11	86
06/08/2000	1000	0.12	35	06/08/2000	1300	0.10	32	06/08/2000	1900	0.11	67
09/08/2000	0400	0.15	84	09/08/2000	1000	0.11	29	09/08/2000	1600	0.10	38
11/08/2000	0700	0.12	58	11/08/2000	1600	0.15	62	11/08/2000	1900	0.14	86
13/08/2000	2200	0.15	78	13/08/2000	0100	0.15	86	13/08/2000	0400	0.16	90
14/08/2000	0100	0.18	96	14/08/2000	0400	0.15	99	14/08/2000	0700	0.13	67
15/08/2000	1000	0.14	48	15/08/2000	1300	0.12	38	15/08/2000	1600	0.15	42
16/08/2000	1000	0.13	24	16/08/2000	1300	0.08	23	16/08/2000	1600	0.10	24
17/08/2000	0700	0.14	56	17/08/2000	1300	0.10	36	17/08/2000	1900	0.15	71
18/08/2000	1600	0.11	46	18/08/2000	1900	0.10	56	19/08/2000	0100	0.11	82
19/08/2000	0700	0.12	65	19/08/2000	1000	0.10	48	19/08/2000	1900	0.15	42
22/08/2000	1300	0.15	50	22/08/2000	1600	0.15	50	22/08/2000	1900	0.10	62
23/08/2000	1000	0.11	49	23/08/2000	1900	0.12	80	24/08/2000	2200	0.10	82
24/08/2000	1000	0.10	58	24/08/2000	1300	0.10	52	24/08/2000	1600	0.10	44
28/08/2000	1000	0.13	48	28/08/2000	1300	0.10	41	28/08/2000	1900	0.13	69
08/09/2000	1900	0.11	62	09/09/2000	2200	0.11	75	09/09/2000	0400	0.13	80
09/09/2000	0700	0.12	55	09/09/2000	1000	0.12	44	09/09/2000	1600	0.11	48
09/09/2000	1900	0.13	59	10/09/2000	0400	0.11	66	12/09/2000	2200	0.22	97
09/09/2000	1900	0.13	59	10/09/2000	0700	0.12	51	12/09/2000	2200	0.22	97
09/09/2000	1900	0.13	59	10/09/2000	1300	0.10	32	12/09/2000	2200	0.22	97
09/09/2000	1900	0.13	59	10/09/2000	1900	0.11	56	12/09/2000	2200	0.22	97
09/09/2000	1900	0.13	59	11/09/2000	2200	0.13	58	12/09/2000	2200	0.22	97
09/09/2000	1900	0.13	59	11/09/2000	1600	0.12	76	12/09/2000	2200	0.22	97
18/09/2000	1300	0.13	56	18/09/2000	1600	0.12	57	18/09/2000	1900	0.15	70
21/09/2000	1000	0.13	63	21/09/2000	1600	0.10	61	21/09/2000	1900	0.11	92
23/09/2000	1600	0.10	45	24/09/2000	2200	0.10	83	24/09/2000	0700	0.10	71
23/09/2000	1600	0.10	45	24/09/2000	0100	0.13	78	24/09/2000	0700	0.10	71
25/09/2000	0700	0.11	64	25/09/2000	1300	0.10	41	26/09/2000	0700	0.13	62
25/09/2000	0700	0.11	64	25/09/2000	1600	0.08	50	26/09/2000	0700	0.13	62
25/09/2000	0700	0.11	64	25/09/2000	1900	0.11	80	26/09/2000	0700	0.13	62
25/09/2000	0700	0.11	64	26/09/2000	2200	0.14	84	26/09/2000	0700	0.13	62
25/09/2000	0700	0.11	64	26/09/2000	0100	0.10	89	26/09/2000	0700	0.13	62
27/09/2000	0700	0.13	92	27/09/2000	0700	0.12	83	28/09/2000	1000	0.15	98
27/09/2000	0700	0.13	92	27/09/2000	1900	0.10	72	28/09/2000	1000	0.15	98

(a)

(b)

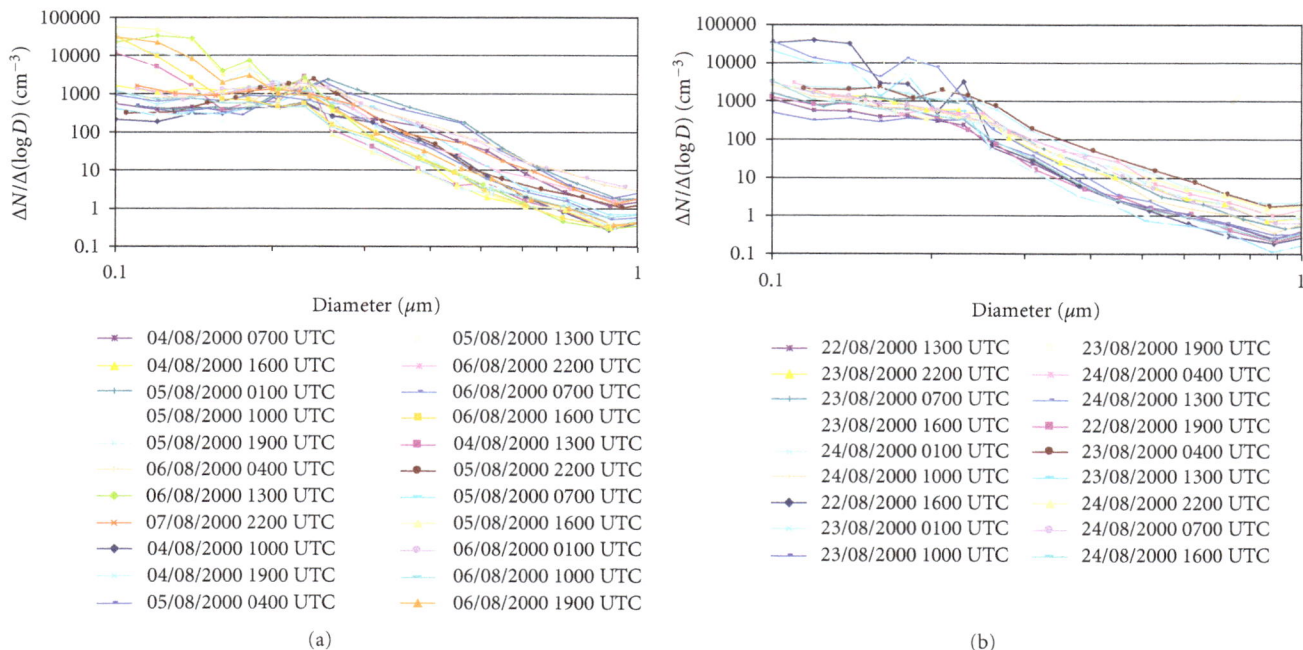

FIGURE 7: Aerosol size distributions of two time intervals: (a) between 4th August 2000 at 0700 UTC and 7th August 2000 at 2200 UTC and (b) between 22nd August 2000 at 1300 UTC and 24th August 2000 at 1600 UTC, representing measurements contaminated by wildfires. Only the sizes between 0.1 and 1 μm are shown.

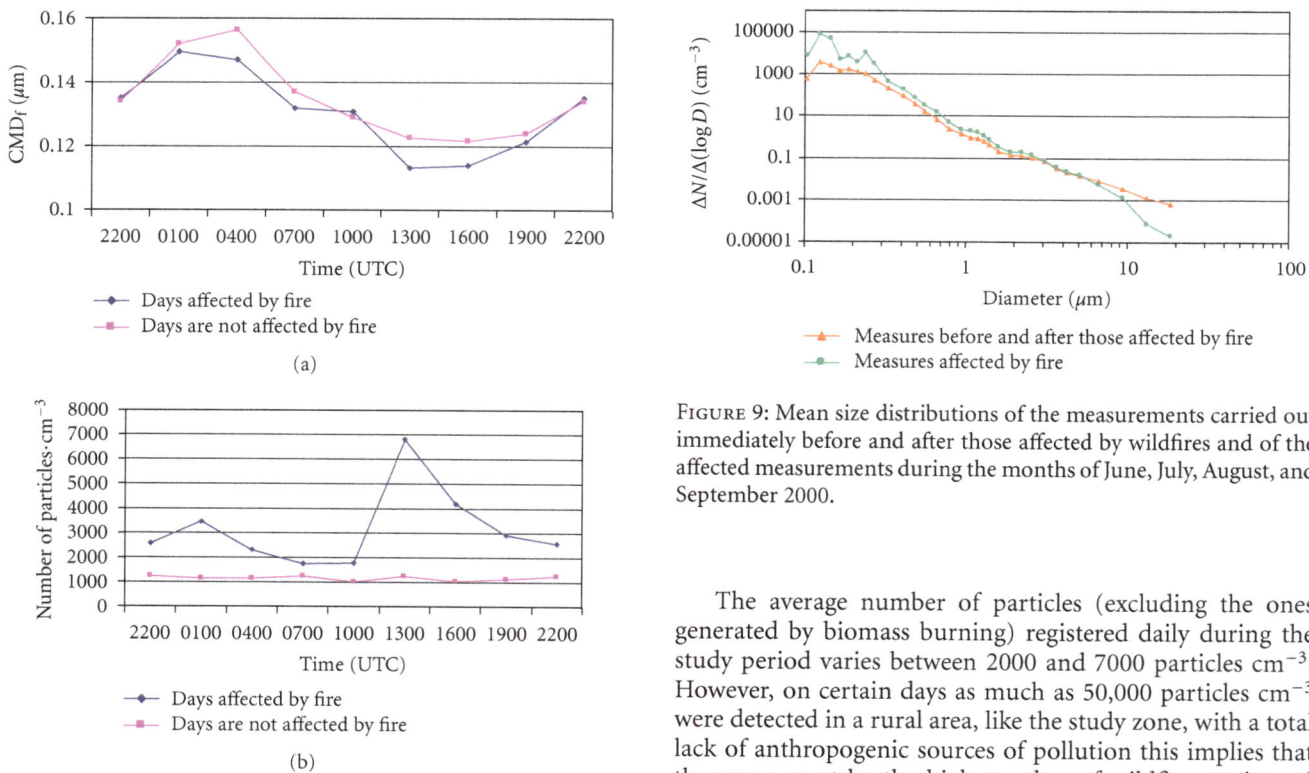

(a)

(b)

FIGURE 8: Comparative analysis of the evolution of (a) the geometric mean diameter (CMD$_f$) and (b) the total number of particles in the fine mode in the 8 daily registers, for data including contaminated and noncontaminated measurements.

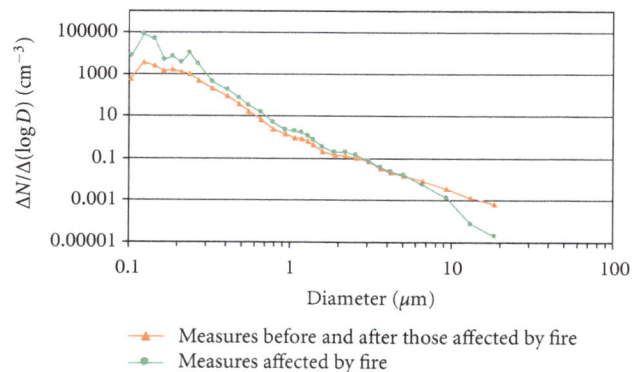

FIGURE 9: Mean size distributions of the measurements carried out immediately before and after those affected by wildfires and of the affected measurements during the months of June, July, August, and September 2000.

The average number of particles (excluding the ones generated by biomass burning) registered daily during the study period varies between 2000 and 7000 particles cm^{-3}. However, on certain days as much as 50,000 particles cm^{-3} were detected in a rural area, like the study zone, with a total lack of anthropogenic sources of pollution this implies that the cause must be the high number of wildfires registered every summer. Part of this increase may be attributed to the stable atmospheric situations typical of the summer season, as well as to the occurrence of deep radiative and/or subsidence thermal inversions. However, in this study, because of the sudden increases observed in the number of particles, the most plausible explanation is the transport of smoke plumes at a regional scale full of particulate matter.

In the province of León, wildfires burnt 20,636 ha between June and September 2000. We claim that wildfires and other large fires in the surrounding provinces are contaminating the air in rural areas during the summer, reaching average values of PM10 concentrations of $86\,\mu m \cdot m^{-3}$ clearly exceeding the threshold value of $50\,\mu m \cdot m^{-3}$ established by the Spanish and European Regulation on Air Quality.

The wildfires cause not only a considerable increase in the number of particles in the atmosphere but also changes in particle size distributions. Average increases of around 1000% have been found in the number of particles in the fine mode. In particles between 0.10 and $0.14\,\mu m$, the increases reach 2000%, revealing that wildfires clearly contribute particles with these sizes. The air affected by the fires presents aerosols with average values of the CMD of the fine mode around $0.11\,\mu m$ versus average values of $0.14\,\mu m$ in clean rural air.

When the relative humidity increases in the measurements contaminated by the fires when compared with the non-contaminated ones (RH around 80%), the CMD of the fine mode increases by 50%. This may be due to the hygroscopic growth of the aerosol, which absorbs water vapor from ambient air, from the river Órbigo, and from the irrigation system in the study zone.

Summing up, in a rural area, with few anthropogenic sources of aerosols, the considerable increases registered in the number of particles cannot be explained by high pressures leading to atmospheric stability, nor by thermal inversions, but must be attributed to wildfires, large or small, with smoke plumes covering sometimes great distances and causing important ambient pollution of nonanthropogenic origin.

The authors believe this research line must be pursued because the devastating consequences of wildfires cause immediate changes in atmospheric composition at local and regional scales, and the increased number of aerosols produce changes in the Earth's global radiation balance, causing both direct and indirect radiative forcing. Moreover, exposure to atmospheric particulate matter may have adverse effects on human health.

Acknowledgments

The authors wish to thank Miguel Angel Olguín, from the Department for the Environment of the Junta de Castilla y León, for his help and cooperation whenever it was needed. The authors are also grateful to Dr. Laura López Campano for her collaboration and to Dr. Noelia Ramón for translating the paper into English. This study was partially supported by the Regional Government of Castilla y León (Grants LE13/00B and LE039A10-2).

References

[1] T. Suzuki and S. Tsunogai, "Daily variation of aerosols of marine and continental origin in the surface air over a small island, Okushiri, in the Japan Sea," *Tellus*, vol. 40, no. 1, pp. 42–49, 1988.

[2] T. Ito, "Size distribution of Antarctic submicron aerosols," *Tellus*, vol. 45, no. 2, pp. 145–159, 1993.

[3] A. Castro, J. L. Marcos, J. Dessens, J. L. Sánchez, and R. Fraile, "Concentration of ice nuclei in continental and maritime air masses in León (Spain)," *Atmospheric Research*, vol. 47-48, pp. 155–168, 1998.

[4] R. Fraile, A. I. Calvo, A. Castro, D. Fernández-González, and E. García-Ortega, "The behavior of the atmosphere in long-range transport," *Aerobiologia*, vol. 22, no. 1, pp. 35–45, 2006.

[5] J. S. Reid, T. F. Eck, S. A. Christopher et al., "A review of biomass burning emissions part III: intensive optical properties of biomas burning particles," *Atmospheric Chemistry and Physics*, vol. 5, pp. 827–849, 2005.

[6] A. I. Calvo, V. Pont, A. Castro et al., "Radiative forcing of haze during a forest fire in Spain," *Journal of Geophysical Research—Atmospheres*, vol. 115, Article ID D08206, 10 pages, 2010.

[7] H. Cachier, C. Liousse, P. Buatmenard, and A. Gaudichet, "Particulate content of savanna fire emissions," *Journal of Atmospheric Chemistry*, vol. 22, no. 1-2, pp. 123–148, 1995.

[8] L. A. Remer, Y. J. Kaufman, B. N. Holben, A. M. Thompson, and D. McNamara, "Biomass burning aerosol size distribution and modeled optical properties," *Journal of Geophysical Research*, vol. 103, no. 24, pp. 31879–31891, 1998.

[9] J. S. Reid and P. V. Hobbs, "Physical and optical properties of young smoke from individual biomass fires in Brazil," *Journal of Geophysical Research*, vol. 103, no. 24, pp. 32013–32030, 1998.

[10] D. R. Cocker, N. E. Whitlock, R. C. Flagan, and J. H. Seinfeld, "Hygroscopic properties of pasadena, California aerosol," *Aerosol Science and Technology*, vol. 35, no. 2, pp. 637–647, 2001.

[11] M. Vakeva, K. Hameri, T. Puhakka, E. D. Nilsson, H. Hohti, and J. M. Makela, "Effects of meteorological processes on aerosol particle size distribution in an urban background area," *Journal of Geophysical Research*, vol. 105, no. 8, pp. 9807–9821, 2000.

[12] Y. Aklilu, M. Mozurkewich, A. J. Prenni et al., "Hygroscopicity of particles at two rural, urban influenced sites during Pacific 2001: comparison with estimates of water uptake from particle composition," *Atmospheric Environment*, vol. 40, no. 15, pp. 2.650–2.661, 2006.

[13] V.-M. Kerminen, "The effects of particle chemical character and atmospheric processes on particle hygroscopic properties," *Journal of Aerosol Science*, vol. 28, no. 1, pp. 121–132, 1997.

[14] J. Warner and S. Twomey, "The production of cloud nuclei by cane fires and the effect on cloud drop concentrations," *Journal of Atmospheric Sciences*, vol. 24, no. 6, pp. 704–713, 1967.

[15] R. C. Eagan, P. V. Hobbs, and L. F. Radke, "Measurements of cloud condensation nuclei and cloud droplet size distributions in the vicinity of forest fires," *Journal of Applied Meteorology*, vol. 13, no. 5, pp. 553–557, 1974.

[16] G. C. Roberts, P. Artaxo, J. Zhou, E. Swietlicki, and M. O. Andreae, "Sensitivity of CCN spectra on chemical and physical properties of aerosol: a case study from the Amazon Basin," *Journal of Geophysical Research*, vol. 107, no. 20, pp. 1–37, 2002.

[17] M. D. Petters, C. M. Carrico, S. M. Kreidenweis et al., "Cloud condensation nucleation activity of biomass burning aerosol," *Journal of Geophysical Research*, vol. 114, no. 22, Article ID D22205, 2009.

[18] A. I. Calvo, F. J. Olmo, H. Lyamani et al., "Chemical composition of wet precipitation at the background EMEP station in

Víznar (Granada, Spain) (2002–2006)," *Atmospheric Research*, vol. 96, no. 2-3, pp. 408–420, 2010.

[19] C. F. Bohren and D. R. Huffman, *Absorption and Scattering of Light by Small Particles*, Wiley, New York, NY, USA, 1983.

[20] Y. J. Kim and J. F. Boatman, "Size calibration corrections for the active scattering aerosol spectrometer probe (ASASP-100X)," *Aerosol Science and Technology*, vol. 12, no. 3, pp. 665–672, 1990.

[21] A. F. Jenkinson and F. P. Collison, *An Initial Climatology of Gales over the North Sea. Synoptic Climatology Branch Memorándum*, vol. 62, Meteorological Office, London, UK, 1977.

[22] P. D. Jones, M. Hulme, and K. R. Briffa, "A comparison of Lamb circulation types with an objective classification scheme," *International Journal of Climatology*, vol. 13, no. 6, pp. 655–663, 1993.

[23] H.H. Lamb, *British Isles Weather Types and a Register of Daily Sequence of Circulation Patterns. 1861–1971*, Geophysical Memoir 116, HMSO, London, UK, 1972.

[24] R. M. Trigo and C. C. DaCamara, "Circulation weather types and their influence on the precipitation regime in Portugal," *International Journal of Climatology*, vol. 20, no. 13, pp. 1559–1581, 2000.

[25] C. Tomás, F. de Pablo, and L. R. Soriano, "Circulation weather types and cloud-to-ground flash density over the Iberian Peninsula," *International Journal of Climatology*, vol. 24, no. 1, pp. 109–123, 2004.

[26] M. Fernández-Raga, R. Fraile, J. J. Keizer et al., "The kinetic energy of rain measured with an optical disdrometer: an application to splash erosion," *Atmospheric Research*, vol. 96, no. 2-3, pp. 225–240, 2010.

[27] A. Castro, E. Alonso-Blanco, M. González-Colino, A. I. Calvo, M. Fernández-Raga, and R. Fraile, "Aerosol size distribution in precipitation events in León, Spain," *Atmospheric Research*, vol. 96, no. 2-3, pp. 421–435, 2010.

[28] S. Fernández-González, S. del Río, A. Castro et al., "Connection between NAO, weather types and precipitation in León, Spain (1948–2008)," *International Journal of Climatology*. In press.

[29] L. Alonso, G. Gangoiti, M. Navazo, M. M. Millán, and E. Mantilla, "Transport of tropospheric ozone over the Bay of Biscay and the Eastern Cantabrian Coast of Spain," *Journal of Applied Meteorology*, vol. 39, no. 4, pp. 475–486, 2000.

[30] G. Gangoiti, L. Alonso, M. Navazo et al., "Regional transport of pollutants over the Bay of Biscay: analysis of an ozone episode under a blocking anticyclone in west-central Europe," *Atmospheric Environment*, vol. 36, no. 8, pp. 1.349–1.361, 2002.

[31] M. Viana, X. Querol, A. Alastuey, G. Gangoiti, and M. Menéndez, "PM levels in the Basque Country (Northern Spain): analysis of a 5-year data record and interpretation of seasonal variations," *Atmospheric Environment*, vol. 37, no. 21, pp. 2.879–2.891, 2003.

[32] M. Escudero, X. Querol, A. Ávila, and E. Cuevas, "Origin of the exceedances of the European daily PM limit value in regional background areas of Spain," *Atmospheric Environment*, vol. 41, no. 4, pp. 730–744, 2007.

[33] K. H. Whitby, "The physical characteristics of sulfur aerosols," *Atmospheric Environment*, vol. 12, no. 1–3, pp. 135–159, 1978.

[34] W. A. Hoppel, G. M. Frick, J. W. Fitzgerald, and R. E. Larson, "Marine boundary layer measurements of new particle formation and the effects nonprecipitating clouds have on aerosol size distribution," *Journal of Geophysical Research*, vol. 99, no. 7, pp. 14.443–14.459, 1994.

[35] J. M. Mäkelä, I. K. Koponen, P. Aalto, and M. Kulmala, "One-year data of submicron size modes of tropospheric background aerosol in Southern Finland," *Journal of Aerosol Science*, vol. 31, no. 5, pp. 595–611, 2000.

[36] W. Birmili, A. Wiedensohler, J. Heintzenberg, and K. Lehmann, "Atmospheric particle number size distribution in central Europe: statistical relations to air masses and meteorology," *Journal of Geophysical Research*, vol. 106, no. 23, pp. 32.005–32.018, 2001.

[37] T. Hussein, K. Hämeri, P. P. Aalto, P. Paatero, and M. Kulmala, "Modal structure and spatial-temporal variations of urban and suburban aerosols in Helsinki-Finland," *Atmospheric Environment*, vol. 39, no. 9, pp. 1655–1668, 2005.

[38] J. S. Reid and P. V. Hobbs, "Physical and optical properties of young smoke from individual biomass fires in Brazil," *Journal of Geophysical Research*, vol. 103, no. 24, pp. 32.013–32.030, 1998.

[39] C. Orr, F. K. Hurd, and W. J. Corbett, "Aerosol size and relative humidity," *Journal of Colloid Science*, vol. 13, no. 5, pp. 472–482, 1959.

18

Applying Econometrics to the Carbon Dioxide "Control Knob"

Timothy Curtin

Emeritus Faculty, Australian National University, Canberra, ACT 0200, Australia

Correspondence should be addressed to Timothy Curtin, tcurtin@bigblue.net.au

Academic Editor: Donald H. Stedman

This paper tests various propositions underlying claims that observed global temperature change is mostly attributable to anthropogenic noncondensing greenhouse gases, and that although water vapour is recognized to be a dominant contributor to the overall greenhouse gas (GHG) effect, that effect is merely a "feedback" from rising temperatures initially resulting *only* from "non-condensing" GHGs and not at all from variations in preexisting naturally caused atmospheric water vapour (i.e., [H₂O]). However, this paper shows that "*initial radiative forcing*" is not exclusively attributable to forcings from noncondensing GHG, both because atmospheric water vapour existed before there were any significant increases in GHG concentrations or temperatures and also because there is no evidence that such increases have produced measurably higher [H₂O]. The paper distinguishes between forcing and feedback impacts of water vapour and contends that it is the *primary* forcing agent, at much more than 50% of the total GHG gas effect. That means that controlling atmospheric carbon dioxide is unlikely to be an effective "control knob" as claimed by Lacis et al. (2010).

1. Introduction: Previous Econometric Modelling

The main technique used in this paper is econometric least squares regression analysis, which enables computation of the relative strength of proposed alternative and independent causal factors in determination of the dependent variable, temperature change. This procedure is not used in Solomon et al. [1] or by Schmidt et al. [2] and Lacis et al. [3]. Instead, they all rely on computer models of the climate system in which parameterized expressions for the main variables under consideration are first used to generate a simulation of the global climate, and when the average of an ensemble of such models generates some conformity with observations, the expressions for one or other of the noncondensing and condensing GHGs are removed in turn from their composite model, and thereby they estimate the relative strength of individual GHGs. However, the claims that only the noncondensing GHGs are the "forcing" agents, and that condensable water vapour has just a feedback role, are built into the models' alternate simulations, and do not constitute confirmatory evidence validating their hypothesis that the only role of water vapour and clouds is to "amplify

the initial [sic] warming provided by the noncondensing GHGs, and in the process, account for the bulk of the total terrestrial greenhouse effect" [3–9]. For that, in the absence of controlled physical experiments like those of Tyndall [10], which are not possible at the global or regional levels with or without computer models, econometrics is essential.

Dessler and Davis [11, page 1] state that the water vapour feedback "is the process whereby an initial warming of the planet, caused, for example, by an increase in long-lived greenhouse gases, leads to an increase in the humidity of the atmosphere. Because water vapour is itself a greenhouse gas, this increase in humidity causes additional warming. This is the most powerful feedback in the climate system, with the capacity by itself to double [sic] the warming from carbon dioxide alone." That claimed positive feedback is what explains how the IPCC's predicted global temperature increase for a doubling in [CO₂] from the c.280 ppm in 1900 of 3°C (central value) to 560 ppm implies an increase of 2.3°C from the extra 60 percent in [CO₂] from 2010, despite the observed only 0.83°C associated with the nearly 40 percent increase in [CO₂] between 1900 and 2010 (Gistemp). This paper's regression analysis tests for the relative importance of changes in [CO₂] and [H₂O] and also as to which comes first, the former according to Dessler and Davis [11], or the latter, in "forcing" temperature changes.

Not many researchers have used time domain econometrics methods to analyze climate change. Stern and Kaufmann [12, page 412], Tol and de Vos [13], and Tol [14], are amongst the few that explicitly use econometric multi-variate regression analysis of time series data to investigate the causes of climate change...[1]

None of these papers addresses the respective proportions of condensing and noncondensing GHGs to the overall greenhouse effect, and none mention [H2O] as an independent variable with potential explanatory value for changes in temperature. Kaufmann et al. [15, 16] have made further use of econometric methods, and comment how "statistical models of the relationship between surface temperature and radiative forcing that are estimated from the observational temperature record often are viewed skeptically by climate modelers. One reason is uncertainty about what statistical models measure. Because statistical models do not represent physical linkages directly, it is difficult to assess the time scale associated with statistical estimates for the effect of a doubling in CO_2 on surface temperature." These papers' database regressions (Section 4) use a wide range of "physical linkages," and the derived coefficients provide an ample resource for "assessing the time scale... for the effect of a doubling in CO_2," which could be more than a hundred years if their analysis is correct.[2]

Hegerl et al. [17], in AR4, [1] claimed that they would attempt to differentiate between climate changes "that result from anthropogenic and natural external forcings" (p.667). However, they do not report any regression results estimating the relative values of those forcings. They concede (p.668) that attribution studies seek to "assess whether the response to a key forcing, such as greenhouse gas increases, is distinguishable from that due to other forcings (Appendix 9A) and add that "these questions are typically investigated using a multiple regression of observations onto several fingerprints [sic] representing climate responses to different forcings... see Section 9.2.2." However, there is no trace of the results of any such analysis anywhere in Hegerl et al. 2007, least of all in either their referenced Section 9.2.2 or their Appendix 9A. The latter (pp.744–745) does have a textbook account of multivariate regression but reports no results. Thus Hegerl et al. [12, p.666] provide no evidence for their assertion "greenhouse gas forcing has very likely caused *most* of the global warming over the last 50 years" where "very likely" means "more than 90 percent probability" [1, page 121], and "most" must mean at least more than 50 percent when only two independent variables are considered.[3] Had these authors done some regression analysis, they could have been more precise, but they never did, nor do they report any by others.

Instead, for both Hegerl and Allen [18] and the many co-authors of the 11 papers cited by Hegerl et al. [17] of which Hegerl was the lead author, "attribution" consists of model outputs with imposed parameters of radiative forcing arising from [CO_2] and other greenhouse gases.[4] In practice, none of these papers perform any regression analysis of both natural and nonnatural forcings and ignore primarily "natural

external forcings" like that from [H2O]. Hegerl and Allen [18] deal only with greenhouses gases and sulphur dioxide, and the latter is even more of anthropogenic origin (mainly comprising emissions from combustion of hydrocarbon fuels) than the former. It is true that sulphate aerosols are usually assumed to have a cooling effect, see Charlson and Wigley [19], but most sulphate aerosols (hereafter [SO_2]) are of the same anthropogenic origin in time and place as emissions of CO_2 although from time to major volcanic eruptions increase both [CO_2] and [SO_2], with only local effects in the case of the latter. The other papers cited by Hegerl et al. [17] adopt much the same approach. For example 7Hegerl et al. [20, page 632] consider only [CO_2] and [SO_2] with just this mention of solar irradiation at the top of the atmosphere (TOA): "We used only a greenhouse gas and a greenhouse gas-plus-aerosol signal pattern, since the solar response pattern could not be sufficiently separated from noise and the greenhouse gas pattern," a curious conclusion in the light of the title of that paper.

Stott et al. [21, page 2][5] use what they call "optimal detection technology" to conclude that "increases in temperature observed in the latter half of the century have been caused by increases in anthropogenic greenhouse gases offset by cooling from tropospheric sulphate aerosols rather than natural variability..." They claim that their "technology" is simply "just least squares regression in which we estimate the amplitude in observed data of prespecified [i.e., modelled] patterns of climate change in space and time" [21, page 1], yet at no point does their paper report adjusted R^2 or any other standard regression statistics (e.g., sum of squares, F, coefficients, standard errors, t-statistics, or P-values arising from their regressions). Nor does their paper report any of the standard tests (Durbin-Watson, Dickey-Fuller) for serial autocorrelation and thereby for spurious correlations, and least of all, any of the normal tests for multicollinearity. Moreover, these authors' "control simulation, in which *external* climate forcings...are kept constant to simulate [sic] natural *internal* variability, has been run for over 1700 years [sic] (our emphasis)" is contradictory.[6]

This paper uses overlooked NOAA-ESRL site-specific databases of statistics on a wider range of both human and natural climatic variables than is analyzed in any of the "detection and attribution" papers noted above. We show that a comprehensive analysis results in relegating [CO_2] to insignificance as a determinant of climate change, and that atmospheric water vapour arising almost exclusively from nonhuman sources is by far the largest source of radiative forcing and temperature change. We thereby hope to achieve a better response to the Kaufmann et al. [15, 16] challenge noted above, however incompletely. Section 2 provides an assessment of the appropriate specifications to be adopted for multivariate regression analysis of various models' climatic variables, while Section 3 outlines the paper's data sources. Section 4 reports its regression results, and the concluding Section 5 provides discussion of the implications of these results.

2. Methodology

Unlike mainstream climate science, which relies wholly on "general circulation models" (GCM) [4, page 749], few of which successfully hindcast the observational record without retrospective fine tuning of parameters, we seek to evaluate the following climate change models using only the observational record. That is represented by measures of monthly or annual temperatures (T), such as minimum, maximum, and mean, at various locations between 1960 and 2006, as potentially mostly determined by one of the following, including rising atmospheric concentration of greenhouse gases in general, represented by x_1, [CO$_2$] (following [13, page 96]), by variations in x_2, solar surface radiation (SSR, in Watt hours per square meter, Wh/m^2), and x_3, atmospheric water vapour ([H$_2$O], in cm.)

$$T = a + bx_1 + cx_2 + dx_3 + ex \ldots_n + u[x]. \qquad (1)$$

Variable $u[x]$ is an error or "noise" term that represents any failure of the linear combination of x_1, x_2, and x_3 to account fully for T. However, because of substantial evidence of spurious correlations when regressing T on the independent variables in (1), we assess the similar hypothesis, that year on year *changes* in temperature are determined by year on year *changes* in those independent variables (see (4) below).

It is important to establish that the RHS variables in (1) are indeed independent of each other, so I run regressions of each of $x_{1,\ldots n}$ on each other in turn; for example, if x_1 represents atmospheric water vapour [H$_2$O] and x_2 is [CO$_2$], then we need to know if x_1 is a function of T and x_2:

$$x_1 = a + fT + cx_2 \ldots . ex_{\ldots n}. \qquad (2)$$

Clearly, total independence of the T and [CO$_2$] variables is questionable, as on colder/hotter days offices and households are likely to use more heating/cooling, and if that involves burning of more hydrocarbon fuels, then large changes in T from ambient levels will affect additions to [CO$_2$]. I have done tests (not reported here) which show that changes in [CO$_2$] appear to have no impact on changes in [H$_2$O]. The outcomes of the regression analysis of (1) and (2) are discussed below. Additional regression results may be found in the Supplementary Material available on line at doi: 10.1100/3012/761473. Although there has been general agreement that the T and [H$_2$O] variables are independent, as "most of (the water vapour in the atmosphere) originates through evaporation from the ocean surface and is not influenced directly by human activity" ([22, page 23] see also IPCC, TAR, [23]), the view in IPCC AR4 [1] is that atmospheric water vapour is increasing because of the rises in temperature attributed to increasing [CO$_2$] (see below for assessment of that claim).

There has been considerable debate since Granger and Newbold [24] on how best to ensure that OLS regression of the variables in (1) does not produce spurious correlations between the temperature and the independent variables x_1, x_2, and x_3. Various tests have been devised to determine whether the variables are "stationary" or have "unit roots." The presence of a unit root in a time series is considered to invalidate standard regression analyses because that series is no longer stationary, this being a necessary condition to ensure avoidance of spurious correlation[7]. For example, many time series in economics have a steady upward trend similar to that of the concentration of carbon dioxide in the atmosphere [CO$_2$]—numbers of television sets, mobile phones, computers, and their broadband connections all show steady upward trends worldwide, but none of these trends can plausibly imply either direct or inverse causal relationships with [CO$_2$] despite no doubt striking correlation coefficients between them and rising [CO$_2$].

One widely applied solution to the problem of nonstationarity in time series is first to difference the series in question, by subtracting the present value of a variable from the previous value, and so on for all values in the series.[8] A simple regression model is merely a *straight line* fitted to a scatter-plot of one variable versus another. So when there are debates as in Kaufmann and Stern [25] and Kaufmann et al. [15, 16] as to whether various statistics, such as local or global temperatures and other climate variables, have a unit root and thereby require cointegration, or are trend stationary, this means only that there is a problem in *system identification*. That means we have to determine whether we are looking at the output of a first order low pass filter (what the statisticians call $I(0)$ or at the output of an integrator—as in $I(1)$). In the former, the variance is a constant (although the distribution may be around a linear trend), in the latter the variance is itself expanding (or inflating).

In this paper's *in situ* (local) model and its data these considerations are irrelevant. All that matters is that the data on changes in [CO$_2$] and [H$_2$O] and any other causative variables should be linearly independent. A key requirement—spelt out in rule (5) in the list below—is that this noise must have a constant variance over the distribution of samples; it must be $I(0)$. We need only to take first differences if we have some reason to suppose that this noise is $I(1)$, and because we do find evidence of multicollinearity when regressing the absolute values of the independent and dependent variables of interest, we focus here mostly on the results of regressions taking the first differences of both the dependent and the independent variables. However, "stationarizing" in this manner is not necessarily one of the general rules for successful application of regression analysis (and calculation of meaningful statistics subsequently).

In general, the various rules or conditions that must be satisfied for a valid regression are the following:

(1) the predictor samples $x_{t1,2\ldots n}$ and y_t must be representative of the population that they are sampling;

(2) the unknown u_t must have zero mean;

(3) the predictors must be linearly independent;

(4) the unknown u_t must be uncorrelated;

(5) the unknown u_t must be samples from a random variable population with constant variance, or *homoscedastic*.

Evidently, there is no particular requirement that the vectors x and y of the respective data should conform to a time series with specific statistical properties. The noise variables u_t in (1) appear to be I(0), with uncorrelated zero mean and with no expansion of the variance, at least there is no evidence that they are not.

The aim is to establish if the level of $[CO_2]$ is or is not—the main explanatory variable of average global or local temperature—in some quasimonotonic relation. For simplicity we stick to basic linear regression.

The Mauna Loa Slope Observatory in Hawaii has provided a test range of CO_2 from 315.71 ppm in April 1958 to 393.39 ppm in April 2011 and such current levels are confirmed by other measurements that started some years later, like those at Pt. Barrow in Alaska and elsewhere, including Cape Grim in Tasmania[9]. We may call this the independent "x_1" variable. Let y represent the averaged annual temperature at either the global or some specific location. Is there a dependence $y \gg f(x)$—or linearized about some operating point, does $y \gg a + bx$?[10] Perhaps so, but it makes no difference whatsoever in the testing whether x itself should exhibit a consistent rising trend (what engineers call a "ramp") or whether it is noise-like. The condition to satisfy rule (1), that x should cover with reasonable uniformity the given range, is clearly satisfied. Where this paper seeks to make a useful advance is in proposing a multiple regression to include all the potential causes of "weather."

One obvious candidate for determining mean maximum (i.e., day) temperature in addition to $[CO_2]$ has to be localized solar surface radiation SSR in Watt hours/sq. meter which I call here x_2. If the sun shines on any given day of successive years more or less "vertically" (or with less albedo) at any one place, subject to the level of atmospheric water vapour at the same place, then the temperature is likely to vary with the respective variations in solar radiation at that place. Similarly, the level of $[H_2O]$ at any given time and place, closely related to the relative humidity (RH) that is well known to make any given temperature level seem "hotter" than otherwise, has a no more evident relationship with $[CO_2]$ than the level of solar surface radiation. That is because $[CO_2]$ is invariant across the globe, at all given times and places, while $[H_2O]$ varies enormously at any given latitudes and times.

Again for simplicity let us introduce this one further possible explanation as $z \gg f(x_1, x_2)$ or linearized $z \gg a + bx_1 + cx_2$. Rule (3) says that formally there should be no linear dependence of x_1 and x_2, as that could produce multicollinearity and spurious correlations with temperatures y. There seems little risk of that with these variables. There is no reason why atmospheric water vapour and total watt hours of sun at any one location during any year would be coupled and connected to the level of $[CO_2]$ at that location in that year. Whether time series $x_1, x_2 \ldots x_n$ and time series y exhibit nonstationarity or not is irrelevant and incidental when they are independent of each other, but, to be on the safe side, we provide standard tests for the presence or not of multicollinearity and show that there is no such presence in any of the regressions of our first differenced data.

What a first differencing exercise may usefully show is a better exhibition of a rising trend in temperature since the "noise" in the measurements hopefully has been reduced by introducing the additional independent variables using $x_2 \ldots x_n$. Thus our multiple regression analysis seeks to remove or at least mitigate the scatter in annual temperature by testing if and when that scatter is linked to changes in solar surface radiation and other climatic variables such as $[H_2O]$, in the hope of revealing a better measurement of a linear trend in temperature (which would be otherwise nondiscernible for the data assembly in our selected sites).

Again, what really matters is the statistical property of the error sequence u_t. We assume that this is an I(1) sequence, because there is evidence for autocorrelation and multicollinearity of the absolute data, and that is why we rely on first differenced data. In general we find that $[CO_2]$ plays at best a marginal role—and one that is usually statistically insignificant—in explaining the temperature changes at various locations in USA over the 47 years inclusive between 1960 and 2006 (when the NOAA discontinued reporting the data sets used here, although a similar but less comprehensive series with data from 1948 to 2011 for locations defined by their latitude and longitude is available from ESRL-NASA).[11]

3. Data Sources

The "BEST" data sets [26] are the latest attempts to "homogenize" the most widely used global temperature sets, namely, Gistemp, HadleyCRU, and NCDC, but exclude the ESRL-NOAA data that have attempted the same task since 1996 [27]. In Section 4 below I use the BEST data set for global assessment since 1990, and in Supplementary Material, the NASA-GISS Gistemp series since 1958. For *in situ* (local) analysis I use the ESRL-NOAA database which covers some 1,200 locations across the whole of the USA since 1990, and of these more than 200 have data extending back to 1960. However I report detailed results of such analysis for just Point Barrow in Arctic Alaska and Hilo in equatorial Hawaii (at the foot of Hawaii's Mauna Loa) and at Mauna Loa itself, where Keeling set up his $[CO_2]$ observatory in 1958. In Supplementary Material I also report regressions of data from various other locations.

Table 7 presents a specimen of the NOAA-ESRL raw data from Point Barrow, in the arctic circle at the northernmost tip of Alaska, where if $[CO_2]$ is to be significant anywhere, it has to be there, given mean temperatures that have always been negative since 1960, despite $[CO_2]$ levels there that are almost identical to those at Mauna Loa in Hawaii and elsewhere on the globe.[12] Not only that, Barrow being in the Arctic Circle is a pristine site, far removed from confusing elements such as the urban heat island (UHI) effect, which is why it was selected as one of the gold standard locations for measurement of $[CO_2]$. That is also why Keeling selected Mauna Loa for his first $[CO_2]$ measurement station, as it too is far aways from other anthropogenic influences, at an altitude of 3,500 meters above sea level.

4. Regression Results

4.1. Is Most of Observed Temperature Change due to Anthropogenic GHGs? I first regress the global mean temperature (GMT) anomalies against the global annual values of the main climate variable evaluated by the IPCC Hegerl et al. [17] and Forster et al. [28] based on Myhre et al. [29], namely, the total radiative forcing of all the noncondensing greenhouse gases [RF]

$$\text{Annual } (T\text{mean}) = a + b[\text{RF}] + u(x)\dots \quad (3)$$

The results appear to confirm the findings of Hegerl et al. [17] with a fairly high R^2 and an excellent t-statistic (>2.0) and P-value (<0.01) but do not pass the Durbin-Watson test (>2.0) for spurious correlation (i.e., serial autocorrelation), see Table 1. This result validates the null hypothesis of *no statistically significant* influence of radiative forcing by noncondensing GHGs on global mean temperatures.[13]

Modifying (3) to represent first differences in both the dependent and independent variable,

$$\Delta\text{Annual } (T\text{mean}) = a + b(\Delta[\text{RF}]) + u(x)\dots \quad (4)$$

regression of year-on-year *changes* in GMT against those in [RF] passes the Durbin-Watson test statistic, but the adjusted R^2 statistic is now far below 0.5, so does not confirm the Hegerl et al. assertion (in Solomon et al. [1]) that "most" (at least more than 50 percent) of changes in GMT result from changes in [GHG] attributable to human causation (see Table 2 and Figure 1). The failure of the regression to reveal *any* contribution of changes in [GHG] to changes in Gistemp's GMT anomalies is obvious both from Figure 1 and from Table 2, which shows *total statistical insignificance* because with $t < 2.0$, and $P > 0.05$, the critical values are not attained. These results validate the null hypothesis from Hegerl et al. [17] that there is no discernible and statistically significant causation of global temperature change attributable to the radiative forcing from anthropogenic changes in noncondensing GHGs.

The minimal level of R^2 indicates serious omitted variable bias, and this could be addressed by using the ESRL-NOAA data for precipitable water [H_2O], with results shown in Table 3. Unfortunately, unlike the NOAA data sets for hundreds of locations in USA from 1960 to 2006, the ESRL-NOAA global reanalysis data sets exclude critical variables like solar surface radiation, and that explains why the minimal R^2 in Table 3 again indicates the absence of such variables and plausibly explains why the radiative forcing from [CO_2] and [H_2O] has such minimal statistical significance and can in no sense be described as "control knobs."

Next, I use first differences regressions to include NOAA data on nonanthropogenic variables at various locations of atmospheric water vapor [H_2O], and solar surface radiation in addition to [CO_2] as the main atmospheric GHG[14]

$$\text{Annual}(T\text{max}_{t1-t0}, T\text{min}_{t1-t0}, \text{AvD}T_{t1-t0})$$

$$= a + b([CO_2]_{t1-t0}) + c(\text{AVGLO}_{t1-t0}) \quad (5)$$

$$+ d(H_2O_{t1-t0})\dots.$$

TABLE 1: Regression of Gistemp anomalies on total noncondensing GHG-radiative forcings.

(a)

Regression statistics	
Multiple R	0.814
R square	0.662
Adjusted R Square	0.651
Standard error	12.384
Observations	31
Durbin Watson	1.749

(b) ANOVA

	Df	SS	MS	F
Regression	1	8718.49	8718.49	56.85
Residual	29	4447.51	153.36	
Total	30	13166		

(c)

	Coefficients	Standard error	t stat	P value
Intercept	−80.581	16.278	−4.950	0.000
Total radiative Forcings	53.584	7.107	7.540	0.000

TABLE 2: First-differenced regression of Gistemp temperature anomalies on total noncondensing GHG-radiative forcing.

(a)

Regression statistics	
Multiple R	0.183
R square	0.033
Adjusted R square	−0.001
Standard error	16.467
Observations	30
Durbin Watson	2.760

(b) ANOVA

	df	SS	MS	F
Regression	1	262.454	262.454	0.968
Residual	28	7592.513	271.161	
Total	29	7854.967		

(c)

	Coefficients	Standard error	t stat	P value
Intercept	−10.073	12.601	−0.799	0.431
dTotalRF	339.775	345.366	0.984	0.334

I first run this model for mean *minimum* temperatures at Point Barrow in Alaska from 1960 to 2006 with results reported in Table 4. This model passes the autocorrelation (D-W > 2.0) and collinearity tests. The nonanthropogenic [H_2O] variable is highly statistically significant in regard to what are night temperatures, at the better than 99 percent

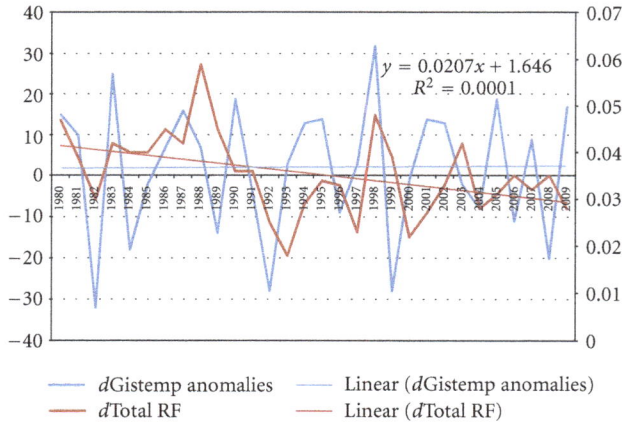

FIGURE 1: Plot of first differences in temperature anomalies and total radiative forcing (by all noncondensing GHGs). Source: Muller et al. [26].

TABLE 3: Regression of temperature change against radiative forcing of $[CO_2]$ and year-on-year changes in $[H_2O]$.

(a)

Regression statistics	
Multiple R	0.088
R square	0.008
Adjusted R square	−0.034
Standard error	0.169
Observations	51

(b) ANOVA

	df	SS	MS	F
Regression	2	0.011	0.005	0.188
Residual	48	1.372	0.029	
Total	50	1.383		

(c)

	Coefficients	Standard error	t stat	P value
Intercept	−0.030	0.084	−0.360	0.721
RF CO_2	0.038	0.071	0.544	0.589
$\Delta[H_2O]$	0.017	0.072	0.230	0.819

Sources: ESRL-NOAA and CDIAC.

level, while the coefficient on the differenced $[CO_2]$ variable is barely positive and remains statistically insignificant.

The adjusted R^2 in Table 4 is somewhat lower at 0.41 than the 0.63 in Table 2, so clearly there is still at least one omitted explanatory variable. As there is virtually no sunshine at Barrow for most of the winter, solar surface radiation is not a serious candidate, but obvious candidates include temperature variation arising from the ocean currents offshore of Pt Barrow, Arctic Ocean heat content, and decadal wind variability (see [30–33]). These variables are beyond the scope of this paper. However, if we now consider mean maximum temperatures at Barrow, then net

TABLE 4: Determinants of temperature change at Pt. Barrow. Mean minimum temperature 1960–2006.

(a)

Regression statistics	
Multiple R	0.660
R square	0.435
Adjusted R square	0.409
Standard error	1.227
Observations	46

(b) ANOVA

	df	SS	MS	F
Regression	2	49.913	24.956	16.566
Residual	43	64.779	1.506	
Total	45	114.692		

(c)

	Coefficients	Standard error	t stat	P value
Intercept	0.001	0.473	0.002	0.998
$\Delta[H_2O]$	17.225	3.076	5.600	0.000
$\Delta[CO_2]$	0.007	0.311	0.021	0.983

Sources: http://rredc.nrel.gov/solar/old_data/nsrdb/1961-1990/dsf/ and http://rredc.nrel.gov/solar/old_data/nsrdb/1991-2005/ For CO_2: http://www.esrl.noaa.gov/gmd/ccgg/trends/.

solar surface radiation ("AVGLO") and opacity of the sky ("OPQ") (using their absolute values) can be added to the regression, with results shown in Supplementary Material. There the unadjusted R^2 is 0.39, and only the $[H_2O]$ variable is statistically significant, accounting for more than 90 per cent of the changes in mean maximum temperature over the period 1960–2006, thereby going beyond the assertion by Schmidt et al. 2010 cited above that atmospheric water vapor accounts for only 50 per cent of the total greenhouse effect [2].

The conclusion from the limited model used in Tables 4 and 5 is that there can be a high degree of confidence that the increasing radiative forcing due to rising $[CO_2]$ at Pt Barrow attributable to anthropogenic emissions plays no role in explaining the climate there since 1960. We are also able to show that which proves to be the case for all the other locations where the same regression analysis is possible (see examples in the Supplementary Material).

I now provide here one further site-specific regression analysis, for the Slope Laboratory at Mauna Loa itself (Table 6), close to the Equator while Barrow is in the Arctic Circle. The Durbin-Watson and collinearity tests are satisfactory, and the R^2 (0.399) along with the t statistics and P values for the coefficient on $\Delta[CO_2]$ do not imply that changes in $[CO_2]$ at the foot of Mauna Loa have had "most" (Hegerl et al. [17]) to do with temperature changes there since 1960. Instead, the coefficients on the variables in Table 6 indicate that $[H_2O]$ accounts for more than 90 per cent of temperature change near where C. D. Keeling began

TABLE 5: Determinants of temperature change at Pt. Barrow. Determinants of *year-on-year* differences in mean *maximum* annual temperatures point Barrow 1960–2006.

(a) Summary output. Dependent variable: year-on-year changes in mean maximum temperatures

Regression statistics	
Multiple R	0.593
R square	0.351
Adjusted R square	0.321
Standard error	1.314
Observations	46

(b) ANOVA

	df	SS	MS	F
Regression	2	40.162	20.081	11.633
Residual	43	74.228	1.726	
Total	45	114.390		

(c)

	Coefficients	Standard error	t stat	P value
Intercept	−0.075	0.743	−0.101	0.920
ΔH_2O	15.456	3.207	4.820	0.000
RF abs	0.065	0.648	0.100	0.921

Sources: http://rredc.nrel.gov/solar/old_data/nsrdb/1961-1990/dsf/ and http://rredc.nrel.gov/solar/old_data/nsrdb/1991-2005/ For CO_2: http://www.esrl.noaa.gov/gmd/ccgg/trends/.

TABLE 6: Regression analysis of *changes* in annual mean temperature at Mauna Loa Slope Observatory with respect to *changes* in the annual level of $[CO_2]$ and $[H_2O]$ 1977–2009.

(a) Summary output. Dependent variable: year on year changes in mean maximum temperatures

Regression Statistics	
Multiple R	0.631
R square	0.399
Adjusted R square	0.356
Standard error	0.565
Observations	46

(b) ANOVA

	df	SS	MS	F
Regression	3	8.876	2.959	9.281
Residual	42	13.389	0.319	
Total	45	22.265		

(c)

	Coefficients	Standard error	t stat	P value
Intercept	−0.248	0.219	−1.133	0.264
$\Delta[CO_2]$	0.195	0.145	1.346	0.186
$\Delta[H_2O]$	2.564	0.548	4.676	0.000
ΔAVGLO	0.001	0.000	3.721	0.001

Durbin-Watson: 2.834

his measurements of the atmospheric concentration of CO_2 back in 1958.[15]

I noted above that water vapor is the most potent greenhouse gas because it absorbs strongly in the infra-red region of the light spectrum, first demonstrated by Tyndall [10], despite the conventional view [2] that because the water vapor content of the atmosphere will increase in response to warmer temperatures, water vapor is only a feedback that merely amplifies the climate warming effect due to increased carbon dioxide alone. In reality, the $[H_2O]$ variable in the NOAA's database proves to be a remarkably powerful determinant of climate variability over the period from 1960 to 2006 not only at Barrow but across all USA, as it is always highly statistically significant at better than the 95% level of confidence for both annual mean minimum and maximum annual temperatures. This is hardly surprising, if only because in reality, as Tans has noted[16], "global annual evaporation equals ~500,000 billion metric tons. Compare that to fossil CO_2 emissions of ~8.5 billion ton C/year," and even the total level of $[CO_2]$ is only 827 billion tonnes of carbon equivalent. It would seem to be a case of the tail wagging the dog if the additions to $[CO_2]$ from human burning of hydrocarbon fuels have raised global temperatures enough (just 0.0125°C p.a. since 1950) to generate annual evaporation of 500,000 billion tonnes of $[H_2O]$, especially when as I have shown here, its role in

explaining temperature changes is much less than claimed by the IPCC's Hegerl et al. [17] (see also [5]) (Figure 2).

5. Conclusion

This paper has used basic econometric (multivariate least squares regression) analysis of observational evidence to falsify or confirm two null hypotheses, first that "most" of observed global warming since around 1950 has *not* been "very likely" caused by emissions of noncondensing anthropogenic GHGs [17], and, second, that the noncondensing GHGs do *not* constitute a "control knob" enabling manipulation of global climate. The regression results in the previous Section confirm the first null, as there is no statistically significant evidence to show that increases in anthropogenic GHGs account for any, let alone "most," of observed global temperature change.

The second null derives from this statement by Lacis et al. [3].

This assessment comes about as the result of climate modeling experiments which show that it is the noncondensing greenhouse gases such as carbon dioxide, methane, ozone, nitrous oxide, and chlorofluorocarbons that provide the necessary atmospheric temperature structure that ultimately determines the sustainable range for atmospheric water vapor and

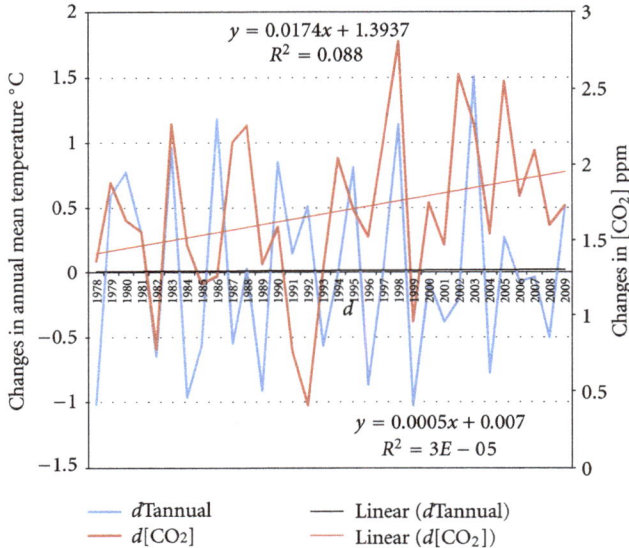

FIGURE 2: Trends in annual *changes* in annual mean temperatures and average annual $[CO_2]$ Mauna Loa Slope Observatory 1977–2009. Notes: neither of the trends has good linear fits, with $R^2 < 0.1$, and there is in fact no discernible trend in changes in the annual mean temperature at Mauna Loa. Temperature data at Mauna Loa are not included in any of the Hadley-CRU, GCHN, and Gistemp data sets.

cloud amounts and thus controls their radiative contribution to the terrestrial greenhouse effect. From this it follows that these noncondensing greenhouse gases provide the temperature environment that is necessary for water vapor and cloud feedback effects to operate, without which the water vapor dominated greenhouse effect would inevitably collapse and plunge the global climate into an icebound Earth state.

Schmidt et al. [2] make a similar claim: "a model simulation performed with zero CO_2 gives a global mean temperature changes of about $-35°C$ and produces an ice-covered planet (A. Lacis, pers. communication)." These paper's regressions do not invalidate the null that *none* of the Schmidt-Lacis effects is evident when econometric analysis is applied to observations of the most relevant climate variables and instead indicate that the planet's slow warming is mainly associated with the much larger primary rather than feedback changes in atmospheric water vapor, which along with rising $[CO_2]$ have major social benefits in terms of supporting the rising food production needed to feed a global population now at 7 billion and projected to reach 9 billion by 2050 [6–9]. This may imply the demonization of atmospheric CO_2 by Hegerl et al. [17] and Schmidt et al. [2], as the alleged primary source of rising temperature could be because of the obvious political difficulty in countries like Australia of blaming increasing rainfall for the observed slow increases in global temperatures evident since 1950.

The basic physical science underlying the results above is very straight forward, despite the misleading claims

in Solomon et al. [1] and Trenberth and Fasullo [34].[17] These and others distinguish between so-called "long-lived" noncondensing GHGs and the certainly short-lived nature of $[H_2O]$ arising from evaporation created by solar energy, since it is true that condensation and precipitation generally follow evaporation within at most around ten days. But that does not eliminate nonanthropogenic evaporation, for as Lim and Roderick show [35, page 14], the average daily level of basic $[H_2O]$ is around 3-4 litres per square meter throughout the year[18]. That is a result of the solar radiative forcing of 342 W/sq. meter [1, page 96]. meter. In contrast the total radiative forcing attributable to noncondensing anthropogenic GHGs is only c. 2.6 W/sq. meter [28]. The annual increase in GMT attributable to up to that level of forcing since 1950 has been only 0.0125°C p.a. But the Clausius-Clapeyron relation which defines the maximum partial pressure of water vapour that can be present in a volume of atmosphere in thermodynamic equilibrium implies that would have only trivial effect on $[H_2O]$. The maximum is known as the saturation vapour pressure, e_s:[19]

$$e_s(T) = 6.1094 \exp\left(\frac{17.625T}{T + 243.04}\right). \qquad (6)$$

This formula suggests that the increase in $[H_2O]$ attributable to rising GMT of 0.0125°C p.a. that could be accommodated in the atmosphere is only 0.047 per cent p.a., not enough to have any measurable effect on GMT, far less than the 2°C to even 3°C and more claimed by Solomon et al. 2007 or Schmidt et al. 2010 for a doubling of $[CO_2]$ from the preindustrial level of 280 ppm.

The data underlying my regressions showing that in general variations in $[H_2O]$ account for as much as 90 per cent of observed changes in temperature both globally and *in situ* suggest that such variations are far larger than those indicated here by Clausius Clapeyron. Thus both my regressions and Clausius Clapeyron fail to invalidate the nulls of the hypotheses advanced by Hegerl et al. [17], Lacis et al. [3], and Schmidt et al. [2]. Consequently it is far from certain that managing the level of atmospheric carbon dioxide concentration really is a meaningful "control knob."

Appendix

AVGLO. A joule is the unit of energy. The Watt is the unit of power and equal to a joule per second, or, equivalently, a joule of energy equals a wattsecond, and a watthour, Wh, equals 3.6 kilojoules of energy. The daily average solar radiation at Barrow in June 1991 (per square meter), 3918 Wh, was 14.1 MJ (mega or million joules)—which lies in the range noted here. The IPCC's [1, page 141] radiative forcing power of 2.6 W (per square meter) is equivalent to 0.225 MJ daily of additional energy in the climate system (per square meter). Thus, we might conclude that, according to the IPCC, atmospheric CO_2 accumulated since preindustrial times exerts 1.6% of the power of the sun on a summer's day in Barrow. More precisely, AVGLO is the average daily total radiation for the "Global" horizontal component of solar radiation (Wh/m²). "Global" solar exposure is the total

Table 7: Specimen of NOAA Data Base. Point Barrow 1960–2006 (selected solar and atmospheric variables, data on average windspeed and relative humidity, and so forth are also available). 700260 BARROW W POST-W ROGERS AK -9 N71 19 W156 37 10 1012.

1960	AVGLO	AVDIR	AVDIF	AVETR	AETRN	TOT	OPQ	H_2O	TAU	MAX_T	MIN_T	AVG_T
January	1	28	1	9	399	4.7	3.4	0.31	0.07	−21.89	−28.5	−25.2
February	259	879	174	692	8541	5.1	3.9	0.29	0.08	−24.36	−30.95	−27.66
March	1568	3422	767	2980	15482	4.5	3.1	0.27	0.09	−22.79	−29.52	−26.17
April	3672	5181	1819	6387	21863	5.1	3.8	0.32	0.11	−15.19	−22.82	−19.01
May	4661	2925	3367	9870	29980	8.1	7.4	0.58	0.12	−4.33	−9.8	−7.05
June	4898	3687	3131	11824	31777	7.9	7.1	1.02	0.14	3.49	−1.26	1.13
July	4456	3878	2627	10926	31671	7.7	6.8	1.38	0.14	7.24	0.89	4.08
August	2624	1576	1962	7760	24588	8.9	8.3	1.26	0.13	5.75	0.75	3.26
September	1338	715	1125	4262	17865	9.2	8.7	0.81	0.11	1.01	−2.76	−0.86
October	478	451	413	1450	11513	8.5	7.7	0.46	0.09	−7.74	−12.89	−10.3
November	25	92	21	110	2665	7	6.1	0.32	0.08	−15.85	−21.59	−18.72
December	0	0	0	0	0	0	0	0.29	0	−20.68	−27.31	−24.01

Source: http://rredc.nrel.gov/solar/old_data/nsrdb/1961-1990/dsf/ and http://rredc.nrel.gov/solar/old_data/nsrdb/. AVGLO/DIR/DIF: Average daily total solar radiation for the GLObal horizontal, DlRect normal, and DlFfuse horizontal elements (Wh/m^2). SDGLO/DIR/DIF: Standard deviation of daily total global, direct, and diffuse solar radiation (see note (2) below) (Wh/m^2). AVETR & AETRN: Average dally total global horizontal (AVETR) and direct normal (AETRN) extraterrestrial solar radiation (Wh/m^2). TOT, OPQ, H2O, TAU: Average TOTal and OPaQue sky cover (tenths), precipitable water (cm), and aerosol optical depth (unitless). MAX_T, MIN_T, AVG_T: Average maximum, minimum, and 24-hour temperatures (°C).

amount of solar energy falling on a horizontal surface. The daily global solar exposure is the total solar energy for a day. Typical values for daily global solar exposure range from 1 to 35 MJ/m^2 (megajoules per square meter). The values are usually highest in clear sun conditions during the summer, and lowest during winter or very cloudy days... Irradiance is a measure of the rate of energy received per unit area and has units of Watts per square meter (W/m^2), where 1 Watt (W) is equal to 1 Joule (J) per second. Radiant exposure is a time integral (or sum) of irradiance. Thus a 1 minute radiant exposure is a measure of the energy received per square meter over a period of 1 minute. Therefore, a 1-minute radiant exposure equals mean irradiance (W/m^2) × 60(s) and has units of joule(s) per square meter (J/m^2)" (see *Solar Radiation Definitions*, Australian Bureau of Meteorology, 2009). The NOAA/NREL data are hourly averages in Wh/m^2, so "AVGLO" of 3918 Wh/m^2 at Point Barrow in June 1991 is equivalent to 94,032 W/m^2 per day or 65 W/m^2 per minute. Thus at Barrow the SSR for the year 1960 amounted to a daily average of 2006 Wh/m^2, or 17,572,560 Watts/m^2 over the year. The AVGLO data in Table 7 are net of outgoing albedo (reflection of sunlight), and the NOAA's database therefore also includes the gross "Direct" and "Diffuse" components of incoming solar radiation. The source data includes standard deviations for these variables. Note that the IPCC uses W/m^2 (not Wh/m^2) for its measures of the (net) *addition* to solar radiation by "radiative forcing" of greenhouse gases, estimated at 2.6 W/m^2 (per minute), that is, 156 Wh/m^2, in 2005 [1, page 141]. That may be compared but never be in Solomon et al. 2007, with the AVGLO (*total net solar radiation*) of 3918 Wh/m^2 at Barrow in June 1991. However, the former is the IPCC's radiative forcing, a change in the balance of the energy fluxes at top of the atmosphere measured in watts per square meter. The latter is the total

energy incident on an average day in say June 1991 on a square meter of Barrow measured in watt hours per square meter, but modified by [H_2O] and other factors like aerosols and albedo. The IPCC's radiative forcing is gross, that is, without taking into account any of the other factors affecting the level of solar radiation actually reaching the surface at Pt Barrow or anywhere else.

H_2O. It is Precipitable water vapour (cm.) The total atmospheric water vapour contained in a vertical column of unit cross-sectional area is extending between any two specified levels, commonly expressed in terms of the height (cm. in Table 7) to which that water substance would stand if completely condensed and collected in a vessel of the same unit cross-section. See also Solomon et al. [1, pages 271–273], where it is stated "global, local, and regional studies all indicate increases in moisture in the atmosphere near the surface."

TOT and OPQ. Opaque sky cover is the amount of sky completely hidden by clouds or obscuring phenomena, while total sky cover includes this plus the amount of sky covered but not concealed (transparent). Sky cover, for any level aloft, is described as thin if the ratio of transparent to total sky cover at and below that level is one-half or more. Sky cover is reported in tenths, so that 0.0 indicates a clear sky and 1.0 (or 10/10) indicates a completely covered sky (excerpt from *Meteorological Glossary*, AMA, accessed 29 September 2010). *En passant*, we note that the presence of more molecules of CO_2 in the atmosphere could be expected to *decrease* TOT and OPQ, and this could help to explain why rising [CO_2] in the local data sets examined here tends to have a *negative*, rather than positive, impact on local temperatures (see Supplementary Material).

Tau—Aerosol Optical Depth (AOD or, in NOAA Data Sets, "Tau"). Aerosol optical depth is a quantitative measure of the extinction of solar radiation by aerosol scattering and absorption between the point of observation and the top of the atmosphere. It is a measure of the integrated columnar aerosol load and the single most important parameter for evaluating direct radiative forcing. The optical depth expresses the quantity of light removed from a beam by scattering or absorption during its path through a medium. If I_0 is the intensity of radiation at the source and I is the observed intensity after a given path, then optical depth τ is defined by the following equation:

$$\frac{I}{I_0} = e^{-\tau}. \qquad (A.1)$$

(From Wikipedia articles "Aerosol Optical Depth" and "Optical Depth," accessed 28 September 2010.)

Acknowledgments

The author is grateful to M. S. Hodgart for his methodological insights and to many others for invaluable comments on early versions of this paper but is responsible for all views expressed and for any remaining errors.

Endnotes

1. Tol and Vellinga [36] used econometric analysis to separate the enhanced greenhouse effect from the influence of the sun at the top of the atmosphere (TOA), while Tol and de Vos [13] used Bayesian analysis. Neither paper considers the role of atmospheric water vapour. Most common is the fingerprint method (e.g. Hegerl and Allen [18]) which claims to produce a human signal by using Global Circulation Models (GCM). But "the fingerprint approach is only applicable for detection of (dis)similarities between patterns; it seems impossible to use it to derive a probability distribution of the climate sensitivity. We use time series analysis. We do not rely on GCM results—at the expense of using an (overly) simple representation of the climate—and show that this allows to estimate a probability distribution of the climate sensitivity" [13, pages 88-89].

2. The textbook by von Storch and Zwiers *Statistical Analysis in Climate Research* [37] offers an advanced treatment but apart from using climate data for illustrative examples does not itself undertake systematic analysis using its own methods. Those methods are also absent from [17].

3. If only two independent variables are specified, "most" must mean more than 50%; if there are three or more, then "most" means that the preferred variable, in this case [CO_2], must have greater potency than the sum of the others; for example, 40% is not sufficient if the others sum to 60%. If the criterion for "most" is only that [CO_2] be the single most potent (and significant) of all the variables, that could be only 1% if all the others *individually* each contribute less than 1% but *jointly* account for 99%. In that case the claims by the IPCC's Solomon et al. that increases in [CO_2] account for "most" observed global warming are consistent with all states of the world.

4. The basic formula for the radiative forcing attributed to rising [CO_2] is RF = 5.35(LN(CO_{2t}/280)), where 280 is the preindustrial level of [CO_2] in ppm., and CO_{2t} is the level in the year in question [29].

5. Stott has 11 citations as a lead author in Hegerl et al. [17].

6. Muller et al. [38] provide data showing very large declines in SO_2 emissions of the US electric power sector between 1990 and 2008. While coal-fired power's CO_2 emissions increased by nearly 200 billion tons over that period, SO_2 emissions by the coal-fired power industry fell from 14.28 million tonnes to 5.5 million tonnes, and similarly for NO_x (7.1 million tonnes to 2.4). These falls should have enhanced the radiative forcing from rising levels of atmospheric CO_2, but there is no evidence to support that. Kaufmann et al. [39] attribute the "hiatus" in global warming since 2000 to SO_2 emissions in China, but their estimate of such global emissions at 65 million tonnes p.a. seems trivial relative to annual emissions of over 30 $GtCO_2$.

7. "In the mathematical sciences, a stationary process is a stochastic process whose joint probability distribution does not change when shifted in time or space. As a result, parameters such as the mean and variance, if they exist, also do not change over time or position." (Wikipedia, October 2010).

8. A stationary series *ab initio* is denoted $I(0)$, the first differenced as $I(1)$, and the second as $I(2)$.

9. The US data are available from http://www.esrl.noaa.gov/gmd/ccgg/iadv/.

10. In the Supplementary Material, I report basic regressions of Gistemp's "global temperatures" as a function of the radiative forcing of the level of [CO_2].

11. Graphing highly autocorrelated time series data showing rising CO_2 concentrations and rising temperatures is not enough to "prove" that the data support the theory that the former is responsible for the latter.

12. Point Barrow is also an ideal test of Arrhenius' model [40], since he himself claimed that the temperature effects of doubled [CO_2] would be significantly higher (6.05°C) at Barrow's latitude (71°N) than at the equator (4.95°C) [40, Table VII] for a doubling of [CO_2], and 3.52°C for just a 50 percent increase in [CO_2], compared with 3.15°C at the equator [40, Table VII]. Arrhenius' prediction of a 3.52°C rise in temperature at Point Barrow given a 50 percent increase in [CO_2] implies that its mean annual temperature would have warmed by more than 3.52°C from the actual minus 12.54°C in 1960 to around minus 9.02°C in 2006, whereas the actual so far is minus 10.2°C, a warming of only 2.52°C. At Hilo near the equator, the predicted "warming" from

1960 to 2006 actually turned out to be a cooling of 0.12°C for the same near 40 percent increase in [CO_2], but consistent with Arrhenius' prediction that warming would be greater at higher latitudes than lower.

13. If the Durbin-Watson statistic is substantially less than 2, there is an evidence of positive serial correlation, "Durbin-Watson statistic," *Wikipedia*, accessed 26th October 2010.

14. Kaufmann et al. [15, page 257] also adopt first differencing: "To avoid statistical problems associated with the lack of cointegration, we take the first difference of equation (7)

$$\Delta xt = \rho \Delta xt - 1 + \Delta \varepsilon t + \theta \Delta tempt + \_vt. \qquad (A.2)$$

Specifying the concentration equation in first differences eliminates all stochastic trends and therefore allows us to avoid the effects of carbon uptake by the unknown carbon sink(s) and measurement error on statistical estimates for the effect of temperature on concentrations."

15. See also Curry at al. [30] and Liu et al. [31].

16. Pers. Comm. See also Tans [41]. For more detailed estimates of global evaporation rates, see Lim and Roderick [35].

17. Trenberth [42] states "Water has a short lifetime in the atmosphere of 9 days on average before it is rained out. Carbon dioxide on the other hand, has a long lifetime, over a century, and therefore plays the most important role in climate change while water vapor provides a positive feedback or amplifying effect; the warmer it gets, the more water vapor the atmosphere can hold by about 4% per degree Fahrenheit". This claim that atmospheric (CO_2) has a long lifetime, over a century...," is at variance with Houghton et al. [24] and Houghton [23], which indicate only about 5 years, because around 20 per cent of atmospheric CO_2 is continuously recycled between the earth's surface and the atmosphere. Moreover, if atmospheric water vapor arising from solar-based evaporation is "rained out" within 9 days, as claimed by Trenberth and Fasullo [34] that must also be true of [H_2O] attributable to rising temperature via the Clausius-Clapeyron relation (see below).

18. One millimetre of measured precipitation is the equivalent of one litre of rainfall per metre squared. The estimates in Lim and Roderick [35] of average global evaporation/precipitation of up to 1,187 mm. p.a. in 1970–1999 imply evaporation/precipitation of c.4 mm. per day. Thus, the average daily evaporation implies average [H_2O] of c. 4 litres per day per metre squared. See also Kelly [43].

19. Wikipedia, "Clausius-Clapeyron", accessed 30 October 2011, and Pierrehumbert et al. [44, page 145].

References

[1] S. Solomon, D. Qin, M. Manning et al., *Climate Change 2007. The Physical Science Basis. Working Group 1, International Panel on Climate Change, Fourth Assessment Report*, Cambridge University Press, Cambridge, UK, 2007.

[2] G. A. Schmidt, R. A. Ruedy, R. L. Miller, and A. A. Lacis, "Attribution of the present-day total greenhouse effect," *Journal of Geophysical Research D*, vol. 115, no. 20, Article ID D20106, 2010.

[3] A. A. Lacis, G. A. Schmidt, D. Rind, and R. A. Ruedy, "Atmospheric CO_2: principal control knob governing earth's temperature," *Science*, vol. 330, no. 6002, pp. 356–359, 2010.

[4] G. A. Meehl, T. F. Stocker, W. D. Collins et al., "Global Climate projections," in *Climate Change 2007: The Physical Science Basis. Contribution of Working Group I to the Fourth Assessment Report of the Intergovernmental Panel on Climate Change*, S. Solomon, D. Qin, M. Manning et al., Eds., Cambridge University Press, Cambridge, UK, 2007.

[5] R. Pielke Sr., K. Beven, G. Brasseur et al., "Climate change: the need to consider human forcings besides greenhouse gases," *Eos*, vol. 90, no. 45, p. 413, 2009.

[6] J. Lloyd and G. D. Farquhar, "The CO_2 dependence of photosynthesis, plant growth responses to elevated atmospheric CO_2 concentrations and their interaction with soil nutrient status. I. General principles and forest ecosystems," *Functional Ecology*, vol. 10, no. 1, pp. 4–32, 1996.

[7] R. J. Norby and Y. Luo, "Evaluating ecosystem responses to rising atmospheric CO_2 and global warming in a multi-factor world," *New Phytologist*, vol. 162, no. 2, pp. 281–293, 2004.

[8] F. Dyson, *A Many-Colored Glass: Reflections on the Place of Life in the Universe*, University of Virginia Press, Charlottesville, Va, USA, 2007.

[9] T. Curtin, "Climate change and food production," *Energy & Environment*, vol. 20, no. 7, pp. 1099–1116, 2009.

[10] J. Tyndall, "On the absorption and radiation of heat by gases and vapours, and on the physical connection of radiation, absorption, and conduction," *Philosophical Magazine Series 4*, vol. 22, no. 169–194, pp. 273–285, 1861.

[11] A. E. Dessler and S. M. Davis, "Trends in tropospheric humidity from reanalysis systems," *Journal of Geophysical Research D*, vol. 115, Article ID D19127, 5 pages, 2010.

[12] D. I. Stern and R. K. Kaufmann, "Detecting a global warming signal in hemispheric temperature series: a structural time series analysis," *Climatic Change*, vol. 47, no. 4, pp. 411–438, 2000.

[13] R. S. J. Tol and A. F. de Vos, "A Bayesian statistical analysis of the enhanced greenhouse effect," *Climatic Change*, vol. 38, no. 1, pp. 87–112, 1998.

[14] R. S. J. Tol, "Greenhouse statistics—time series analysis: part II," *Theoretical and Applied Climatology*, vol. 49, no. 2, pp. 91–102, 1994.

[15] R. K. Kaufmann, H. Kauppi, and J. H. Stock, "Emissions, concentrations, & temperature: a time series analysis," *Climatic Change*, vol. 77, no. 3-4, pp. 249–278, 2006.

[16] R. K. Kaufmann, H. Kauppi, and J. H. Stock, "The relationship between radiative forcing and temperature: what do statistical analyses of the instrumental temperature record measure?" *Climatic Change*, vol. 77, no. 3-4, pp. 279–289, 2006.

[17] G. Hegerl, F. W. Zwiers, P. Braconnot et al., "Understanding and attributing climate change," in *Climate Change 2007: The Physical Science Basis. Contribution of Working Group I to the*

Fourth Assessment Report of the Intergovernmental Panel on Climate Change, S. Solomon, D. Qin, M. Manning et al., Eds., Cambridge University Press, Cambridge, UK, 2007.

[18] G. Hegerl and M. Allen, "Origins of model-data discrepancies in optimal fingerprinting," *Journal of Climate*, vol. 15, no. 11, pp. 1348–1356, 2002.

[19] R. J. Charlson and T. M. L. Wigley, "Sulfate aerosol and climatic change," *Scientific American*, vol. 270, no. 2, pp. 28–35, 1994.

[20] G. C. Hegerl, K. Hasselmann, U. Cubasch et al., "Multi-fingerprint detection and attribution analysis of greenhouse gas, greenhouse gas-plus –aerosol and solar forced climate change," *Climate Dynamics*, vol. 13, Article ID 613D634, 22 pages, 1997.

[21] P. A. Stott, S. F. B. Tett, G. S. Jones, M. R. Allen, W. J. Ingram, and J. F. B. Mitchell, "Attribution of twentieth century temperature change to natural and anthropogenic causes," *Climate Dynamics*, vol. 17, no. 1, pp. 1–21, 2001.

[22] J. Houghton, *Global Warming: The Complete Briefing*, Cambridge University Press, Cambridge, UK, 2004.

[23] J. T. Houghton, Y. Ding, D. J. Griggs et al., *Climate Change 2001. The Scientific Basis. Working Group 1, Third Assessment Report*, Cambridge University Press, Cambridge, UK, 2001.

[24] C. Granger and P. Newbold, "Spurious regressions in econometrics," *Journal of Econometrics*, vol. 2, no. 2, pp. 111–120, 1974.

[25] R. K. Kaufmann and D. I. Stern, "Cointegration analysis of hemispheric temperature relations," *Journal of Geophysical Research D*, vol. 107, no. 1-2, pp. 8–10, 2002.

[26] R. A. Muller, J. A. Curry, D. Groom et al., *Earth Atmospheric Land Surface Temperature and Station Quality*, University of California, Los Angeles, Berkeley, Calif, USA, 2011.

[27] E. Kalnay, M. Kanamitsu, R. Kistler et al., "The NCEP/NCAR 40-year reanalysis project," *Bulletin of the American Meteorological Society*, vol. 77, no. 3, pp. 437–471, 1996.

[28] P. Forster, V. Ramaswamy, P. Artaxo et al., "Changes in atmospheric constituents and in radiative forcing," in *Climate Change 2007: The Physical Science Basis. Contribution of Working Group I to the Fourth Assessment Report of the Intergovernmental Panel on Climate Change*, S. Solomon, D. Qin, M. Manning et al., Eds., Cambridge University Press, Cambridge, UK, 2007.

[29] G. Myhre, E. J. Highwood, K. P. Shine, and F. Stordal, "New estimates of radiative forcing due to well mixed greenhouse gases," *Geophysical Research Letters*, vol. 25, no. 14, pp. 2715–2718, 1998.

[30] J. A. Curry, J. L. Schramm, M. C. Serreze, and E. E. Ebert, "Water vapor feedback over the Arctic Ocean," *Journal of Geophysical Research*, vol. 100, no. 7, pp. 14223–14229, 1995.

[31] J. Liu, J. A. Curry, and Y. Hu, "Recent Arctic sea ice variability: connections to the Arctic Oscillation and the ENSO," *Geophysical Research Letters*, vol. 31, Article ID L09211, 4 pages, 2004.

[32] A. H. Lynch, J. A. Curry, R. D. Brunner, and J. A. Maslanik, "Toward an integrated assessment of the impacts of extreme wind events on Barrow, Alaska," *Bulletin of the American Meteorological Society*, vol. 85, no. 2, pp. 209–141, 2004.

[33] J. S. Turner, "The melting of ice in the arctic ocean: the influence of Double-Diffusive transport of heat from below," *Journal of Physical Oceanography*, vol. 40, no. 1, pp. 249–256, 2010.

[34] K. E. Trenberth and J. T. Fasullo, "Tracking earth's energy," *Science*, vol. 328, no. 5976, pp. 316–317, 2010.

[35] W. H. Lim and M. L. Roderick, *An Atlas of the Global Water Cycle*, ANU E Press, Canberra, Australia, 2009.

[36] R. S. J. Tol and P. Vellinga, "Climate change, the enhanced greenhouse effect and the influence of the sun: a statistical analysis," *Theoretical and Applied Climatology*, vol. 61, no. 1-2, pp. 1–7, 1998.

[37] H. Von Storch and F. W. Zwiers, *Statistical Analysis in Climate Research*, Cambridge University Press, Cambridge, UK, 1999.

[38] N. Muller, R. Mendelsohn, and W. Nordhaus, "Environmental accounting for pollution in the united states economy," *American Economic Review*, vol. 101, no. 5, pp. 1649–1675, 2011.

[39] R. K. Kaufmann, H. Kauppi, M. L. Mann, and J. H. Stock, "Reconciling anthropogenic climate change with observed temperature 1998–2008," *Proceedings of the National Academy of Sciences*, vol. 108, no. 29, pp. 11790–11793, 2011.

[40] S. Arrhenius, "On the influence of carbonic acid in the air upon the temperature of the ground," *Philosophical Magazine*, vol. 41, pp. 237–276, 1896.

[41] P. Tans, "An accounting of the observed increase in oceanic and atmospheric CO_2 and an outlook for the future," *Oceanography*, vol. 22, no. 4, pp. 26–35, 2009.

[42] K. E. Trenberth, "Tracking Earth's Energy—a key to climate variability and change," *Skeptical Science*, July 2011.

[43] A. Kelly, "A materials scientist ponders global warming/cooling," *Energy & Environment*, vol. 21, no. 6, pp. 611–632, 2010.

[44] R. T. Pierrehumbert, H. Brogniez, and R. Roca, "On the relative humidity of the atmosphere," in *The Global Circulation of the Atmosphere*, T. Schneider and A. Sobel, Eds., chapter 6, Princeton University Press, Princeton, NJ, USA, 2007.

19

Source Apportionment of Particle Bound Polycyclic Aromatic Hydrocarbons at an Industrial Location in Agra, India

Anita Lakhani

Department of Chemistry, Dayalbagh Educational Institute, Agra 282005, India

Correspondence should be addressed to Anita Lakhani, anitasaran2003@yahoo.co.in

Academic Editor: Xavier M. Querol

16 US EPA priority polycyclic aromatic hydrocarbons (PAHs) were quantified in total suspended ambient particulate matter (TSPM) collected from an industrial site in Agra (India) using gas chromatography. The major industrial activities in Agra are foundries that previously used coal and coke as fuel in cupola furnaces. These foundries have now switched over to natural gas. In addition, use of compressed natural gas has also been promoted and encouraged in automobiles. This study attempts to apportion sources of PAH in the ambient air and the results reflect the advantages associated with the change of fuel. The predominant PAHs in TSPM include high molecular weight (HMW) congeners BghiP, DbA, IP, and BaP. The sum of 16 priority PAHs had a mean value of $72.7 \pm 4.7 \, \mathrm{ng \, m^{-3}}$. Potential sources of PAHs in aerosols were identified using diagnostic ratios and principal component analysis. The results reflect a blend of emissions from diesel and natural gas as the major sources of PAH in the city along with contribution from emission of coal, coke, and gasoline.

1. Introduction

Polycyclic aromatic hydrocarbons (PAHs) are comprised of carbon and hydrogen atoms in two or more aromatic rings [1]. They are a group of ubiquitous persistent organic pollutants possessing carcinogenic, mutagenic, and immunotoxic properties [2]. They occur in the atmosphere due to emissions from gasoline- and diesel-powered vehicles and other sources such as coal, biomass, gas, and oil combustion [3–7]. The strainless arrangement of sp^2-hybridized carbon atoms in six membered rings and, above all, the energy gain from the delocalized p-electrons, makes PAHs thermally very stable and therefore PAH are abundant molecules in the combustion zone. Therefore, they are always formed in combustion processes and are attributed to unburned, pyrolyzed, or partially oxidized fuel and lubricant oil that are transferred from the gas phase to the particulate phase by adsorption and condensation onto the existing particles or by nucleation of new particles when the exhaust cools [2].

PAHs occur in gaseous form and are adsorbed to particles in the atmosphere, depending on the volatility of the PAH species. Higher condensed molecules with four and more rings are particle-bound, whereas smaller PAHs mainly remain in the gas phase [8]. Once PAHs are released into the atmosphere, they are subjected to various atmospheric processes through which their dispersion, removal, transport and degradation can occur. Dispersion of atmospheric PAH is dependent not only on the magnitude of the emissions but also on the stability of PAH in the atmosphere [8]. PAH concentration is highly dependent upon the size of airborne particulate matter, with the greatest concentration being in the respirable size range [9].

PAHs have attracted much attention in the studies on air pollution recently because some of them are highly carcinogenic or mutagenic. In particular, benzo(a)pyrene has been identified as a highly carcinogenic [10]. The occurrence of PAH in urban air has caused particular concern because of the continuous nature of the exposure and the size of the population at risk. Humans can be affected by direct inhalation of polluted air, tobacco smoke, ingestion of contaminated and processed food and water, and through dermal contact.

PAHs have been less monitored in Asia, specifically in India, in comparison to its western counterpart. India is a developing nation which has experienced an increase in population and industrial expansion that has been accompanied by drastic increase in vehicular transportation. Information on the airborne concentration of toxic pollutants in developed countries is relatively abundant. However, for less developed nations as India, data are not routinely collected. In India, studies on measurements of PAHs in the ambient air of major cities are limited. Total PAH concentrations in ambient aerosols in Ahmedabad, Mumbai, Nagpur, Calcutta, and Kanpur show that PAH concentrations are 10–50 times higher than those reported internationally and range between 23 and 190 ng m^{-3} [11–15]. In an earlier study, it was reported that the annual average total PAH concentrations at four locations in New Delhi, which ranged from 150 to 1,800 ng m^{-3}, were dominated by vehicular traffic [16]. Recently atmospheric concentrations of total PAHs were found to range from 1,049 to 1,344 ng m^{-3} at three sites in Delhi with higher concentrations in winter. Diesel- and gasoline-driven vehicles were identified as the principal sources of PAHs [17]. In addition, in the Indian urban environment, cooking fuel combustion is also a likely source of PAH. High concentrations of PAH have been measured in smoke from solid-fuel stoves burning wood, coal, and dried cattle manure [11], which along with kerosene stoves [18] are used as the primary cooking device by urban slum residents.

Agra, located in the state of Uttar Pradesh of northern India, is known for Taj Mahal, one of the wonders of the world. In addition to Taj Mahal, there are three more world heritage sites, namely, Agra Fort, Fatehpur Sikri, and the bird sanctuary at Bharatpur National Park. In addition, the Akbar's tomb at Sikandra in Agra (in proximity to Agra) and the Imtad-ud-Daulahs tomb in Agra are proposed as world heritage sites. The majority of industries in Agra comprise of foundries and several diesel generator manufacturing units. In foundries, the principal source of emission is cupola. The volume of gas exhausted and its concentration depends on the cupola, operations, melting rates, characteristic of charging material, and the coke [19]. Gases escape while drawing the hot metal and during casting. In the pit type of cupola, emissions are fugitive type. Besides a number of petha (a sweet) industries are operating in the city, which mainly use coal as fuel. There are also several sweet shops, pottery units that use coal, cow dung, and wood [19]. The vehicle fleet in Agra comprises of trucks, buses, delivery vans, taxis, private cars, and autorickshaws that are mainly diesel driven. The number of registered on-road diesel vehicles in Agra till January 2008 was approximately 51,938 [20]. Diesel-driven generators are almost exclusively used in the industrial units as well as in the households as a stand by to meet power requirements during periods of load shedding. The diesel quality used earlier was poor containing a high S content, to the extent of 0.25%, and a large amount of aromatics [19]. Moreover, a large population of the slum residents of the area use coal, coke, wood, kerosene, and cow dung cakes as the domestic fuel, which is burnt in small home made cook stoves.

The area in the immediate vicinity of the Taj Mahal in the form of a trapezoid is known as the Taj Trapezium Zone. It extends between 27°30′ N and 77°30′ E to 27°45′ N

and 77°15′ E and 26°45′ N and 77°15′ E to 27°00′ N and 78°30′ E. In 1999, the Ministry of Environment and Forest, Government of India, has notified the Taj Trapezium Zone (Pollution Prevention and Control) Authority for protection and improvement of the environment in the Trapezium. In this connection several actions have been taken to control industrial and vehicular pollution. Industries have been encouraged to use natural gas instead of coal and there is strict vigilance on these industries not to use coal or coke; the brick kilns that were previously operational around the city have been closed. Likewise stringent actions have also been taken to control vehicular pollution. These include phasing out of grossly polluting vehicles plying within the city area and age limits for different categories of vehicles were fixed by Road Transport Authority (RTA). Diesel-driven autos fitted with particle traps are only given license and registration and to control emissions low sulfur (0.05%) diesel is made available. More and more CNG/LPG (compressed natural gas/liquefied natural gas) and battery-operated vehicles are being promoted. Similarly gas-driven electricity generator sets have been developed and their use has been encouraged. Simultaneously for the in-use vehicles strict checking of vehicular emissions is performed [19].

In the recent past we have conducted PAH measurement in total suspended particulate matter at Agra for short durations at different locations [7, 21]. These studies also revealed higher concentrations of PAH in TSPM that were mainly attributed to originate from vehicular traffic and combustion of coal and biomass comprising wood, crop residues, and agricultural wastes burnt in the domestic and industrial sector. In the present study results of PAH measurements carried out from May 2006 to December 2008 at a location close to the industrial area as well as the major highway (NH-2) crossing in the city are presented. An attempt has also been made to see the influence of the use of cleaner fuels through determination of sources of PAH by principal component analysis.

2. Materials and Methods

2.1. Description of Sampling Site. Agra is situated in the extreme southwest corner of Uttar Pradesh and stretches across 26°44′ N to 27°25′ N and 77°26′ E to 78°32′ E. Its borders touch Rajasthan to its west and south, the district of Firozabad to its East, and the districts of Mathura and Etah to its North. It is situated on the banks of river Yamuna, and it has a limited forest area supporting mainly deciduous trees. According to census 2001, the area of Agra district is 4027 km^2 with a total population of about 1,316,177 and density about 21,148 of km^2 with 386,635 vehicles registered and 32,030 generator sets. In Agra, 60% of NOx pollution is due to vehicles [20]. Three national highways (NH-2, NH-11, and NH-3) pass through the city. Like most cities of north India, the weather and climate of Agra is extreme and tropical. Agra suffers from extremities of climate with scorching hot summers and chilly winters. It is about 169 m above the mean sea level (msl) and has been reported [19] as having semiarid climate with atmospheric temperature in the range of 11–48°C (max) 0.7–30°C (min), relative humidity

25–95%, light intensity 0.7–5.6 oktas (cloudiness), and rainfall 650 mm per year. The climate of Agra has been broadly divided into three seasons: winter (October to February), summer (March to June), and monsoon (July to September). During monsoon period the temperature ranges between 26 and 39°C and relative humidity varies between 70 and 100%, respectively [19]. The winter months are cool and temperature ranges from 2 to 15°C while relative humidity is 60 to 90%. Summers are characterized by high temperature ranging from 23 to 45°C and low relative humidity 25 to 40%.

Sampling was conducted on the roof of a single storeyed building situated in proximity to the industrial area as well as about 1.5 km from a busy roadway intersection situated on the National Highway Number2 (NH-2). NH-2 connects Delhi to Kolkata via Agra and is one of the busiest highways. On an average six to seven thousand vehicles ply on this highway and comprise about 3500 light motor vehicles (LMV), 500 light commercial vehicles (LCVs), and 2500 heavy commercial vehicles (HCVs) [22]. In this study total suspended particles (TSPs) were collected during May 2006 to December 2009 using a high-volume sampler. The samples were collected on predesiccated and preweighed glass fibre filters (Whatman, EPM 2000). Air was drawn at a flow rate of $1.1 \, m^3 \, min^{-1}$ for 24 hr on an average. The accurate flow rate of the sampling device was determined by averaging the flow rates measured at the beginning and the end of the sampling period. Before bringing the samples to the laboratory, the filters were packed in aluminium foils to protect them from dust and were stored at low temperature to prevent the volatilization of low molecular weight PAHs.

The particulate matter contained in the filter was extracted thrice by using ultrasonic agitation with 200 mL dichloromethane for 1 hr after intervals of 10 min after every 15 min to ensure maximum recovery. All the three extracts were mixed together to make a composite sample. The extract was purified by means of cleanup in silica gel column to remove elements interfering in the analysis. The extract was transferred to the top of a glass column (10 cm × 1.0 cm i.d.) slurry packed with 2 g of silica gel (Fluka, 230 mesh). The column was eluted with dichloromethane to give fraction enriched with PAH. The PAH containing fraction was concentrated to 1.5 mL by a rotary evaporator and stored in Teflon vials at low temperature till analysis.

PAHs were analyzed in the splitless mode using a temperature gradient program [23] by gas chromatograph, (Shimadzu 17AATF, version 3.0) equipped with a FID detector and capillary column (25 m length, 0.3 mm internal diameter: BP) with dimethyl polysiloxane as stationary phase. Nitrogen was the carrier gas at a flow rate of $12.7 \, mL \, min^{-1}$. The oven temperature was held at 40°C for 5 minutes and programmed to rise to 179°C at $10°C \, min^{-1}$, held for 2 minutes, and then elevated to 300°C at $9°C \, min^{-1}$. The temperature of injector and detector was maintained at 210°C and 310°C, respectively. The GC was calibrated with a standard solution of 16 PAH compounds (Supelco EPA 610 PAH mixture). The procured PAH mixture contained the following 16 EPA priority PAHs in mixed solvent (methanol: dichloromethane; vol/vol, 1 : 1); Naphthalene (Nap), Acen-aphthylene (Acy), Acenaphthene (Ace), Fluorene (Flu), Phenanthrene (Phen), Anthracene (Anth), Fluoranthene (Fla), Pyrene (Pyr), Benzo(a)anthracene (BaA), Chrysene (Chr), Benzo(b)fluroanthene (BbF), Benzo(k)fluoranthene (BkF), Benzo(a) pyrene (BaP), Dibenzo(a,h)anthracene (DbA), Benzo(ghi)perylene (BghiP), and Indeno (1,2,3-cd)pyrene (IP) in the range of 100 to 2000 $\mu g \, mL^{-1}$. Five point calibration curves for all the target analytes ranging were obtained by analysis of serial dilution of PAHs standard. Calibration curves were plotted by regression analysis. 1 μL of the extracted sample was injected into GC and the program was set to run for 40 minutes. Individual PAH were identified by comparing their retention time with the standard chromatogram. Each PAH compound was quantified by plotting its peak area on the regression curve of standard. The limit of detection of the chromatographic method determined through serial dilution of the PAH standard varied between 0.007 and 0.16 ng for the different compounds. The limit of quantification (LOQ) defined as the limit of detection divided by the sampling volume [24] was in the range of 1.8×10^{-7} and $4.10 \times 10^{-5} \, ng \, m^{-3}$ [7]. The recovery efficiency of the method was evaluated by the analysis of filters spiked with known concentration of standard PAH compounds [7]. Most of the compounds provided high recoveries with mean values ranging between 70 and 80%. Field and laboratory blank were routinely analyzed for quality control. Blanks levels of individual analytes were normally very low and in most cases not detectable. Some PAH compounds like BaA and Chr, BbF and BkF, and DbA and IP coelute in the method followed in this study; hence these coeluting pairs have been reported as their sum [25].

3. Results and Discussion

3.1. PAH Concentrations and Seasonal Variations. Both TSP and PAH concentration presented a large variability with a skewed distribution. The geometric mean concentrations of TSP, individual PAH, and total PAH (i.e., sum of all determined PAH in each sample, TPAH) are listed in Table 1.

The mean concentration of individual PAH ranged from 0.9 to 20.5 $ng \, m^{-3}$. PAH can be classified into low molecular weight compounds (Nap, Acy, Ace Flu, Phen, Anth, Fla and Pyr) and high molecular weight compounds (BaA, Chr, BbF, BkF, BaP, DbA, IP, and BghiP). The characteristic mass distribution pattern of PAH shows that the mass distribution in air was dominated by high molecular weight PAH with 5 to 6 rings such as BbF, BkF, DbA, IP, BaA, Chr, BaP and BghiP. PAHs are semivolatile organic compounds and occur in both gaseous and particulate phases in the atmosphere. The vapor-particle partition exhibits a strong dependence on molecular weight. High molecular weight compounds dominate in the particulate phase while low molecular weight PAHs reach higher concentrations in the vapor phase. The HMW PAHs are mainly derived from the vehicular emissions and are known to exist adsorbed on particulate matter owing to their low vapor pressure. Moreover, some of the higher molecular weight PAHs in the air are short lived; half life is about 2.4 h for BaP, 1–13 h for DbA and BaA, and 0.31–10 h for BghiPs, therefore, they are easily detected at

TABLE 1: Mean and standard deviations for all PAHs ($ng\,m^{-3}$).

	Average	Standard deviation ($ng\,m^{-3}$)
Naphthalene (NaP)	0.9	1.7
Acenaphthylene (Acy)	1.7	4.7
Acenaphthene (Ace)	1.8	1.2
Fluorene (Flu)	1.0	0.7
Phenanthrene (Phen)	0.9	0.2
Anthracene (Anth)	2.2	1.3
Fluoranthene (Fla)	0.9	1.3
Pyrene (Pyr)	3.1	2.3
Benzo(a)anthracene + Chrysene (BaA + Chr)	14.1	4.7
Benzo(b)fluoranthene + Benzo(k)fluoranthene (BbF + BkF)	20.5	6.8
Benzo(a)pyrene (BaP)	12.3	6.0
Dibenz(ah)anthracene + Indeno (123-cd) pyrene (DbA + IP)	17.3	3.1
B(ghi)perylene (BghiP)	11.1	6.2
TPAH	72.7	4.7
TSPM*	348	31.5
CPAH	47.6	—
CPAH/TPAH	0.66	—
Total Carcinogenic PAH	64%	—

*$\mu g\,m^{-3}$.

sites where sampling is carried out close to emission sources [26]. Among the low molecular weight compounds Acy, Ace, and Anth are in higher concentrations than Pyr, Fla, Flu, and Phen. The sum of the concentration of nine major combustion PAHs (CPAH) (Fla, Pyr, BaA, Chr, $B(b+k)F$, BaP, IP, and BghiP) accounted for 77% of the total PAH mass. The ratio of CPAH/TPAH was 0.66 and is higher than the ratio reported for noncatalyst (0.41), catalyst (0.51) automobiles, and heavy duty diesel trucks (0.30) [27]. Noncatalyst vehicles emit 27 times more PAHs especially HMW PAHs than catalyst-equipped vehicles [27]. Less efficient emission control system in the vehicle fleet and more extensive combustion activities in the city may account for this high CPAH/TPAH ratio. Sum of the carcinogenic PAHs including BaA, BbF, BkF, BaP, DbA, and IP accounted for 64% of TPAH. IARC has classified total carcinogenic PAH as "probably carcinogenic" to humans (namely, BaA, BaP, and DbA) and as "possibly carcinogenic" to humans (namely, BbF, BkF, and IP) [28]. In the present study the "probably car-cinogenic" compounds accounted for 38% while "possibly carcinogenic" compounds accounted for 26% of total PAH. The carcinogenic PAH have high molecular weights and are especially bound to suspended particles. BaP has been the most extensively measured PAH in urban areas around the world due to its high carcinogenic property. Average BaP concentration at the site was $12.3\,ng\,m^{-3}$, that is, less than the permissible exposure limit ($0.2\,mg\,m^{-3}$) of PAH set by the Occupational Safety and Health Administration (OSHA) [29].

Atmospheric PAH levels can differ to an extent of 40% depending on the sampling system, extraction solvents, analytical techniques, and detectors [30]. PAHs determined in airborne particles are also influenced by regional climatic and source characteristics [31]. A comparison has been made with other Indian sites in Table 2.

PAH concentrations varied inversely with temperature ($r = -0.65$, $P = 0.05$), that is, higher concentrations in the cold months and lower ones in the warm months. The winter-to-summer ratio of individual compounds varied between 1.14 (DbA) and 2.65 (IP). The increase in particulate PAH concentration during the winter and the dependence of PAH concentration on atmospheric temperature have been reported in a number of studies [35–38]. Source emissions and meteorological conditions as well as gas particle partitioning may result in winter and summer difference of PAHs concentrations [39]. Reduced atmospheric dispersion resulting from lower mixing height as well as reduced atmospheric reaction can lead to higher pollutant concentrations in ambient air during winter. Low atmospheric temperature can affect the distribution of PAHs between the gas and particle phases and result in a relatively larger portion of PAH partitioning to the particle phase in winter. In contrast during the summer a higher ambient temperature could change the distribution of PAHs between the gaseous and particulate phases by increasing the vapor pressure of pollutants that adhered to atmospheric aerosols and favoring the volatilization of PAHs from the particulate to gaseous phase. It is also well known that PAHs in the atmosphere are subjected to photochemical or thermal reactions with ozone, nitric oxides, and hydroxyl radicals that lead to PAHs degradation, especially in the warmer seasons [40]. High solar radiation could enhance the reaction of volatile organic compounds in forming ozone active hydroxyl radicals that in turn react with PAHs and reduce their concentrations [41]. In addition to temperature effects on the physicochemical property of atmospheric PAHs, anthropogenic factors can also lead to

TABLE 2: Comparison of total PAHs ($\mathrm{ng\,m^{-3}}$) with various cities of India.

Total PAHs	Range	City	Substrate	Extraction	Analysis	Reference
$\sum11$ PAHs (PM$_{2.5}$)	326–791 (Mean at 4 Sites)	Chennai	PTFE membrane Filters	DCM : Methanol (60 : 40); Ultrasonication	HPLC (Fluorescence Detector)	[32]
$\sum9$ PAHs (PM$_{2.5}$)	202.6–333.7 (Mean at 4 Sites)	Tiruchirappalli	PTFE membrane Filters	DCM : Methanol (60 : 40); Ultrasonication	HPLC (Fluorescence Detector)	[32]
$\sum11$ PAHs (PM$_{10}$)	8–97.9 (Mean at 4 Sites)	Agra	Glass fibre filter	Methylene Chloride; Soxhlet Extraction	GC-MS	[33]
$\sum12$ PAHs (TSP)	1049–1344 (Mean at 3 Sites)	Delhi	Glass fibre filter	Toluene; Ultrasonication	GC-FID	[17]
$\sum12$ PAHs (TSP)	672 (Mean)	Delhi	Glass fibre filter	Toluene; Ultrasonication	GC-FID	[34]
$\sum16$ PAHs (TSP)	72.7 (Mean)	Agra	Glass fibre filter	DCM Ultrasonication	GC-FID	Present Study

TABLE 3: Diagnostic ratios.

	Present study	Diesel	Gasoline	Wood	Coal	Other sources	References
IP/(IP + BghiP)	0.72	0.37		0.62	0.56		[54]
		0.35–0.7	0.18				[53, 54]
			0.21–0.22				[27]
					0.33		[49]
							[53]
BaA/(BaA + Chy)	0.44	0.38–0.64					[52]
			0.22–0.55				[52]
					0.50		[53]
BaP/(BaP + Chy)	0.61	>0.73	0.49				[46, 55]
Phen/(Phen + Anth)	0.68	0.65	0.5		0.76	>0.70 (Crude oil)	[51, 55]
BaA/BaP	0.50	0.90–1.70	0.50–0.70	1.0–1.5			[50]
BbF/BkF	1.08	>0.5					[52]

seasonal variation of particulate PAHs. During the cold season, PAH emissions from automobile exhaust are higher because of low ambient temperature and increased cold start impacts [42]. Another reason for the higher concentrations in winter could be due to an increase in the emissions owing to the fossil fuel usage for space heating purposes. Sharma et al. [17] have also found winter/summer ratios of total PAH between 2.41 and 3.69 at three sites in Delhi. However PAHs in winter have been found higher by a factor of 1.5–10 than that in summer in studies carried out in Europe and the USA [8, 43–45]. Similarly, Guo et al. [46] found high winter/summer ratios of 8.6 and 7.5 at two sites in the city of Hong Kong.

The contribution of the combustion-derived PAH (CPAH) to total PAH also varied between the two seasons. Their contributions were 64 and 72%, respectively, in summer and winter. The lower contribution is summer meant that vehicular emissions may be a contribution in summer while in the winter season, emission sources were more complex.

3.2. Diagnostic Analysis and Principal Component Analysis. Studies indicate that particulate organic samples collected in

tunnels are enriched in benzo(ghi)perylene and coronene, which are characteristic of gasoline engines [47]. Diesel exhaust is found to be enriched in Fla, Chr, and Pyr [48] while Anth, Phen, Fla, and Pyr have been identified as source fingerprints of wood combustion and BaA and Chr as markers of coal combustion [49]. The concentrations of these marker compounds and their ratios can give some indication about the impact of different sources of airborne compounds and can be used in distinguishing emissions [49, 50]. In this study, the following PAH concentration diagnostic ratios characteristic of the anthropogenic emissions were calculated: IP (IP + BghiP), BaA/(BaA + Chr), BaP/(BaP + Chr), BghiP/IP, Phen/(Phen + Anth), BaA/BaP, and BbF/BkF. A comparison between the various diagnostic ratios obtained in this study with the standard values reported in the literature is shown in Table 3. Analysis of the ratios could be associated to various sources. The IP/(IP + BghiP) ratio is found to be 0.72, which is similar to that reported for diesel and gasoline emissions [27, 44, 46]. Similarly the value of BaA/(BaA + Chr) ratio has been reported to vary between 0.38 and 0.64 for diesel emissions and between 0.22 and 0.5 for gasoline emissions [51, 52]. For emissions from coal burning this ratio has been observed to be 0.50 [53].

In the present study this ratio is 0.44 which can be attributed to vehicular (diesel and gasoline) emissions. The vehicular influence can further be assessed from Phen/(Phen + Anth) ratio whose value of 0.68 is comparable to values reported for diesel, coal, and crude oil burning and gasoline emissions [49, 51]. The ratio BaP/(BaP + Chr) with its value 0.61 again indicates towards contributions from diesel emissions. The BaA/BaP and BghiP/IP ratios of 0.50 at this site may be inferred to the contribution from gasoline emissions [50]. Further BbF/BkF ratio with a value >0.5 may be attributed to diesel emissions [10]. These results suggest the influence of multiple sources like the burning of fuels, such as diesel oil, gasoline, and coal. Thus diagnosis of ratios contributes quantitatively to the identification of the main emission sources in the studied area. It is to be noted that the diagnostic ratios method should be used with caution because it is often difficult to discriminate between some sources [4]. For example, a ratio of 0.62 for IP/IP + BghiP has been suggested for wood burning [54] whereas a value between 0.35 and 0.70 indicates diesel emissions [55]. Hence, it would be difficult to differentiate diesel emission from biomass emission based only on one proposed diagnostic ratio. Recently Galarneau [56] has reviewed the caveats associated with the use of diagnostic ratios in source apportionment of PAHs in ambient air. The validity of this approach rests on the assumption that each source or source type is associated with relative proportions of the compounds that are unique and secondly the relative proportions of the compounds are assumed to be conserved between the emission source and the receptor or the points of measurement. These assumptions hold for PAH under limited set of conditions [56]. These ratios can also be altered due to the reactivity of some PAH species with other atmospheric species like ozone and/or oxides of nitrogen or by the volatility and solubility of some PAH [4, 57]. To minimize this bias, the diagnostic ratio of PAH with similar physicochemical properties should be used. Degradation processes during the sampling process may also modify PAH concentrations and thus the ratios between PAHs.

In addition to the diagnostic ratios between PAH, Principal Component Analysis (PCA) was used as an exploratory tool as it enhances the accuracy of emission source identification by selecting statistically independent source tracers [46]. This method segregates measured ambient concentrations according to groups of covarying compounds and comparing these groups to suspected source profiles. Thus using PCA, it is possible to simplify the interpretation of complex systems and to reduce the set of variables to few new ones, called factors. Each of these factors can be identified as either an emission source or a chemical interaction by the most representative compounds. Some PAH compounds and their combinations are frequently used as source markers such as Flu, Chr, and Pyr for diesel exhaust [46, 58], Flu, Pyr, BbF, and BkF for heavy duty diesel vehicles [4, 59], BghiP for gasoline vehicles [4, 58, 60], Flu and BaP [31] or Ant, Phe, Flu, and Pyr [4, 5, 31, 49] or NaP and Ant [49] for wood combustion, Ant, Phe, Flu, Pyr, BaA, and Chr for coal combustion [4, 5, 49, 60], Phe and Flu [4, 49] for coke production, Flu and Pyr [4, 61] for oil burning, and Ant, Phe, BaP, BghiP, and Chr [4] for steel industry emissions.

Before subjecting the data to PCA, a pretreatment involving an autoscaling of the data was performed. In this autoscaling procedure all variables were mean centered and scaled to unit variance, by subtracting the mean x_j of each variable j from the result x_{ij} of each sample i for that variable and dividing the result by the standard deviation s_j of that variable. The resulting autoscaled z_{ij} were used instead of the original x_{ij}:

$$z_{ij} = \frac{(x_{ij} - x_j)}{s_j}. \tag{1}$$

As a result of this autoscaling, each variable has about the same range and it is avoided that some variables would be more important than others because of scale effects. In the present study, PCA was attempted by SPSS (version 10.0) by adopting Varimax procedure for rotation of factor matrix into one that was easier to interpret. Deciding the number of factors to retain for each data set is a critical issue. Retaining too few factors results in combining different sources, while retaining too many factors splits sources among factors in a physically unreasonable manner. The number of factors to be retained was decided on the basis of the breaks in the slopes of scree plots that are plots between eigen values and PCs as shown in Figure 1(a). PCs with eigen values greater than 1were retained as they provided a reasonable physical interpretation of sources. The PCs were interpreted on the basis of their loadings. Factor loadings greater than 0.5 were considered statistically significant. The factor profiles were constructed on the basis of factor loadings. These are shown in Figure 1(b). Three factors were obtained. They explained approximately 80% of the total variance of data. Factor 1 was associated with BaA + Chr, Fla, BbF + BkF, BaP, Flu, and Pyr and accounted for 44% of the total variance.

This factor was attributed to industrial emissions (stationary source). BbF and BkF are indicators of diesel exhaust emissions mainly emitted from diesel generators that are frequently operated during power cuts in the various industrial units located in this area and partly also contributed by heavy duty diesel vehicles plying around the area. Fla, Pyr, BaP, and BaA + Chr in this factor can be associated to emissions from natural gas combustion. BaA, Chr, and BaP are associated with combustion of natural gas [16, 27, 52] and BaA has been considered as a tracer for this source [52]. The presence of Chr with BaA, Fla, and Pyr in natural gas combustion has also been reported by Daisey et al. [62]. Factor 2 explains 22% of the total variance and includes Acy, Ace, Flu, Phen, and Anth. This factor can be attributed to coal and wood combustion. Khalili et al. [49] identified that Ant, Phe, Fla, and Pyr were source fingerprints of wood combustion, while the third factor explaining 14% of the variance can be associated to emissions from gasoline vehicles showing higher loadings on BghiP and DbA + IP.

3.3. Health Risk Assessment. An essential prerequisite for determining the human health risk associated with PAH in the environment is accurately characterizing the toxicities of individual PAH. As the toxicities of individual PAH differ considerably, toxicity assessments of PAH are complex [63].

TABLE 4: Proposed Toxic Equivalency Factors (TEFs) for individual PAHs and calculated TEF-adjusted concentrations of measured samples.

PAH	TEF (data from [69])	TEF-adjusted concentrations in ng m^{-3} (Present Study)
Naphthalene (NaP)	0.001	0.0009
Acenaphthylene (Acy)	0.001	0.0017
Acenaphthene (Ace)	0.001	0.0018
Fluorene (Flu)	0.001	0.0010
Phenanthrene (Phen)	0.001	0.0009
Anthracene (Anth)	0.01	0.0223
Fluoranthene (Fla)	0.001	0.0002
Pyrene (Pyr)	0.001	0.0031
Benzo(a)anthracene (BaA)	0.1	0.616
Chrysene (Chr)	0.01	0.079
Benzo(b)fluoranthene Benzo(k)fluoranthene (BbF + BkF)	0.1	1.064
Benzo(k)fluoranthene (BkF)	0.1	0.984
Benzo(a)pyrene (BaP)	1	12.27
Indeno(123-cd)pyrene (IP)	0.1	1.479
Dibenz(ah)anthracene (DbA)	1	0.023
B(ghi)perylene (BghiP)	0.01	11.13

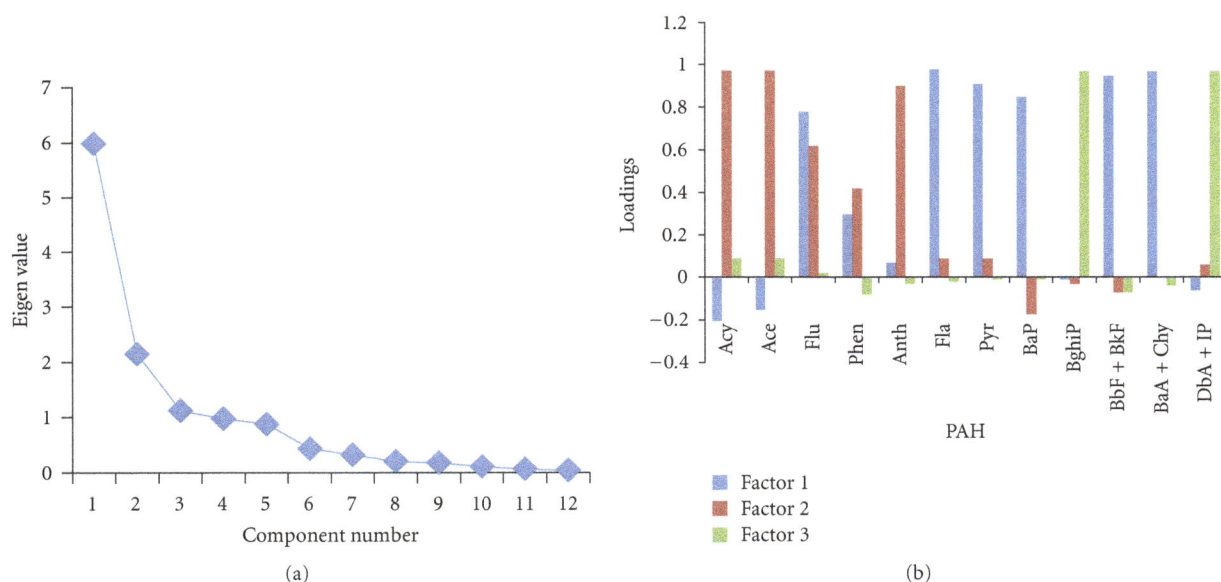

FIGURE 1: (a) Scree plot. (b) Factor profile.

One approach commonly referred as the Toxic equivalence Factor (TEF) calculates the inhalation risk for excess lung cancer over the risk posed by BaP for each of its copollutant carcinogenic PAH in the polluted ambient air [64]. TEFs are toxicity potency factors that are used by scientists and regulators globally as a consistent method to evaluate the toxicities of highly variable mixtures of PAH compounds. TEF was initially developed for estimation of the risk caused by complex mixtures, such as dioxins, and could help to characterize more precisely the carcinogenic properties of complex mixtures. The method relates individual compound toxic potency to a compound, often the most toxic compound in the mixture [65]. A few proposals for TEFs are available [66–70]; among these the list of TEFs completed by Nisbet and LaGoy [69] reflects well the actual knowledge of the toxic potency of each individual PAH compound [70]. In the PAH family, BaP the most toxic member, is assigned a TEF of one. According to this approach the health risk assessment of PAHs can be assessed based on its BaP equivalent concentration (BaPeq). BaPeq concentration for each PAH is calculated by multiplying its concentration with the corresponding TEF. The carcinogenic potency of total PAHs can be assessed by the sum of the BaPeq concentration of each PAHs. The TEFs-adjusted concentrations of each PAH along with their proposed TEFs are presented in Table 4. The total BaPeq concentration was found to be 27.6 ng m^{-3}. The LMW compounds show only a negligible contribution towards carcinogenicity. The major contributor is BaP followed by BghiP. This underlines and confirms the importance of BaP and BghiP as a surrogate compound for PAH mixture in ambient air. The carcinogenic PAHs (BbF, BkF, BaP, BghiP, and IP) accounted for 97% of the total BaPeq concentrations which is similar to 97% and 93% in Ho Chi Minh city, Vietnam, and Osaka, Japan, respectively [58]. These results imply that particle bound PAHs with high molecular weight play an important role on total BaP eq concentration from health risk point of view. Other compounds like Acy, Ace, Anth, Pyr, BaA, and Chr which also dominate in the PAH mixture obviously play a minor role in the carcinogenicity of the PAH mixture in ambient air.

4. Conclusion

In this study, TSP samples were collected and quantified for 16 PAH compounds. Mean total PAH concentration was found to be 72.7 ± 4.7 ng m^{-3} that is lower than the permissible exposure limit (0.2 mg m^{-3}) of PAH set by the Occupational Safety and Health Administration (OSHA). The high molecular weight PAHs, namely, BghiP, DbA, IP, and BaP were the most abundant. Average BaP concentration was 12.3 ng m^{-3}. Concentrations of Total PAHs as well as the individual compounds were lower in summer than in winter. Diagnostic ratios and principal component analysis indicated that the influences of vehicular sources and stationary sources involving both diesel and natural gas combustion were the principal sources while emissions from coal-, coke-, and gasoline-driven vehicles had a minor contribution.

Acknowledgment

The author is grateful to the Director, Dayalbagh Educational Institute Agra, Head, Department of Chemistry and Department of Science and Technology, Project no. SR/S4/AS:282/07, New Delhi, for financial assistance.

References

[1] E. Velasco, P. Siegmann, and H. C. Siegmann, "Exploratory study of particle-bound polycyclic aromatic hydrocarbons in different environments of Mexico City," *Atmospheric Environment*, vol. 38, no. 29, pp. 4957–4968, 2004.

[2] D. Golomb, E. Barry, G. Fisher, P. Varanusupakul, M. Koleda, and T. Rooney, "Atmospheric deposition of polycyclic aromatic hydrocarbons near New England coastal waters," *Atmospheric Environment*, vol. 35, no. 36, pp. 6245–6258, 2001.

[3] K. Ravindra, L. Bencs, E. Wauters et al., "Seasonal and site-specific variation in vapour and aerosol phase PAHs over Flanders (Belgium) and their relation with anthropogenic activities," *Atmospheric Environment*, vol. 40, no. 4, pp. 771–785, 2006.

[4] K. Ravindra, R. Sokhi, and R. Van Grieken, "Atmospheric polycyclic aromatic hydrocarbons: source attribution, emission factors and regulation," *Atmospheric Environment*, vol. 42, no. 13, pp. 2895–2921, 2008.

[5] X. Wang, H. Cheng, X. Xu, G. Zhuang, and C. Zhao, "A wintertime study of polycyclic aromatic hydrocarbons in PM2.5 and PM2.5-10 in Beijing: assessment of energy structure conversion," *Journal of Hazardous Materials*, vol. 157, no. 1, pp. 47–56, 2008.

[6] K. F. Chang, G. C. Fang, J. C. Chen, and Y. S. Wu, "Atmospheric polycyclic aromatic hydrocarbons (PAHs) in Asia: a review from 1999 to 2004," *Environmental Pollution*, vol. 142, no. 3, pp. 388–396, 2006.

[7] N. Rajput and A. Lakhani, "Measurements of polycyclic aromatic hydrocarbons in an urban atmosphere of Agra, India," *Atmosfera*, vol. 23, no. 2, pp. 165–183, 2010.

[8] S. O. Baek, R. A. Field, M. E. Goldstone, P. W. Kirk, J. N. Lester, and R. Perry, "A review of atmospheric polycyclic aromatic hydrocarbons: sources, fate and behavior," *Water, Air, and Soil Pollution*, vol. 60, no. 3-4, pp. 279–300, 1991.

[9] N. Ré-Poppi and M. Santiago-Silva, "Polycyclic aromatic hydrocarbons and other selected organic compounds in ambient air of Campo Grande City, Brazil," *Atmospheric Environment*, vol. 39, no. 16, pp. 2839–2850, 2005.

[10] S. S. Park, Y. J. Kim, and C. H. Kang, "Atmospheric polycyclic aromatic hydrocarbons in Seoul, Korea," *Atmospheric Environment*, vol. 36, no. 17, pp. 2917–2924, 2002.

[11] C. V. Raiyani, J. P. Jani, N. M. Desai, J. A. Shaha, and S. K. Kashyap, "Levels of polycyclic aromatic hydrocarbons in ambient environment of Ahmedabad," *Indian Journal of Environmental Protection*, vol. 13, pp. 206–216, 1993.

[12] G. G. Pandit, S. Sharma, A. M. Mohan Rao, and T. M. Krishnamoorthy, "Chromatographic methods for the estimation of polycyclic aromatic hydrocarbons in atmospheric particulates," in *Proceeding of the 5th National Symposium on Environment*, pp. 133–136, 1996.

[13] R. Vaishali, K. M. Phadke, R. Thakre, and M. Z. Hasan, "PAHs in respirable particulate matter in Nagpur city," *Journal of the Indian Association for Environmental Management*, vol. 24, pp. 11–16, 1997.

[14] G. Chattopadhyay, G. Samanta, S. Chatterjee, and D. Chakraborti, "Determination of particulate phase polycyclic aromatic hydrocarbons in ambient air of Calcutta for three years during winter," *Environmental Technology*, vol. 19, no. 9, pp. 873–882, 1998.

[15] P. Kulkarni, *Chemical Mass Balance for source apportionment of particulate polycyclic aromatic hydrocarbons in Indian Cities*, M.S. thesis, Centre for Environmental Science and Engineering, Indian Institute of Technology, Bombay, India, 1997.

[16] P. Kulkarni and C. Venkataraman, "Atmospheric polycyclic aromatic hydrocarbons in Mumbai, India," *Atmospheric Environment*, vol. 34, no. 17, pp. 2785–2790, 2000.

[17] H. Sharma, V. K. Jain, and Z. H. Khan, "Atmospheric polycyclic aromatic hydrocarbons (PAHs) in the urban air of Delhi during 2003," *Environmental Monitoring and Assessment*, vol. 147, no. 1–3, pp. 43–55, 2008.

[18] S. Saksena, *Integrated exposure assessment of airborne pollutants in an urban community using biomass and kerosene cooking fuels*, Ph.D. thesis, Centre for Environmental Science and Engineering, Indian Institute of Technology, Bombay, India, 1999.

[19] NAAQMS, *National Ambient Air Quality Monitoring Series: NAAQMS/28/2006-2007. Urban Air Monitoring: A Case Study*

in Agra 2002–2006, Central Pollution Control Board, New Delhi, India, 2006.

[20] RTO. Road Traffic Office. Transport Nagar Agra, India. Personal Communication, 2008.

[21] N. Rajput and A. Lakhani, "Measurements of polycyclic aromatic hydrocarbons at an industrial site in India," *Environmental Monitoring and Assessment*, vol. 150, no. 1–4, pp. 273–284, 2009.

[22] NHAI, National Highway Authority of India, Data from Shipping, Ministry of Road Transport and Highways, Delhi to Agra (km 108 to 199.63) at Madhuvan, India, 2007.

[23] N. Rajput, L. D. Khemani, and A. Lakhani, "Determination of polycyclic aromatic hydrocarbons in atmospheric particulate matter using gas chromatography," *Pollution Research*, vol. 26, no. 4, pp. 541–549, 2007.

[24] P. C. Vasconcellos, D. Zacarias, M. A. F. Pires, C. S. Pool, and L. R. F. Carvalho, "Measurements of polycyclic aromatic hydrocarbons in airborne particles from the metropolitan area of São Paulo City, Brazil," *Atmospheric Environment*, vol. 37, no. 21, pp. 3009–3018, 2003.

[25] N. Rajput and A. Lakhani, "PAHs and their carcinogenic potencies in diesel fuel and diesel generator exhaust," *Human and Ecological Risk Assessment*, vol. 15, no. 1, pp. 201–213, 2009.

[26] Parivesh, *Polycyclic Aromatic Hydrocarbons in Air and Their Effects on Human Health*, Central Pollution Control Board, Ministry of Environment and Forests, 2003.

[27] W. F. Rogge, L. M. Hildemann, M. A. Mazurek, G. R. Cass, and B. R. T. Simoneit, "Sources of fine organic aerosol. 2. Non-catalyst and catalyst-equipped automobiles and heavy-duty diesel trucks," *Environmental Science and Technology*, vol. 27, no. 4, pp. 636–651, 1993.

[28] A. D. Pereira Netto, I. F. Cunha, F. C. Muniz, and E. C. P. Rego, "Polycyclic aromatic hydrocarbons in street Dust of Niteroi City, RJ, Brazil," *Bulletin of Environmental Contamination and Toxicology*, vol. 72, no. 4, pp. 829–835, 2004.

[29] ATSDR (Agency for Toxic Substances and Disease Registry), *Toxicological Profile for Polycyclic Aromatic Hydrocarbons (PAHs)*, US Department of Health and Human Services, Atlanta, Ga, USA, 1995.

[30] A. M. Mastral, M. S. Callén, J. M. López, R. Murillo, T. García, and M. V. Navarro, "Critical review on atmospheric PAH. Assessment of reported data in the Mediterranean basin," *Fuel Processing Technology*, vol. 80, no. 2, pp. 183–193, 2003.

[31] C. Bourotte, M. C. Forti, S. Taniguchi, M. C. Bícego, and P. A. Lotufo, "A wintertime study of PAHs in fine and coarse aerosols in São Paulo city, Brazil," *Atmospheric Environment*, vol. 39, no. 21, pp. 3799–3811, 2005.

[32] R. Mohanraj, G. Solaraj, and S. Dhanakumar, "PM 2.5 and PAH concentrations in urban atmosphere of Tiruchirappalli, India," *Bulletin of Environmental Contamination and Toxicology*, vol. 87, no. 3, pp. 330–335, 2011.

[33] A. Masih, R. Saini, R. Singhvi, and A. Taneja, "Concentrations, sources, and exposure profiles of polycyclic aromatic hydrocarbons (PAHs) in particulate matter (PM10) in the north central part of India," *Environmental Monitoring and Assessment*, vol. 163, no. 1–4, pp. 421–431, 2010.

[34] H. Sharma, V. K. Jain, and Z. H. Khan, "Characterization and source identification of polycyclic aromatic hydrocarbons (PAHs) in the urban environment of Delhi," *Chemosphere*, vol. 66, no. 2, pp. 302–310, 2007.

[35] M. Tsapakis and E. G. Stephanou, "Occurrence of gaseous and particulate polycyclic aromatic hydrocarbons in the urban atmosphere: study of sources and ambient temperature effect on the gas/particle concentration and distribution," *Environmental Pollution*, vol. 133, no. 1, pp. 147–156, 2005.

[36] E. G. Sanderson, A. Raqbi, A. Vyskocil, and J. P. Farant, "Comparison of particulate polycyclic aromatic hydrocarbon profiles in different regions of Canada," *Atmospheric Environment*, vol. 38, no. 21, pp. 3417–3429, 2004.

[37] A. Eiguren-Fernandez, A. H. Miguel, J. R. Froines, S. Thurairatnam, and E. L. Avol, "Seasonal and spatial variation of polycyclic aromatic hydrocarbons in vapor-phase and PM2.5 in Southern California urban and rural communities," *Aerosol Science and Technology*, vol. 38, no. 5, pp. 447–455, 2004.

[38] J. Li, G. Zhang, X. D. Li, S. H. Qi, G. Q. Liu, and X. Z. Peng, "Source seasonality of polycyclic aromatic hydrocarbons (PAHs) in a subtropical city, Guangzhou, South China," *Science of the Total Environment*, vol. 355, no. 1–3, pp. 145–155, 2006.

[39] J. H. Tan, X. H. Bi, J. C. Duan, K. A. Rahn, G. Y. Sheng, and J. M. Fu, "Seasonal variation of particulate polycyclic aromatic hydrocarbons associated with PM10 in Guangzhou, China," *Atmospheric Research*, vol. 80, no. 4, pp. 250–262, 2006.

[40] G. Dörr, M. Hippelein, H. Kaupp, and O. Hutzinger, "Baseline contamination assessment for a new resource recovery facility in Germany: part VI: levels and profiles of polycyclic aromatic hydrocarbons (PAH) in ambient air," *Chemosphere*, vol. 33, no. 8, pp. 1569–1578, 1996.

[41] D. W. M. Sin, Y. C. Wong, Y. Y. Choi, C. H. Lam, and P. K. K. Louie, "Distribution of polycyclic aromatic hydrocarbons in the atmosphere of Hong Kong," *Journal of Environmental Monitoring*, vol. 5, no. 6, pp. 989–996, 2003.

[42] D. Ludykar, R. Westerholm, and J. Almén, "Cold start emissions at +22, -7 and -20∘C ambient temperatures from a three-way catalyst (TWC) car: regulated and unregulated exhaust components," *Science of the Total Environment*, vol. 235, no. 1–3, pp. 65–69, 1999.

[43] R. M. Harrison, D. I. T. Smith, and L. Luhana, "Source apportionment of atmospheric polycyclic aromatic hydrocarbons collected from an urban location in Birmingham, U.K," *Environmental Science and Technology*, vol. 30, no. 3, pp. 825–832, 1996.

[44] A. M. Caricchia, S. Chiavarini, and M. Pezza, "Polycyclic aromatic hydrocarbons in the urban atmospheric particulate matter in the city of Naples (Italy)," *Atmospheric Environment*, vol. 33, no. 23, pp. 3731–3738, 1999.

[45] T. Ohura, T. Amagai, T. Sugiyama, M. Fusaya, and H. Matsushita, "Characteristics of particle matter and associated polycyclic aromatic hydrocarbons in indoor and outdoor air in two cities in Shizuoka, Japan," *Atmospheric Environment*, vol. 38, no. 14, pp. 2045–2054, 2004.

[46] H. Guo, S. C. Lee, K. F. Ho, X. M. Wang, and S. C. Zou, "Particle-associated polycyclic aromatic hydrocarbons in urban air of Hong Kong," *Atmospheric Environment*, vol. 37, no. 38, pp. 5307–5317, 2003.

[47] A. H. Miguel, T. W. Kirchstetter, R. A. Harley, and S. V. Hering, "On-road emissions of particulate polycyclic aromatic hydrocarbons and black carbon from gasoline and diesel vehicles," *Environmental Science and Technology*, vol. 32, no. 4, pp. 450–455, 1998.

[48] P. Masclet, G. Mouvier, and K. Nikolaou, "Relative decay index and sources of polycyclic aromatic hydrocarbons," *Atmospheric Environment*, vol. 20, no. 3, pp. 439–446, 1986.

[49] N. R. Khalili, P. A. Scheff, and T. M. Holsen, "PAH source fingerprints for coke ovens, diesel and gasoline engines, highway tunnels, and wood combustion emissions," *Atmospheric Environment*, vol. 29, no. 4, pp. 533–542, 1995.

[50] C. K. Li and R. M. Kamens, "The use of polycyclic aromatic hydrocarbons as source signatures in receptor modeling," *Atmospheric Environment*, vol. 27, no. 4, pp. 523–532, 1993.

[51] M. A. Sicre, J. C. Marty, A. Saliot, X. Aparicio, J. Grimalt, and J. Albaiges, "Aliphatic and aromatic hydrocarbons in different sized aerosols over the Mediterranean Sea: occurrence and origin," *Atmospheric Environment*, vol. 21, no. 10, pp. 2247–2259, 1987.

[52] M. F. Simcik, H. Zhang, S. J. Eisenreich, and T. P. Franz, "Urban contamination of the Chicago/Coastal Lake Michigan atmosphere by PCBs and PAHs during AEOLOS," *Environmental Science and Technology*, vol. 31, no. 7, pp. 2141–2147, 1997.

[53] N. Tang, T. Hattori, R. Taga et al., "Polycyclic aromatic hydrocarbons and nitropolycyclic aromatic hydrocarbons in urban air particulates and their relationship to emission sources in the Pan-Japan Sea countries," *Atmospheric Environment*, vol. 39, no. 32, pp. 5817–5826, 2005.

[54] G. Grimmer, J. Jacob, and K. W. Naujack, "Profile of the polycyclic aromatic compounds from crude oils. Part 3. Inventory by GCGC/MS.—PAH in environmental materials," *Fresenius Zeitschrift fur Analytische Chemie Labor und Betriebsverfahren*, vol. 314, no. 1, pp. 29–36, 1983.

[55] I. G. Kavouras, P. Koutrakis, M. Tsapakis et al., "Source apportionment of urban particulate aliphatic and polynuclear aromatic hydrocarbons (PAHs) using multivariate methods," *Environmental Science and Technology*, vol. 35, no. 11, pp. 2288–2294, 2001.

[56] E. Galarneau, "Source specificity and atmospheric processing of airborne PAHs: implications for source apportionment," *Atmospheric Environment*, vol. 42, no. 35, pp. 8139–8149, 2008.

[57] J. Mantis, A. Chaloulakou, and C. Samara, "PM10-bound polycyclic aromatic hydrocarbons (PAHs) in the Greater Area of Athens, Greece," *Chemosphere*, vol. 59, no. 5, pp. 593–604, 2005.

[58] T. T. Hien, P. P. Nam, S. Yasuhiro, K. Takayuki, T. Norimichi, and B. Hiroshi, "Comparison of particle-phase polycyclic aromatic hydrocarbons and their variability causes in the ambient air in Ho Chi Minh City, Vietnam and in Osaka, Japan, during 2005-2006," *Science of the Total Environment*, vol. 382, no. 1, pp. 70–81, 2007.

[59] M. D. R. Sienra, N. G. Rosazza, and M. Préndez, "Polycyclic aromatic hydrocarbons and their molecular diagnostic ratios in urban atmospheric respirable particulate matter," *Atmospheric Research*, vol. 75, no. 4, pp. 267–281, 2005.

[60] F. Tian, J. Chen, X. Qiao et al., "Sources and seasonal variation of atmospheric polycyclic aromatic hydrocarbons in Dalian, China: factor analysis with non-negative constraints combined with local source fingerprints," *Atmospheric Environment*, vol. 43, no. 17, pp. 2747–2753, 2009.

[61] G. Wang, L. Huang, Xin Zhao, H. Niu, and Z. Dai, "Aliphatic and polycyclic aromatic hydrocarbons of atmospheric aerosols in five locations of Nanjing urban area, China," *Atmospheric Research*, vol. 81, no. 1, pp. 54–66, 2006.

[62] J. M. Daisey, J. L. Cheney, and P. J. Lioy, "Profiles of organic particulate emissions from air pollution sources: status and needs for receptor source apportionment modeling," *Journal of the Air Pollution Control Association*, vol. 36, no. 1, pp. 17–33, 1986.

[63] N. T. Kim Oanh, L. H. Nghiem, and Y. L. Phyu, "Emission of polycyclic aromatic hydrocarbons, toxicity, and mutagenicity from domestic cooking using sawdust briquettes, wood, and kerosene," *Environmental Science and Technology*, vol. 36, no. 5, pp. 833–839, 2002.

[64] A. Papageorgopoulou, E. Manoli, E. Touloumi, and C. Samara, "Polycyclic aromatic hydrocarbons in the ambient air of Greek towns in relation to other atmospheric pollutants," *Chemosphere*, vol. 39, no. 13, pp. 2183–2199, 1999.

[65] Z. Wang, J. Chen, P. Yang, X. Qiao, and F. Tian, "Polycyclic aromatic hydrocarbons in Dalian soils: distribution and toxicity assessment," *Journal of Environmental Monitoring*, vol. 9, no. 2, pp. 199–204, 2007.

[66] M. Chu and C. Chen, "Evaluation and estimation of potential carcinogenic risks of polynuclear aromatic hydrocarbons," in *Proceedings of the Symposium on Polycyclic Aromatic Hydrocarbons in the Workplace, Pacific Rim Risk Conference*, Honolulu, Hawaii, USA, December 1984.

[67] D. Krewski, T. Thorslund, and J. Withey, "Carcinogenic risk assessment of complex mixtures," *Toxicology and Industrial Health*, vol. 5, no. 5, pp. 851–867, 1989.

[68] T. Thorslund and D. Farrer, *Development of Relative Potency Estimates for PAHs and Hydrocarbon Combustion Product Fractions Compared to Benzo(a)Pyrene and Their Use in Carcinogenic Risk Assessments*, United States Environmental Protection Agency, 1991.

[69] I. C. T. Nisbet and P. K. LaGoy, "Toxic equivalency factors (TEFs) for polycyclic aromatic hydrocarbons (PAHs)," *Regulatory Toxicology and Pharmacology*, vol. 16, no. 3, pp. 290–300, 1992.

[70] T. Petry, P. Schmid, and C. Schlatter, "The use of toxic equivalency factors in assessing occupational and environmental health risk associated with exposure to airborne mixtures of polycyclic aromatic hydrocarbons (PAHs)," *Chemosphere*, vol. 32, no. 4, pp. 639–648, 1996.

Permissions

The contributors of this book come from diverse backgrounds, making this book a truly international effort. This book will bring forth new frontiers with its revolutionizing research information and detailed analysis of the nascent developments around the world.

We would like to thank all the contributing authors for lending their expertise to make the book truly unique. They have played a crucial role in the development of this book. Without their invaluable contributions this book wouldn't have been possible. They have made vital efforts to compile up to date information on the varied aspects of this subject to make this book a valuable addition to the collection of many professionals and students.

This book was conceptualized with the vision of imparting up-to-date information and advanced data in this field. To ensure the same, a matchless editorial board was set up. Every individual on the board went through rigorous rounds of assessment to prove their worth. After which they invested a large part of their time researching and compiling the most relevant data for our readers. Conferences and sessions were held from time to time between the editorial board and the contributing authors to present the data in the most comprehensible form. The editorial team has worked tirelessly to provide valuable and valid information to help people across the globe.

Every chapter published in this book has been scrutinized by our experts. Their significance has been extensively debated. The topics covered herein carry significant findings which will fuel the growth of the discipline. They may even be implemented as practical applications or may be referred to as a beginning point for another development. Chapters in this book were first published by Hindawi Publishing Corporation; hereby published with permission under the Creative Commons Attribution License or equivalent.

The editorial board has been involved in producing this book since its inception. They have spent rigorous hours researching and exploring the diverse topics which have resulted in the successful publishing of this book. They have passed on their knowledge of decades through this book. To expedite this challenging task, the publisher supported the team at every step. A small team of assistant editors was also appointed to further simplify the editing procedure and attain best results for the readers.

Our editorial team has been hand-picked from every corner of the world. Their multi-ethnicity adds dynamic inputs to the discussions which result in innovative outcomes. These outcomes are then further discussed with the researchers and contributors who give their valuable feedback and opinion regarding the same. The feedback is then collaborated with the researches and they are edited in a comprehensive manner to aid the understanding of the subject.

Apart from the editorial board, the designing team has also invested a significant amount of their time in understanding the subject and creating the most relevant covers. They scrutinized every image to scout for the most suitable representation of the subject and create an appropriate cover for the book.

The publishing team has been involved in this book since its early stages. They were actively engaged in every process, be it collecting the data, connecting with the contributors or procuring relevant information. The team has been an ardent support to the editorial, designing and production team. Their endless efforts to recruit the best for this project, has resulted in the accomplishment of this book. They are a veteran in the field of academics and their pool of knowledge is as vast as their experience in printing. Their expertise and guidance has proved useful at every step. Their uncompromising quality standards have made this book an exceptional effort. Their encouragement from time to time has been an inspiration for everyone.

The publisher and the editorial board hope that this book will prove to be a valuable piece of knowledge for researchers, students, practitioners and scholars across the globe.

List of Contributors

I. Hůnová, J. Horálek, M. Schreiberová
Ambient Air Quality Department, Czech Hydrometeorological Institute, 14306 Prague, Czech Republic

M. Zapletal
Faculty of Philosophy and Science, Silesian University at Opava, 74601 Opava, Czech Republic
Ekotoxa s.r.o.-Centre for Environment and Land Assessment, 74601 Opava, Czech Republic

Julio Tóta
Universidade do Estado do Amazonas, Manaus, AM, Brazil

David Roy Fitzjarrald
State University of New York, Albany, NY, USA

Maria A. F. da Silva Dias
Universidade de S˜ao Paulo, Sao Paulo, SP, Brazil

Juan Antonio Añel
Smith School of Enterprise and the Environment, University of Oxford, Oxford OX1 2BQ, UK

Laura de la Torre and Luis Gimeno
Ephyslab, Facultad de Ciencias de Ourense, Universidade de Vigo, 32004 Ourense, Spain

Guor-Cheng Fang and Ci-Song Huang
Department of Safety, Health and Environmental Engineering, Hung Kuang University, Sha-Lu, Taichung 433, Taiwan

Vyoma Singla, Tripti Pachauri, Aparna Satsangi, K.Maharaj Kumari and Anita Lakhani
Department of Chemistry, Faculty of Science, Dayalbagh Educational Institute, Dayalbagh, Agra 282110, India

Libor Kukačka and Radka Kellnerová
Department of Meteorology and Environment Protection, Faculty of Mathematics and Physics, Charles University in Prague, VHoleˇsoviˇckˊach 2, 180 00 Prague, Czech Republic
Institute of Thermo mechanics AS CR, v.v.i., Dolejˇskova 1402/5, 182 00 Prague, Czech Republic

Stěpán Nosek, Klára Jurčáková and Zbyněk Jaňour
Institute of Thermo mechanics AS CR, v.v.i., Dolejˇskova 1402/5, 182 00 Prague, Czech Republic

Daniela Dias, Oxana Tchepel, Anabela Carvalho, Ana IsabelMiranda and Carlos Borrego
Centre for Environmental and Marine Studies and Department of Environment and Planning, University of Aveiro, 3810-193 Aveiro, Portugal

Sunil D. Pawar and V. Gopalakrishnan
Thunderstorm Dynamics Program, Indian Institute of Tropical Meteorology, Pune 411 008, India

Pasquale Avino
Air Chemical Laboratory, DIPIA, INAIL Settore Ricerca, Via IV Novembre 144, 00187 Rome, Italy

Geraldo Capannesi and Alberto Rosada
UTFIST-CATNUC, ENEA, Via Anguillarese 301, 00060 Rome, Italy

Francesco Lopez
Dipartimento di Agricoltura, Ambiente Alimenti (DIAAA), University of Molise, Via De Sanctis, 86100 Campobasso, Italy

Seehyung Lee, Jeongwoo Lee and Eui-Chan Jeon
Department of Earth and Environmental Sciences, Sejong University, Seoul 143-747, Republic of Korea

Jinsu Kim
Cooperate Course for Climate Change, Sejong University, Seoul 143-747, Republic of Korea

Armineh Barkhordarian
Helmholtz-Zentrum Geesthacht Centre for Materials and Coastal Research, Max-Planck-Straße, 21502 Geesthacht, Germany

Jaber Almedeij
Civil Engineering Department, Kuwait University, P.O. Box 5969, 13060 Safat, Kuwait

Mabrouk Chaâbane and Chafai Azri
Facult´e des Sciences, Universit´e de Sfax, B.P. 1171, Sfax 3000, Tunisia

KhaledMedhioub
Institut Pr´eparatoire aux Etudes d'Ing´enieurs de Sfax, Universit´e de Sfax, BP 805, Sfax 3018, Tunisia

S. Saavedra,1 A. Rodr´ıguez, J. A. Souto and J. J. Casares
Department of Chemical Engineering, University of Santiago de Compostela, 15782 Santiago de Compostela, Spain

J. L. Berm´ udez and B. Soto
Department of Environment, As Pontes Power Plant, Endesa Generaci´on S.A., 15320 La Coruna, Spain

Sujata Pattanayak, U. C.Mohanty, and Krishna K. Osuri
Centre for Atmospheric Sciences, Indian Institute of Technology, Delhi, Hauz Khas, New Delhi 110016, India

NattasutMantananont and Savitri Garivait
The Joint Graduate School of Energy and Environment, KingMongkut's University of Technology Thonburi, Bangkok 10140, Thailand

Suthum Patumsawad
Department of Mechanical Engineering, King Mongkut's University of Technology North Bangkok, Bangkok 10800, Thailand

E. Alonso-Blanco, R. Fraile, and A. Castro
Department of Physics (IMARENAB), University of Le´on, Le´on 24071, Spain

A. I. Calvo
Centre for Environmental and Marine Studies (CESAM), University of Aveiro, Aveiro 3810-193, Portugal

Timothy Curtin
Emeritus Faculty, Australian National University, Canberra, ACT 0200, Australia